Neuroplasticity Unleashed: Innovations and Applications in Brain Rewiring

Selen Gür Özmen and Aygül Köseoğlu Kurt (eds)

Neuroplasticity Unleashed: Innovations and Applications in Brain Rewiring

PETER LANG
Berlin · Bruxelles · Chennai · Lausanne · New York · Oxford

Bibliographic Information published by the Deutsche Nationalbibliothek
The German National Library lists this publication in the German National Bibliography; detailed bibliographic data is available on the Internet at http://dnb.d-nb.de.

Library of Congress Cataloging-in-Publication Data
Names: Gür Özmen, Selen editor | Köseoğlu Kurt, Aygül editor
Title: Neuroplasticity unleashed : innovations and applications in brain rewiring / Selen Gür Özmen and Aygül Köseoğlu Kurt (eds).
Description: Berlin ; New York : Peter Lang, [2026] | Includes bibliographical references.
Identifiers: LCCN 2026002571 (print) | LCCN 2026002572 (ebook) | ISBN 9783631930182 print | ISBN 9783631950982 ePDF | ISBN 9783631950999 ePub
Subjects: LCSH: Neuroplasticity
Classification: LCC QP363.3 .N488 2026 (print) | LCC QP363.3 (ebook)
LC record available at https://lccn.loc.gov/2026002571
LC ebook record available at https://lccn.loc.gov/2026002572

Cover image: Yasin Akşit, “Innovations and Applications”
Cover design by Peter Lang Group AG

ISBN 978-3-631-93018-2 (Print)
ISBN 978-3-631-95098-2 (ePDF)
ISBN 978-3-631-95099-9 (ePUB)
DOI 10.3726/b23673

Published by Peter Lang GmbH, Berlin, Germany

info@peterlang.com

This publication has been peer reviewed.

www.peterlang.com

Contact for General Product Safety Regulation (GPSR): gpsr@peterlang.com

Contents

Figures

Nutritional Neuroscience: Dietary Influences on Brain Plasticity

Exercise-Induced Neuroplasticity: Mechanisms and Benefits

Motor Learning and Neuroplasticity: From Theory to Practice

Neuroplasticity in Sports Medicine: Injury Prevention and Performance Enhancement

Table

Nutritional Neuroscience: Dietary Influences on Brain Plasticity

Disclosures

Conflict of interest: The author declares no conflicts of interest related to this manuscript.

Use of AI for Writing Assistance: AI tools were utilized to edit the manuscript language. The author independently developed all intellectual contributions, critical content, and interpretations.

Financial Disclosure: No financial support was received to prepare this manuscript.

Peer review: This manuscript has been reviewed internally and externally by peers to ensure the accuracy and quality of the content.

Neuroplasticity and Nutrition: Neuro Foods

Ecenur Özkul Erdoğan

1. Introduction

Nutrition is vital for learning and memory. It supports nerve cell development and function (Georgieff, 2023). Key dietary elements like vitamins and minerals boost brain health. They enhance communication between neurons and increase synaptic plasticity. A healthy diet can improve cognitive functions, learning, and reduce neurological disease risk. It also fosters neuroplasticity, aiding learning and memory (Bekdash, 2024).

Deficiencies in vitamins and minerals can harm neuroplasticity, leading to cognitive issues like learning difficulties and memory loss. However, better nutrition may help mitigate these deficits and support cognitive functions. Adopting healthy eating habits supports neuron development, enhances learning, and improves neurological health. This chapter highlights the positive impact of diet on neuroplasticity, detailing how nutrition affects the brain through biochemical and physiological processes.

2. Neuroplasticity and Nutrition

Nutrition is vital for health and development. It regulates many biological processes. The first 1,000 days of life, from pregnancy to infancy, are crucial for brain development. During this time, brain cells grow, differentiate, and connect. These processes are linked to neuroplasticity, the brain's ability to adapt. Nutrients significantly influence this capacity (Georgieff, 2023).

Adequate nutrient intake supports brain cell growth, neurotransmitter production, and energy use. Key nutrients include omega-3 fatty acids, iron, zinc, B vitamins, and antioxidants. These are essential for neuroplasticity. Malnutrition can cause permanent damage to brain development. It affects the growth of oligodendrocytes and myelination, crucial for cognitive function. Research shows that poor nutrition harms neuroplasticity. This leads to lasting deficits in cognitive abilities, learning, memory, and social skills (Mansour, 2022).

Nutrition affects brain development directly and through epigenetics. Epigenetics changes gene activity without altering DNA (Bekdash, 2024). Nutrients can influence gene expression. For instance, vitamin B12 deficiency is linked to neural tube defects and early cognitive issues. Iron deficiency can hinder brain cell growth, affecting learning and memory. Thus, nutrition's role in brain development is complex (Bacon & Brinton, 2021).

The "gut–brain axis" is key to linking nutrition and brain development. Gut microbiota significantly influences brain health. These microorganisms affect brain functions, digestion, and immunity. The gut–brain axis connects the nervous and digestive systems. Gut bacteria aid in neurotransmitter production and communicate with the brain. They also influence serotonin levels, impacting mood and behavior (Damiani et al., 2023).

Gut health is linked to nutrition. Malnutrition disrupts gut microbiota, harming brain functions. This imbalance can lead to anxiety and depression. Probiotics and prebiotics may support brain health via gut microbiota (Damiani et al., 2023).

In summary, the link between nutrition and brain development is intricate. It involves neuroplasticity, epigenetic mechanisms, and the gut–brain axis. Good nutrition in early life is vital for brain growth and lifelong cognitive and emotional health. Thus, the importance of nutrition in brain development extends beyond individuals to society.

2.1. Omega-3, Polyunsaturated Fatty Acids and Neuroplasticity

Omega-3 and polyunsaturated fatty acids are vital for brain health and neuroplasticity. These fatty acids are crucial during development to prevent neurological disorders. Omega-3s, particularly DHA and EPA, improve neuron structure and function. They enhance communication between neurons. DHA aids in forming synapses, supports neuron movement, and boosts synaptic plasticity (Miceli et al., 2020). Adequate omega-3 intake during pregnancy and after birth is key for infant brain development. In breast milk, DHA and ARA are essential for neuroplasticity. Studies show that a lack of omega-3s can harm children's brain development and lead to disorders. This includes lower levels in conditions like autism and ADHD (Yook & Lee, 2020).

Omega-3s also affect neuroplasticity through the endocannabinoid system (eCB). This system strengthens synaptic connections and enhances neuron plasticity. Omega-3s trigger the release of BDNF, which supports neuron survival and is crucial for learning and memory (Miceli et al., 2020).

A new study has identified N-docosahexaenoylethanolamide (DHEA), a DHA derivative, as crucial for synapse formation. It interacts with the GPR110 receptor to reduce neuroinflammation. Notably, DHEA may promote synapse creation better than DHA. This suggests DHEA may have therapeutic potential in strategies aimed at brain development and protection by influencing specific genes (Park & Watkins, 2021).

Brain cell membranes act as storage for long-chain polyunsaturated fatty acids (LC PUFAs). These are released by certain enzymes, aiding nerve communication (Miceli et al., 2020). Researchers use two main methods to study LC PUFAs. First, they observe how these fatty acids interact with neurotransmitters like glutamate, GABA, and dopamine, which also affects serotonin. Second, they boost dietary PUFA levels in brain membranes. DHA and ARA then impact brain functions through specific enzymes. Oxidized linoleic acid derivatives also modulate neurotransmission. These processes are vital for brain health (Barón-Mendoza & González-Arenas, 2022).

A diet rich in omega-3 fatty acids boosts brain health by improving neuronal communication. The Mediterranean diet, packed with omega-3s and polyphenols, is a prime example. Olive oil, high in polyphenols, has anti-inflammatory properties and may support cellular maintenance processes (*see Chapter 2: Nutritional Neuroscience: Dietary Influences on Brain Plasticity*). It also enhances brain development, memory, and learning. In studies, rats fed olive oil outperformed others in memory tests and showed increased cellular cleanup (Maharjan et al., 2020). Additionally, olive oil reduces amyloid plaques in the brain. It triggers the release of GLP-1, which lowers blood sugar and protects brain cells from amyloid damage (Georgieff, 2023).

2.2. Flavonoids and Neuroplasticity

Flavonoids are compounds in plants known for health benefits. Notably, anthocyanins and flavones can boost cognitive functions. Eating fruits and vegetables rich in flavonoids, such as blueberries, blackberries, cherries, spinach, kale, and onions, greatly enhances young people's cognitive performance. These compounds protect the brain by improving its plasticity, increasing blood flow, and affecting learning and memory pathways (Maharjan et al., 2020). Flavonoids boost key pathways for memory and synaptic plasticity. They also reduce brain inflammation and oxidative stress. Animal studies show flavonoid-rich diets, including green tea (catechins), turmeric (curcumin), and red grapes or red wine (resveratrol), improve cognition and protect against neurodegenerative diseases. Thus, adding may these foods to diets could enhance cognition and prevent cognitive decline and neurodegenerative diseases (Cichon et al., 2020). In particular, anthocyanins, catechins, curcumin, and resveratrol stand out in these dietary recommendations.

2.2.1. Anthocyanins and Neuroplasticity

Anthocyanins are flavonoids mainly found in red and purple fruits and vegetables, especially berries. Recently, their benefits for brain health, particularly neuroplasticity, have gained attention (Zaa et al., 2023).

Research shows anthocyanins aid synaptic plasticity in the brain. For instance, a study by Fang et al. (2020) found that blueberry anthocyanins boost memory and learning, while also enhancing synaptic plasticity. It showed that consuming anthocyanins raised BDNF levels and improved neuron communication. Additionally, another study found that anthocyanins increased neurotransmitter levels and neurotrophic factor release. Mice given blueberry extract showed better learning and memory, alongside higher BDNF levels. These results suggest that anthocyanins may improve learning and memory by enhancing neuron communication (Vauzour et al., 2021a).

Anthocyanins activate cellular signaling pathways by binding to specific receptors in brain cells. For example, anthocyanins have been shown to activate biochemical pathways such as AMPK (AMP-activated protein kinase) and SIRT1 (Sirtuin 1), regulating energy metabolism and responses to cell stress. These mechanisms modulate synapse formation and synaptic communication (Zaa et al., 2023).

It has also been shown that anthocyanins support neurological health through their ability to combat oxidative stress. Brain cells are sensitive to damage caused by free radicals, and the potent antioxidant properties of anthocyanins help protect these cells. These protective effects activate defense mechanisms against free radicals through the Nrf2 pathway (nuclear factor erythroid 2) (Vauzour et al., 2021a). In a study, anthocyanins were found to activate the Nrf2 pathway, increasing the expression of antioxidant enzymes and supporting neuroplasticity by preventing cellular damage. Adequate dietary intake of anthocyanins can help prevent cognitive decline, particularly in older people, emerging as a supportive strategy for overall brain health. Therefore, a sufficient dietary intake of anthocyanins is vital in preserving and enhancing brain health (Vauzour et al., 2021b).

2.2.2. Resveratrol and Neuroplasticity

Resveratrol, found in grape skins and some fruits, may support brain health. Recent studies highlight its effects on neuroplasticity, as it influences cell signaling and protects neurons (Tu et al., 2023).

It may increase BDNF levels, which aids neuron growth, survival, and synapse formation. This enhances synaptic plasticity in the hippocampus, improving learning and memory (Calado et al., 2024).

Additionally, resveratrol activates pathways like MAPK/ERK and PI3K/Akt. These pathways promote neuron growth and survival. By activating the PI3K/Akt pathway, it enhances cell survival and synaptic plasticity (Terracina et al., 2022).

Research highlights resveratrol's role in preventing age-related brain diseases. Studies in Alzheimer's models suggest that it may enhance cognitive function and reduce amyloid plaque accumulation. This is linked to its capacity to prevent neuron damage and its anti-inflammatory properties (Tu et al., 2023). Moreover, resveratrol reduces brain cell inflammation by modulating proteins like SIRT1 and NF-κB. Its antioxidant and anti-inflammatory traits make it a strong candidate for treating these diseases. Studies confirm its positive impact on brain health and clarify the underlying mechanisms (Calado et al., 2024).

2.2.3. Curcumin and Neuroplasticity

Curcumin, the active compound in turmeric, is a natural polyphenol associated with many health benefits. Its protective effects on brain health are particularly noted for their positive impact on neuroplasticity.

The effects of curcumin on neuroplasticity occur mainly through biochemical mechanisms. One of these mechanisms is the enhancement of BDNF release by curcumin. A study determined that curcumin increased BDNF levels in brain tissue, improving synaptic plasticity. Curcumin also demonstrates its effects on hippocampal neurogenesis through the CREB (cAMP receptor element-binding protein)–BDNF signaling pathway. CREB supports the transcription of BDNF, promoting the formation of new neurons. Curcumin activates this signaling pathway by increasing CREB phosphorylation, promoting BDNF production. This mechanism improves the formation of new neurons, improving memory and learning abilities (Namgyal et al., 2020).

The ability of curcumin to reduce oxidative stress is also significant. Oxidative stress occurs due to the accumulation of free radicals that can damage brain cells. Curcumin reduces oxidative stress by increasing the activity of antioxidant enzymes such as superoxide dismutase (SOD) and glutathione peroxidase (GPx). This aspect supports neuron survival and facilitates neuroplastic changes (Mansour, 2022).

Additionally, the anti-inflammatory properties of curcumin play an essential role in supporting brain health. Curcumin reduces inflammation by inhibiting the NF-kB pathway, contributing to improving neuroplastic processes. A study found that curcumin decreased the production of inflammatory cytokines, resulting in enhanced synaptic communication (González-Granillo et al., 2022).

Another significant effect of curcumin is its interaction with sirtuin proteins. Sirtuins are essential proteins that regulate cellular aging, energy metabolism, and gene expression. Curcumin activates SIRT1. This increases BDNF and other neurotrophic factors. This is one way curcumin supports cognitive functions (Namgyal et al., 2020).

Finally, curcumin also has positive effects on glucose metabolism. Glucose is an energy source for neurons, and proper glucose metabolism supports brain health. Curcumin improves insulin sensitivity by activating the AMPK pathway (AMP-activated protein kinase), facilitating cell glucose uptake. This mechanism is critical for regulating energy metabolism and preserving neuronal health (Kandezi et al., 2020).

Curcumin shows notable effects against age-related issues. It combats neurodegeneration, which often leads to reduced cognitive function. In studies, curcumin improved cognitive function in rats of all ages and eased age-related changes (González-Granillo et al., 2022). In young people, it boosted neuroplasticity, learning, and memory. In older adults, it helped prevent neurodegeneration. These results highlight curcumin's protective role against cognitive decline with age. Its ability to enhance cognition and reduce neurodegeneration makes it a promising treatment for neurological disorders (Kandezi et al., 2020).

2.2.4. Catechins and Neuroplasticity

Catechins, found in plants and fruits like cocoa and green tea, are vital for brain health. They are rich in flavonoids. Specifically, catechins in cocoa and green tea help prevent neurological conditions and enhance cognitive functions (Unno & Nakamura, 2021).

Catechins boost neuroplasticity and BDNF in several ways. They activate the CREB pathway, which aids synaptic plasticity and BDNF production. This activation improves cognitive functions, making cocoa flavanols effective for long-term memory and learning (Melgar-Locatelli et al., 2024). Catechins also trigger the NADPH oxidase pathway, enhancing BDNF's neurogenic activity. This may supports neuron growth and health. Research shows green tea catechins promote neuron development through this pathway (Unno & Nakamura, 2021).

Moreover, catechins increase long-term memory by boosting synaptic plasticity. A study found that catechin intake enhanced working memory and cognitive flexibility in older adults. This underscores its brain health benefits, particularly for the elderly (Tejada et al., 2024).

Catechins are powerful antioxidants that protect brain cells from oxidative stress. They boost enzymes like SOD and GPx, vital for neuron health. Additionally, catechins reduce inflammatory cytokines, lowering neuroinflammation. This is key in preventing neurodegenerative diseases and cognitive decline, especially in the elderly (Ammar et al., 2020).

Cocoa and green tea are essential in supporting neuroplasticity and enhancing cognitive functions. Catechins boost BDNF, enhance synapses, and reduce oxidative stress. These benefits protect the brain. Thus, catechin-rich foods are key for better brain health and preventing cognitive decline with age. They are especially vital for those at risk of neurodegenerative diseases. Such foods help improve thinking and maintain brain health. (Tejada et al., 2024).

2.3. Vitamins and Neuroplasticity

Vitamins are essential nutrients that significantly affect brain development and neuroplasticity. The following discusses the biochemical mechanisms and their effects on neuroplasticity in detail.

2.3.1. Vitamin A

Vitamin A is obtained from various fruits and vegetables and animal sources such as egg yolks, liver, and dairy products.

Retinoic acid, derived from vitamin A, is vital for gene regulation and brain plasticity in embryos. It also supports neuron health, dopamine function, brain growth, and memory in later life (Kumar et al., 2024b).

Animal studies show retinoic acid affects memory in the hippocampus. It boosts cholinergic activity by raising certain gene expressions. Mice lacking retinoic acid receptors showed memory issues in tests. Additionally, long-term beta-carotene intake lowers cognitive impairment risk. Vitamin A may help protect against neuroinflammation by blocking harmful cytokines. A lack of vitamin A links to sleep and memory problems (Moramarco & McCaffery, 2024).

2.3.2. B Vitamins

Vitamin B1 (thiamine): Thiamine is essential for production and synaptic function. It plays a critical role in the production of glutamate and GABA. Deficiency can negatively affect energy production in nerve cells and neurotransmission, leading to neurodegenerative disorders and loss of neuroplasticity (Zarei et al., 2023).

Vitamin B2 (Riboflavin): Riboflavin is a cofactor for enzymes that reduce oxidative stress. It helps meet nerve cells' energy needs and protects them from oxidative damage. Deficiency of this vitamin can predispose to neurodegeneration and affect neuroplasticity processes.

Vitamin B3 (Niacin): Niacin is a component of coenzymes such as NAD and NADP, which are involved in energy transfer and support the energy metabolism of nerve cells. It also plays a role in DNA repair, contributing to the protection of brain cells from damage. Niacin deficiency can cause impaired cognitive functions and loss of neuroplasticity.

Vitamin B5 (pantothenic acid): Pantothenic acid is vital for making acetylcholine and powering nerve cells. It also aids neurotransmission, which affects learning and memory (Kumar et al., 2024b).

Vitamin B6 (pyridoxine): Vitamin B6 is essential for synthesizing neurotransmitters such as dopamine, serotonin, and GABA. It is essential for serotonin synthesis and brain development. Deficiency can lead to communication disruptions among nerve cells and loss of neuroplasticity (Xu et al., 2021).

Vitamin B7 (Biotin): Biotin is critical in energy metabolism and fatty acids. A biotin deficiency can negatively impact nerve cell energy production, reducing neuroplasticity.

Folic acid (vitamin B9): Folic acid is crucial for DNA synthesis and methylation. Its 5-methyltetrahydrofolate form converts homocysteine to methionine, aiding DNA methylation. This process boosts neuroplasticity through epigenetic changes (Bekdash, 2024). Moreover, folic acid is vital for making DNA and RNA. These are key for developing neural tubes and growing nerve cells. A lack of folic acid can slow cell renewal. This negatively impacts nervous system development and harms learning and memory (Pourié et al., 2020).

Vitamin B12 (Cobalamin): Vitamin B12 is vital for myelin production and nerve health. It also aids in creating methionine, acting as a key methyl donor. This process is crucial for gene regulation and supports brain adaptability to learning and experiences (Pourié et al., 2022).

Vitamin B12 is vital for DNA and nerve repair. It enhances brain flexibility. A deficiency can lead to myelin loss, nerve damage, and memory decline. It also disrupts brain connections. Therefore, maintaining B12 levels is crucial for brain health and resilience (Bekdash, 2024).

2.3.3. Vitamin C

Vitamin C is abundant in brain neurons. It helps make neurotransmitters and converts dopamine to norepinephrine. Too much dopamine, however, can harm neurons by producing hydrogen peroxide (Maharjan et al., 2020). Additionally, vitamin C protects neurons from oxidative stress and prevents excitotoxicity caused by glutamate. It also inhibits TNF-α and NF-κB in human neuroblastoma. Studies on guinea pigs showed that vitamin C deficiency leads to neuronal damage and increased oxidative stress (Jazvinšćak et al., 2021).

2.3.4. Vitamin D

Active vitamin D (calcitriol) is mainly produced in the skin under UV light and can also be ingested in small amounts. It's called a "neglected neurosteroid" because it has a widespread impact on brain function. This is due to the many vitamin D receptors found in the central nervous system (CNS) (Cu I & Eyles, 2022). These receptors are especially abundant in the brain's ventricular walls, key areas for new neuron growth. Vitamin D boosts certain brain cells, hinting at its role in the dopamine system. It also regulates factors vital for brain cell survival and growth. A lack of vitamin D during pregnancy can harm brain function and trigger neurological diseases. Moreover, it may protect against psychiatric disorders such as autism, depression, and schizophrenia (Li et al., 2024). Vitamin D supplementation may reduce neurotoxicity in several ways. They regulate neurotrophic factors, prevent the loss of nerve growth factor, and block harmful calcium activity. Additionally, they combat free radicals and reduce oxidative stress. Furthermore, vitamin D directly affects calcium absorption, which is crucial for synaptic transmission (Kouba & Rodrigues, 2024).

2.3.5. Vitamin E

Vitamin E (α-tocopherol) is found in wheat germ oil, soybeans, rice bran, olives, and coconut. It is critical for various neurological functions. Moreover, it regulates gene expression that encodes proteins necessary for cell cycle regulation, apoptosis, lipoprotein receptors, neuronal development, and synaptic plasticity. Vitamin E has anti-inflammatory properties, which inhibit the expression of cyclooxygenase-2 and 5-lipoxygenase. In a mouse model, vitamin E suppressed the activation of nuclear factor kappa by lipopolysaccharides and cytokines such as IL-1β and IL-6 (da Cunha Germano et al., 2023).

2.4. Minerals and Neuroplasticity

The balance of minerals is critical for brain functions. Psychiatric disorders are characterized by the presence of pathological features such as metabolic dysregulation, oxidative stress, and protein aggregation, all of which are associated with the effects of minerals. Specific minerals have been shown to affect brain functions. Iron, iodine, and zinc are essential minerals for synthesizing norepinephrine, serotonin, and synaptic plasticity.

2.4.1. Zinc (Zn^{2+})

Zinc is found in various brain regions, including the dentate gyrus, granular cell axons, CA3 pyramidal cells, the neocortex, and the hippocampus. It plays an important role in signal modulation within CNS. Zinc is present in the synaptic vesicles of certain neurons in the thalamus, cerebral cortex, hippocampus, and olfactory cortex. Additionally, zinc affects the structure of voltage-gated potassium and calcium channels and regulates the production of biogenic amines. Zn^{2+} is associated with long-term potentiation in the hippocampus and modulates synaptic transmission and excitability by controlling the NMDA, GABA, and glycine receptors (Jafari et al., 2021).

Postnatal zinc deficiency disrupts DNA synthesis, affects brain polysomes, reduces protein synthesis, and decreases cerebellar granular cells. Furthermore, neonatal Zn^{2+} deficiency adversely affects learning, memory, attention, and mood. Zinc is also known to enhance BDNF levels, which is crucial for neuroplasticity and cognitive function. By increasing BDNF expression, zinc can support neuronal growth and resilience, significantly improving conditions such as depression (Mlyniec, 2021). Iron, zinc, choline, and copper influence neurotransmitter levels, receptors, and reuptake. Zinc regulates the electrophysiological potential of neurons by protecting healthy mitochondria and producing sufficient ATP. It also reduces pro-inflammatory molecules such as C-reactive protein, IL-6, and TNF-α (Leal et al., 2020).

2.4.2. Magnesium (Mg^{2+})

Magnesium is essential for nerve transmission and neuromuscular coordination. Magnesium protects cells from excitotoxicity by blocking calcium channels in aspartate receptors. Magnesium deficiency can lead to hyperexcitability in the CNS. This deficiency increases certain excitatory neurotransmitters (catecholamines, acetylcholine, and excitatory amino acids) and a decrease in inhibitory neurotransmitters (adenosine, glycine, GABA, and taurine). It also increases the synthesis of neuropeptides, prostanoids, and pro-inflammatory cytokines while reducing antioxidant defenses. It also supports neuroplasticity and may increase BDNF expression (Afsharfar et al., 2021). Magnesium deficiency has been associated with increased inflammatory and pro-inflammatory markers such as TNF-α, vascular cell adhesion molecule-1, IL-1β, fibrinogen, complement, and plasminogen activator inhibitor-1. Additionally, dietary magnesium intake is inversely related to C-reactive protein levels. Electroencephalogram changes have been observed in rats with magnesium deficiency during auditory stimulation, indicating a relationship between magnesium deficiency and increased excitability (Orhan et al., 2022).

2.4.3. Iron (Fe^{3+})

Iron is vital for neuroplasticity, energy metabolism, neurotransmitter synthesis, and synapse formation. It's crucial in brain development for enzymes like cytochrome oxidase (Leal et al., 2020). Moreover, it aids in producing neurotransmitters such as dopamine and serotonin. Dopamine, important for memory and learning, relies on sufficient iron. A deficiency can reduce the production of these neurotransmitters (Tarnacka et al., 2021).

Furthermore, iron plays a role in myelin synthesis through oligodendrocytes, which facilitates efficient communication between neurons. A loss or deficiency of myelin can alter cognitive functions. In particular, during the early stages of development, iron deficiency can inhibit the maturation and connectivity of nerve cells, reducing synaptic plasticity. This can impair learning capacity and memory (Urrutia et al., 2021). Iron deficiency's effects on the brain can cause lasting changes in cognitive performance, mood, and behavior. Low iron intake during childhood has been linked to impairments in motor performance, cognitive functions, social orientation, and expressive language skills, whereas iron supplementation has shown improvements in these areas (Salami et al., 2021).

2.4.4. Calcium (Ca^{2+})

Calcium (Ca^{2+}) balance is vital for many brain functions. These include action potentials, nerve growth, learning, and memory. It also affects synaptic plasticity (Wang et al., 2020). Calcium ions help neurons communicate. They control neural activity and trigger neurotransmitter release. Moreover, calcium influences synaptic plasticity through NMDA receptors. These receptors are crucial for memory processes like long-term potentiation (LTP). Calcium also affects gene expression. It activates signaling proteins, enhancing learning and memory. Low calcium levels can harm synaptic plasticity and cell renewal. This is linked to cognitive decline and neurological disorders (Mosayebi-Samani et al., 2020).

2.4.5. Copper (Cu^{2+})

Copper (Cu^{2+}) is essential for brain function, activating neuropeptides, forming myelin, and making neurotransmitters. It aids enzymes crucial for producing dopamine and norepinephrine. This regulation boosts learning and memory (Leal et al., 2020).

Moreover, copper is key in fighting oxidative stress. It forms part of the SOD1 enzyme, protecting neurons and enhancing neuroplasticity (Tarnacka et al., 2021). Low copper levels lead to cell damage and cognitive decline.

It also aids in making myelin, which speeds up nerve signal transmission and improves synaptic plasticity. Thus, copper is important for maintaining flexible and adaptable neural networks (Nicoletti et al., 2022).

2.4.6. Iodine

Iodine is vital for making thyroid hormones (T3 and T4) needed for brain growth and flexibility. These hormones control brain cells, myelin, and connections. Hormone imbalances, especially in early life, can harm growth and learning. Iodine is crucial for developing brain structures in fetuses. A lack of iodine disrupts the brain's outer layer and harms cell movement. It also increases cell damage risk, affecting brain flexibility (Fuzi & Loh, 2022).

2.4.7. Manganese

Manganese is crucial for brain functions, especially in glutamatergic neurons. A lack of it can raise epilepsy risk (Tarnacka et al., 2021). Moreover, abnormal manganese in the basal ganglia links to Parkinson's disease. Thus, studying manganese's impact on the brain is key. This research could reveal its protective benefits in psychiatric and neurological disorders (Nicoletti et al., 2022).

2.5. Prebiotics and Neuroplasticity

Prebiotics boost neuroplasticity by improving gut bacteria. These are mainly indigestible fibers. Certain gut bacteria turn prebiotics into short-chain fatty acids (SCFAs). This process helps good bacteria grow (Romo-Araiza & Ibarra, 2020). For instance, *Bifidobacteria* break down prebiotic fibers to make SCFAs like butyrate, propionate, and acetate. Gut cells use these SCFAs for energy. They also strengthen the gut barrier and lower inflammation. Butyrate, in particular, inhibits certain enzymes, affecting gene expression. This reduces pro-inflammatory cytokines and boosts regulatory T cells (Tregs). Tregs are vital for controlling immune responses (Paiva et al., 2024).

SCFAs affect the brain through the gut–brain axis. Butyrate boosts BDNF, aiding neuron survival and growth. Animal studies show prebiotics enhance synaptic plasticity and memory. For instance, B-GOS and FOS supplementation improves cognition by raising BDNF and NMDAR in the hippocampus (Vaghef-Mehrabani et al., 2023). Research on prebiotics in humans is limited but encouraging. One study found B-GOS supplementation improved cortisol awakening in healthy people. This suggests benefits for emotional disorders.

Additionally, inulin-rich oligofructose boosted mood and memory. These results imply prebiotics can enhance mental performance and mood (Romo-Araiza & Ibarra, 2020).

Foods rich in prebiotics include garlic, onions, oats, and bananas. These foods can support intestinal health, thus positively affecting brain health:

Garlic: Supports the gut microbiota due to its fructooligosaccharides content. Garlic consumption increased BDNF levels, thereby improving neuroplasticity.

Onion: Its high inulin content supports gut health and can positively influence cognitive functions.

Oats: Contains beta-glucan; oats can increase SCFA production, support neurotransmitter levels, and improve mental health.

Banana: Contains prebiotic fructooligosaccharides and nourishes gut bacteria, positively affecting brain health.

Regular consumption of these foods may strengthen the gut microbiota and improve brain health. Increasing prebiotic intake can also improve mental functions (Shanmugam et al., 2022).

2.6. Probiotics and Neuroplasticity

Probiotics are live microorganisms that can significantly affect health by balancing the gut microbiota. Recent research has shown that probiotics have positive effects on digestive health, brain development, and neuroplasticity.

Dysbiosis in the intestinal microbiota can adversely affect mental health. Therefore, restoring the gut microbiota can be used to improve mental health. Alteration of the intestinal microbiota can influence the functioning of the hippocampus. Bacterial toxins (lipopolysaccharides; typical consequences of dysbiosis) and amyloid beta (Aβ) can interact with specific pathways. These include the vagus nerve pathway, cytokine-associated immune pathways, and hormone and neurotransmitter pathways. They increase the permeability of the blood-brain barrier and intestinal barrier. This can lead to dysfunction of the hippocampus (Damiani et al., 2023).

Probiotics are key to reducing oxidative stress. They do this by producing antioxidant enzymes (catalase and superoxide dismutase), antioxidants (butyrate, folate, and glutathione), and chelating metal ions. (Peterson, 2020). Probiotics can prevent immune actions, such as inflammatory responses, by inhibiting TLR activation. Reducing inflammation can strengthen the blood-brain barrier and boost brain function. Probiotics, known for their antioxidant and anti-inflammatory effects, also aid cognitive function in depression. They achieve this by minimizing HPA axis problems and boosting monoamines and neuroplasticity. Moreover, probiotics enhance neuroplasticity by balancing gut bacteria and increasing short-chain fatty acids (Kumar et al., 2024a).

Research shows probiotics boost BDNF and enhance cognitive performance. They adjust neurotransmitters like serotonin, acetylcholine, and dopamine. This regulation affects brain activity. Prolonged probiotic use raises tryptophan levels and improves mental health (Rahimlou et al., 2022). Probiotics can help with depression by boosting serotonin-making enzymes. They also affect certain brain receptors and dopamine levels. This might change brain chemistry. For instance, *Rouxiella badensis subsp. acadiensis* raises specific mRNA and serotonin receptors in the hippocampus, potentially improving cognition. Moreover, strains like *Lactobacillus rhamnosus* can reduce anxiety and depression symptoms, thus enhancing learning (Peterson, 2020).

Probiotics support LTP processes, which strengthen synapses in the brain. LTP boosts synapse efficacy, enhancing information transfer. Research shows certain probiotics can improve learning and memory by speeding up this process (Rahimlou et al., 2022).

Probiotics might boost brain development and adaptability by influencing gut bacteria. They increase substances like short-chain fatty acids (SCFAs). These, in turn, modify genes in neurons and enhance synaptic connections by raising neurotrophic factors, such as BDNF (Bacon & Brinton, 2021). Probiotics may improve memory, learning, and neurological growth. Research highlights certain bacterial strains for their effectiveness.

Lactobacillus plantarum: May improve brain health by reducing stress levels and supporting the immune system.

Bifidobacterium longum: Has the potential to enhance memory and learning abilities.

Lactobacillus helveticus: Supports synaptic communication by reducing symptoms of anxiety and depression.

Bifidobacterium breve: May prevent age-related cognitive decline due to its neuroprotective effects (Damiani et al., 2023).

2.7. Choline and Neuroplasticity

Choline, a water-soluble essential nutrient that, while synthesized in small amounts by the body, must primarily be obtained through diet to maintain optimal levels. It is found in a variety of foods, including eggs (particularly the yolk), beef liver, chicken, fish, soybeans, and other legumes. Cruciferous vegetables like broccoli and Brussels sprouts, as well as nuts and seeds, also provide notable amounts of choline. This nutrient is critical for cellular structure and function, particularly in brain development and neuroplasticity. It plays a pivotal role in synthesizing phospholipids like phosphatidylcholine, producing acetylcholine, and regulating gene expression. Acetylcholine, a key neurotransmitter in learning and memory processes, is enhanced by increased choline levels, leading to improved synaptic plasticity and, consequently, neuroplasticity (Prabhu et al., 2021).

Choline operates through multiple mechanisms to bolster neuroplasticity by influencing the growth and connections of neurons. A choline deficiency can detrimentally affect fetal brain development, leading to cognitive impairments. For instance, insufficient choline during gestation has decreased global methylation levels in the hippocampus, thereby altering the expression of specific genes. This underscores the pivotal role of choline in gene regulation and its significance during brain development (Vallianou et al., 2024).

2.8. Coffee and Neuroplasticity

Coffee, rich in caffeine and other compounds, may support brain health. Caffeine stimulates the central nervous system and enhances neuroplasticity. This is the brain's ability to adapt structurally and functionally to environmental changes, learning, and experiences (França et al., 2020).

Caffeine boosts dopamine and serotonin by blocking adenosine receptors. Normally, adenosine slows brain activity. Caffeine lifts this barrier, improving communication and neurotransmission. This process enhances learning, memory, and cognitive functions. Specifically, dopamine drives reward and motivation. Meanwhile, serotonin stabilizes mood and emotion (Zulkifly et al., 2021).

Caffeine has been shown to enhance synaptic plasticity by supporting LTP processes. LTP contributes to strengthening the connections between neurons and strengthening learning and memory. Additionally, caffeine supports the growth and survival of neurons by increasing BDNF levels (França et al., 2020).

Moreover, the antioxidants present in coffee play a significant role in reducing oxidative stress. Oxidative stress is critical in developing neurological diseases; antioxidants protect brain health. Research suggests that regular coffee consumption can reduce the risk of degenerative diseases such as Alzheimer's and Parkinson's (Zulkifly et al., 2021).

Additionally, coffee is believed to positively influence the gut microbiota, which may support neuroplasticity through the gut–brain axis. Healthy gut microbiota is essential for neurotransmitter production and overall brain health. By improving gut health, coffee can indirectly enhance neuroplasticity (Carneiro et al., 2021).

2.9. Whole Grain vs Sugar Consumption and Neuroplasticity

Whole grains, a rich source of fiber, vitamins, minerals, and antioxidants, are crucial in supporting brain health. Micronutrients, particularly B vitamins (folate, B6, and B12), magnesium, and zinc, are involved in methylation processes that support neuroplasticity and are crucial for the methylation of DNA and histones. Increased consumption of whole grains can positively influence brain functions by enhancing insulin sensitivity and reducing inflammation. Furthermore, the impact of whole grains on the gut–brain axis is critical for brain health (Elesawy et al., 2021).

Research indicates that whole grain consumption positively affects the gut microbiome by increasing the production of anti-inflammatory compounds. This enhances the release of components that promote neuroplasticity and signaling from the gut to the brain (Damiani et al., 2023). Polyphenols in whole grains create an environment supporting neuroplasticity by reducing oxidative stress. Consuming whole grains has been shown to improve cognitive functions and support neuroplasticity. Additionally, it helps regulate blood sugar levels while maintaining insulin levels, supporting the energy metabolism of brain cells (Maharjan et al., 2020).

In contrast, excessive consumption of simple sugars can have a detrimental effect on neuroplasticity. While sugar provides a quick energy source, it can lead to fluctuations in blood sugar levels. High blood sugar can cause insulin resistance, affecting the functionality of brain cells. Moreover, high sugar intake can trigger inflammatory responses, harming brain health. Research has shown that excess sugar consumption can impair neuroplasticity in the hippocampus and worsen the negative effects on learning and memory. High sugar intake has been found to decrease BDNF levels and inhibit neuroplasticity (Beecher et al., 2021).

3. Conclusion

In summary, nutrition plays a significant role in neuroplasticity, enhancing learning and memory. Essential nutrients, such as vitamins and minerals, promote BDNF expression, which is vital for brain health. A balanced diet can mitigate deficiencies, strengthen neural connections, and support cognitive well-being.

Future research should investigate the impact of diet on neuroplasticity. Long-term studies focusing on nutritional interventions in at-risk populations are crucial. Additionally, exploring the relationship between gut bacteria and nutrient absorption may provide valuable insights. Understanding these connections will enable us to develop improved dietary strategies for brain health.

4. Takeaways

1. **Importance of Nutrition:** A balanced diet should support brain health and enhance neuroplasticity.
2. **Support for Acetylcholine:** To ensure adequate choline intake, it is recommended that eggs, meat, and dairy products be consumed.
3. **Increase in Dopamine and Serotonin:** To prevent omega-3 fatty acid deficiencies, consume fatty fish (such as salmon and sardines) at least twice a week and consider high-quality omega-3 supplements when necessary.
4. **Increasing BDNF:** Regularly consuming antioxidant-rich fruits and vegetables (such as blueberries and spinach) is recommended to support learning and memory.
5. **Probiotic Use:** It should be ensured that probiotic strains used for specific health benefits are appropriate (for example, *Lactobacillus* and *Bifidobacterium* strains).
6. **Benefits of Whole Grains:** Whole grain products (such as whole wheat, quinoa, and oats) should be included in the diet daily to support brain health.
7. **Reduction of Sugar Intake:** Limiting the high intake of simple sugars is essential to prevent fluctuations in blood sugar and help maintain cognitive functions.
8. **Reduction of Inflammation:** Consumption of healthy fats (such as olive oil, nuts, and avocados) in the diet helps reduce inflammation.

9. **Caffeine Consumption:** Drinking 1–2 cups of coffee daily is suggested as it may increase synaptic plasticity.
10. **Maintaining Gut Health:** Gut health is directly related to brain health, so it is recommended that you adopt a diet rich in fiber (from legümes, whole grains, vegetables and fruits).
11. **Vitamin D:** To maintain adequate vitamin D levels, exposure to sunlight and consumption of fortified foods such as dairy products and fatty fish are recommended. Vitamin D supplementation should be considered for individuals with limited sun exposure during winter.
12. **B Vitamins:** Consuming foods rich in B vitamins (such as leafy greens, legumes, nuts, and whole grains) supports cognitive functions and neuroplasticity.
13. **Magnesium:** Magnesium-rich foods (such as dark chocolate, nuts, seeds, and leafy greens) are recommended to support brain health and enhance learning processes.
14. **Zinc:** To improve memory and cognitive performance, it is necessary to consume zinc-rich foods (such as meat, shellfish, legumes, and seeds).
15. **Antioxidant Vitamins:** To protect the brain from oxidative stress and support neuroplasticity, fruits and vegetables rich in vitamins C and E (such as citrus fruits, blueberries, nuts, and leafy greens) are suggested.

Abbreviations

5-HT1A: 5-Hydroxytryptamine Receptor 1A
AL: Arachidonic Acid
AMPK: AMP-Activated Protein Kinase
ASD: Autism Spectrum Disorder
Aβ: Amyloid Beta
B GOS: Bifidobacterium Galactooligosaccharides
BDNF: Brain-Derived Neurotrophic Factor
CREB: cAMP Response Element-Binding Protein
DHA: Docosahexaenoic Acid
DHEA: Dehydroepiandrosterone
eCB: Endocannabinoids
EPA: Eicosapentaenoic Acid

FOS: Fructooligosaccharides
GPx: Glutathione Peroxidase
HPA: Hypothalamic-Pituitary-Adrenal
IL-6: Interleukin 6
LA: Linoleic Acid
LC PUFA: Long-Chain Polyunsaturated Fatty Acids
LTP: Long-Term Potentiation
MDD: Major Depressive Disorder
NF-κB: Nuclear Factor kappa B cells
NGF: Nerve Growth Factor
Nrf2: Nuclear Factor Erythroid 2-Related Factor 2
PLA2: Phospholipase A2
SCFA: Short-Chain Fatty Acids
SCZ: Schizophrenia
SIRT1: Sirtuin 1
SOD: Superoxide Dismutase
TNF-α: Tumor Necrosis Factor Alpha
Tregs: Regulatory T Cells

Bibliography

1. Afsharfar, Maryam, Mansour Shahraki, Mansour S., Omid Asbaghi, & Alireza Dashipour. (2021). "The effects of magnesium supplementation on serum level of brain-derived neurotrophic factor (BDNF) and depression status in patients with depression." *Clinical Nutrition ESPEN* 42:381–86. https://doi.org/10.1016/j.clnesp.2020.12.022
2. Ammar, Achraf, Khaled Trabelsi, Omar Boukhris, Bassem Bouaziz, Patrick Müller, Jordan M Glenn, Nicholas T. Bott, Notger Müller, Hamdi Chtourou, Tarak Driss, ve Anita Hökelmann. (2020). "Effects of Polyphenol-Rich Interventions on Cognition and Brain Health in Healthy Young and Middle-Aged Adults: Systematic Review and Meta-Analysis." *Journal of Clinical Medicine* 9(5):1598. https://doi.org/10.3390/jcm9051598
3. Bacon, E. R., & Roberta Diaz Brinton. (2021). "Epigenetics of the developing and aging brain: Mechanisms that regulate onset and outcomes of brain reorganization." *Neuroscience & Biobehavioral Reviews* 125:503–16. https://doi.org/10.1016/j.neubiorev.2021.02.040

4. Barón-Mendoza, Isabel, and Aliesha González-Arenas. (2022). "Relationship between the effect of polyunsaturated fatty acids (PUFAs) on brain plasticity and the improvement on cognition and behavior in individuals with autism spectrum disorder." *Nutritional Neuroscience* 25(2):387–410. https://doi.org/10.1080/1028415X.2020.1755793
5. Beecher, K., Joshua Wang, A. Jacques, Nicholas Chaaya, Fatemeh Chehrehasa, Arnauld Belmer, a. E. Bartlett. (2021). "Sucrose Consumption Alters Serotonin/Glutamate Co-Localisation Within the Prefrontal Cortex and Hippocampus of Mice." *Frontiers in Molecular Neuroscience* 14. https://doi.org/10.3389/fnmol.2021.678267
6. Bekdash, Rola A. (2024). "Epigenetics, Nutrition, and the Brain: Improving Mental Health through Diet". *International Journal of Molecular Sciences* 25(7):4036. https://doi.org/10.3390/ijms25074036
7. Calado, Caio Matheus Santos da Silva, Raul Manhães-de-Castro, Sabrina da Conceição Pereira, Vanessa da Silva Souza, Leticia Nicoly Ferreira Barbosa, Osmar Henrique dos Santos Junior, Claudia Jacques Lagranha, Pedro Alberto Romero Juárez, Luz Torner, Omar Guzmán-Quevedo, andAna Elisa Toscano. (2024). "Resveratrol Reduces Neuroinflammation and Hippocampal Microglia Activation and Protects Against Impairment of Memory and Anxiety-Like Behavior in Experimental Cerebral Palsy". *Molecular Neurobiology* 61(6): 3619–40. https://doi.org/10.1007/s12035-023-03772-3
8. Carneiro, Sofia M., M. Beatriz P. P. Oliveira and Rita C. Alves. (2021). "Neuroprotective properties of coffee: An update." *Trends in Food Science & Technology* 113:167–79. https://doi.org/10.1016/j.tifs.2021.04.052
9. Cichon, Natalia, Joanna Saluk-Bijak, Leslaw Gorniak, Lukasz Przyslo, and Michal Bijak. (2020). "Flavonoids as a Natural Enhancer of Neuroplasticity—An Overview of the Mechanism of Neurorestorative Action." *Antioxidants* 9(11):1035. https://doi.org/10.3390/antiox9111035
10. da Cunha Germano, Bianca Caroline, Lara Cristina Carlos de Morais, Francisca Idalina Neta, Amélia Carolina Lopes Fernandes, Francisco Irochima Pinheiro, Amália Cinthia Meneses do Rego, Irami Araújo Filho, Eduardo Pereira de Azevedo, José Rodolfo Lopes de Paiva Cavalcanti, Fausto Pierdona Guzen, and Ricardo Ney Cobucci. (2023). "Vitamin E and Molecular Effects in Experimental Models of Neurodegenerative Diseases". International Journal of Molecular Sciences 24(13):11191. https://doi.org/ 10.3390/ijms241311191

11. Damiani, Francesca, Sara Cornuti, and Paola Tognini. (2023). "The gut-brain connection: Exploring the influence of the gut microbiota on neuroplasticity and neurodevelopmental disorders." *Neuropharmacology* 231:109491. https://doi.org/10.1016/j.neuropharm.2023.109491
12. Elesawy, Basem H., Walaa F. Alsanie, Mubarak Ali Algahtany, Jawaher M. AL- Ashkhari, Aya K. Alyarobi, Hussein F. Sakr. (2021). "Whole and Refined Grains Change Behavior and Reduce Brain-Derived Neurotrophic Factor and Neurotrophin-3 in Rats". *Journal of Food Biochemistry* 45(8):e13867. https://doi.org/10.1111/jfbc.13867
13. Fang, Ji-Li, Yang Luo, Song-Heng Jin, Ke Yuan, and Ying Guo. (2020). "Ameliorati effect of anthocyanin on depression mice by increasing monoamine neurotransmitter and up-regulating BDNF expression." *Journal of Functional Foods* 66:103757. https://doi.org/10.1016/j.jff.2019.103757
14. França, Angela Patricia, Marissa Giovanna Schamne, Bruna Soares de Souza, Débora da Luz Scheffer, Angelica Karina Bernardelli, Thiago Corrêa, Geison de Souza Izídio, Alexandra Latini, José Eduardo da Silva-Santos, Paula M. Canas, Rodrigo A. Cunha, and Rui Daniel Prediger. 2020. "Caffeine Consumption plus Physical Exercise Improves Behavioral Impairments and Stimulates Neuroplasticity in Spontaneously Hypertensive Rats (SHR): An Animal Model of Attention Deficit Hyperactivity Disorder". *Molecular Neurobiology* 57(9):3902–19. https://doi.org/10.1007/s12035-020-02002-4
15. Fuzi, Salma Faeza Ahmad, and Su Peng Loh. 2022. "Iodine: A Critical Micronutrient in Brain Development". Ss. 49–67 *Role of Micronutrients in Brain Health*, editör W. Mohamed and T. Yamashita. Singapore: Springer.
16. Georgieff, Michael K. 2023. "Early Life Nutrition and Brain Development: Breakthroughs, Challenges and New Horizons". *Proceedings of the Nutrition Society* 82(2):104–12. https://doi.org/10.1017/S0029665122002774
17. González-Granillo, Aldo Efrain, Dino Gnecco, Alfonso Díaz, Linda Garcés-Ramírez, Fidel de la Cruz, Ismael Juarez, Julio César Morales-Medina, and Gonzalo Flores. (2022). "Curcumin induces cortico-hippocampal neuronal reshaping and memory improvements in aged mice." *Journal of Chemical Neuroanatomy* 121:102091. https://doi.org/10.1016/j.jchemneu.2022.102091

18. Jafari, F., Hamed Mohammadi, & Reza Amani. (2021). "The effect of zinc supplementation on brain-derived neurotrophic factor: A meta-analysis." *Journal of Trace Elements in Medicine and Biology* 66:126753. https://doi.org/10.1016/j.jtemb.2021.126753
19. Jazvinšćak Jembrek, Maja, Nada Oršolić, Lucija Mandić, Anja Sadžak, and Suzana Šegota. (2021). "Anti-Oxidative, Anti-Inflammatory and Anti-Apoptotic Effects of Flavonols: Targeting Nrf2, NF-κB and P53 Pathways in Neurodegeneration". *Antioxidants* 10(10):1628. https://doi.org/10.3390/antiox10101628
20. Kandezi, Niyoosha, Mahsa Mohammadi, Maryam Ghaffari, Mina Gholami, Majid Motaghinejad, and Sepideh Safari. (2020). "Novel Insight to Neuroprotective Potential of Curcumin: A Mechanistic Review of Possible Involvement of Mitochondrial Biogenesis and PI3/Akt/GSK3 or PI3/Akt/CREB/BDNF Signaling Pathways". *International Journal of Molecular and Cellular Medicine* 9(1):1. https://doi.org/10.22088/IJMCM.BUMS.9.1.1
21. Kouba, Bruna R., and Ana Lúcia S. Rodrigues. (2024). "Neuroplasticity-related effects of vitamin D relevant to its neuroprotective effects: A narrative review." *Pharmacology Biochemistry and Behavior* 245:173899. <https://doi.org/10.1016/j.pbb.2024.173899>
22. Kumar, A., Bhagavathi Sundaram Sivamaruthi, S. D., Yogesh Kumar, R. M., Bhupendra G. Prajapati, & Chaiyavat Chaiyasut. (2024a). "Probiotics as Modulators of Gut-Brain Axis for Cognitive Development." *Frontiers in Pharmacology* 15. https://doi.org/10.3389/fphar.2024.1348297
23. Kumar, S., Kumar Mishra, & Harendra Kumar. (2024b). "Vitamins and Their Role in Brain Function". Ss. 169–81 içinde.
24. Leal, M. F. C., R. I. L. Catarino, A. M. Pimenta, and M. R. S. Souto. (2020). "Roles of Metal Microelements in Neurodegenerative Diseases". *Neurophysiology* 52(1):80–88. https://doi.org/10.1007/s11062-020-09854-5
25. Li, Zefang, Xueqiang Wu, Hanzheng Li, Cong Bi, Can Zhang, Yiqing Sun, and Zhaojun Yan. (2024). "Complex Interplay of Neurodevelopmental Disorders (NDDs), Fractures, and Osteoporosis: A Mendelian Randomization Study." *BMC Psychiatry* 24(1):232. https://doi.org/10.1186/s12888-024-05693-4
26. Maharjan, Reeju, Liliana Diaz Bustamante, K. N. Ghattas, Shahbakht Ilyas, Reham Al-Refai, & Safeera Khan. (2020). "Role of Lifestyle

in Neuroplasticity and Neurogenesis in an Aging Brain." *Cureus* 12(9):e10639. https://doi.org/10.7759/cureus.10639

27. Mansour, Heba M. (2022). "Nutrition and Brain Functions in Health and Disease". Ss. 3–26 içinde *Nutrition and Psychiatric Disorders*, editör W. Mohamed and F. Kobeissy. Singapore: Springer Nature. https://doi.org/10.1007/978-981-19-5021-6_1
28. Melgar-Locatelli, Sonia, M. Carmen Mañas-Padilla, Adriana Castro-Zavala, Patricia Rivera, María del Carmen Razola-Díaz, Francisco J. Monje, Celia Rodríguez-Pérez, and Estela Castilla-Ortega. (2024). "Diet Enriched with High-Phenolic Cocoa Potentiates Hippocampal Brain-Derived Neurotrophic Factor Expression and Neurogenesis in Healthy Adult Micewith Subtle Effects on Memory". *Food & Function* 15(16):8310–29. https://doi.org/10.1039/D4FO01201A
29. Miceli, Mathieu Di, Clémentine Bosch-Bouju, and Sophie Layé. (2020). "PUFA and Their Derivatives in Neurotransmission and Synapses: A New Hallmark of Synaptopathies." *Proceedings of the Nutrition Society* 79(4):388–403. https://doi.org/10.1017/S0029665120000129
30. Mlyniec, K. (2021). "Interaction between Zinc, GPR39, BDNF, and Neuropeptides in Depression". *Current Neuropharmacology* 19(11):2012. https://doi.org/10.2174/1570159X19666210225153404
31. Moramarco, Francesca, and Peter McCaffery. (2024). "Retinoic Acid Regulation of Homoeostatic Synaptic Plasticity and Its Relationship to Cognitive Disorders." https://doi.org/10.1530/JME-22-0177
32. Mosayebi-Samani, M., Lorena Melo, D. A., Michael A. Nitsche, & Min-Fang Kuo. (2020). "Ca^{2+} channel dynamics explain the nonlinear neuroplasticity induction by cathodal transcranial direct current stimulation over the primary motor cortex". *European Neuropsychopharmacology* 38:63–72. https://doi.org/10.1016/j.euroneuro.2020.07.011
33. Namgyal, D., Sher Ali, Rachna Mehta, & Maryam Sarwat. (2020). "The neuroprotective effect of curcumin against Cd-induced neurotoxicity and hippocampal neurogenesis promotion through CREB-BDNF signaling pathway." *Toxicology* 442:152542. https://doi.org/10.1016/j.tox.2020.152542
34. Nicoletti, Vincenzo Giuseppe, Krisztián Pajer, Damiano Calcagno, Gholam Pajenda, and Antal Nógrádi. (2022). "The Role of Metals in the Neuroregenerative Action of BDNF, GDNF, NGF and Other Neurotrophic Factors". *Biomolecules* 12(8):1015. https://doi.org/10.3390/biom12081015

35. Orhan, Cemal, Mehmet Tuzcu, Patrick Brice Deeh Defo, Nurhan Sahin, Sara Perez Ojalvo, Sarah Sylla, James R. Komorowski, and Kazim Sahin. (2022). "Effects of a Novel Magnesium Complex on Metabolic and Cognitive Functions and the Expression of Synapse-Associated Proteins in Rats Fed a High-Fat Diet." *Biological Trace Element Research* 200(1):247–60. https://doi.org/10.1007/s12011-021-02619-z
36. Paiva, Igor Henrique Rodrigues de, Laís Macedo Maciel, Rodrigo Soares da Silva, Ingrid Prata Mendonça, José Roberto Botelho de Souza, and Christina Alves Peixoto. (2024). "Prebiotics modulate the microbiota–gut–brain axis and ameliorate anxiety and depression-like behavior in HFD-fed mice." *Food Research International* 182:114153. https://doi.org/10.1016/j.foodres.2024.114153
37. Park, Y., & Bruce A. Watkins. (2021). "Chapter Seven - Endocannabinoids and aging—Inflammation, neuroplasticity, mood and pain." Ss. 129–72 içinde *Vitamins and Hormones*. C. 115, *Hormones and Aging*, editör G. Litwack. Academic Press.
38. Peterson, C. Tara. (2020). "Dysfunction of the Microbiota-Gut-Brain Axis in Neurodegenerative Disease: The Promise of Therapeutic Modulation With Prebiotics, Medicinal Herbs, Probiotics, and Synbiotics." *Journal of Evidence-Based Integrative Medicine* 25:2515690X20957225. https://doi.org/10.1177/2515690X20957225
39. Pourié, G., Guéant, J. L., & Quadros, E. V. (2022). Behavioral profile of vitamin B12 deficiency: A reflection of impaired brain development, neuronal stress, and altered neuroplasticity. In *Vitamins and Hormones* (Vol. 119, pp. 377–404). Academic Press.
40. Pourié, Grégory, Nicolas Martin, Jean-Luc Daval, Jean-Marc Alberto, Rémy Umoret, Jean-Louis Guéant, and Carine Bossenmeyer-Pourié. 2020. "The Stimulation of Neurogenesis Improves the Cognitive Status of Aging Rats Subjected to Gestational and Perinatal Deficiency of B9–12 Vitamins". *International Journal of Molecular Sciences* 21(21):8008. https://doi.org/10.3390/ijms21218008
41. Prabhu, Gayathri S., Mohandas K. G. Rao, and Kiranmai S. Rai. (2021). "Hippocampal neural cell degeneration and memory deficit in high-fat diet-induced postnatal obese rats– exploring the comparable benefits of choline and DHA or environmental enrichment." *International Journal of Neuroscience* 131(11):1066–77. https://doi.org/10.1080/00207454.2020.1773819

42. Rahimlou, Mehran, Seyed Ahmad Hosseini, Nastaran Majdinasab, Mohammad Hosein Haghighizadeh, and Durdana Husain. (2022). "Effects of long-term administration of Multi-Strain Probiotic on circulating levels of BDNF, NGF, IL-6, and mental health in patients with multiple sclerosis: a randomized, double-blind, placebo-controlled trial". *Nutritional Neuroscience* 25(2):411–22. https://doi.org/10.1080/1028415X.2020.1758887.
43. Romo-Araiza, Alejandra, and Antonio Ibarra. (2020). "Prebiotics and probiotics as a potential therapy for cognitive impairment." *Medical Hypotheses* 134:109410. https://doi.org/10.1016/j.mehy.2019.109410
44. Salami, Alireza, Goran Papenberg, Rouslan Sitnikov, Erika J. Laukka, Jonas Persson, and Grégoria Kalpouzos. (2021). "Elevated neuroinflammation contributes to the deleterious impact of iron overload on brain function in aging." *NeuroImage* 230:117792. https://doi.org/10.1016/j.neuroimage.2021.117792
45. Shanmugam, Haripriya, Subhasish Ganguly, and Badma Priya. (2022). "Plant Food Bioactives and Its Effects on Gut Microbiota Profile Modulation for Better Brain Health and Functioning in Autism Spectrum Disorder Individuals: A Review." *Food Frontiers* 3(1):124–41. https://doi.org/10.1002/fft2.125
46. Tarnacka, Beata, Anna Jopowicz, and Maria Maślińska. (2021). "Copper, Iron, and Manganese Toxicity in Neuropsychiatric Conditions." *International Journal of Molecular Sciences* 22(15):7820. https://doi.org/10.3390/ijms22157820
47. Tejada, Silvia, Fiorella Sarubbo, Manuel Jiménez-García, Margarida R. Ramis, Margalida Monserrat-Mesquida, Maria Magdalena Quetglas-Llabrés, Xavier Capó, Susana Esteban, Antoni Sureda, and David Moranta. (2024). "Mitigating Age-Related Cognitive Decline and Oxidative Status in Rats Treated with Catechin and Polyphenon-60". *Nutrients* 16(3):368. https://doi.org/10.3390/nu16030368
48. Terracina, Sergio, Carla Petrella, Silvia Francati, Marco Lucarelli, Christian Barbato, Antonio Minni, Massimo Ralli, Antonio Greco, Luigi Tarani, Marco Fiore, and Giampiero Ferraguti. (2022). "Antioxidant Intervention to Improve Cognition in the Aging Brain: The Example of Hydroxytyrosol and Resveratrol". *International Journal of Molecular Sciences* 23(24):15674. https://doi.org/10.3390/ijms232415674
49. Tu, W., Meiying Song, & Xiang Fan. (2023). "Does Resveratrol Improve Cognition in Humans? A Scientometric Study to an in-Depth

Review". *CNS Neuroscience & Therapeutics* 29(9):2413–29. https://doi.org/10.1111/cns.14276
50. Unno, Keiko, & Yoriyuki Nakamura. (2021). "Green Tea Suppresses Brain Aging". *Molecules* 26(16):4897. https://doi.org/ 10.3390/molecules26164897
51. Urrutia, Pamela J., Daniel A. Bórquez, and Marco Tulio Núñez. (2021). "Inflaming the Brain with Iron". *Antioxidants* 10(1):61. https://doi.org/10.3390/antiox10010061
52. Vaghef-Mehrabani, Elnaz, Roya Harouni, Maryam Behrooz, Fatemeh Ranjbar, Mohammad Asghari-Jafarabadi, and Mehrangiz Ebrahimi-Mameghani. (2023). "Effects of Inulin Supplementation on Inflammatory Biomarkers and Clinical Symptoms of Women with Obesity and Depression on a Calorie-Restricted Diet: A Randomised Controlled Clinical Trial." *British Journal of Nutrition* 129(11):1897–1907. https://doi.org/10.1017/S000711452200232X
53. Vallianou, Natalia G., Dimitris Kounatidis, Sotiria Psallida, Fotis Panagopoulos, Theodora Stratigou, Eleni Geladari, Irene Karampela, Dimitrios Tsilingiris, and Maria Dalamaga. (2024). "The Interplay Between Dietary Choline and Cardiometabolic Disorders: A Review of Current Evidence." *Current Nutrition Reports* 13(2):152–65. https://doi.org/10.1007/s13668-024-00521-3
54. Vauzour, David, Catarina Rendeiro, Alfonsina D'Amato, Pierre Waffo-Téguo, Tristan Richard, Jean Michel Mérillon, Matthew G. Pontifex, Emily Connell, Michael Müller, Laurie T. Butler, Claire M. Williams, and Jeremy P. E. Spencer. (2021a). "Anthocyanins Promote Learning through Modulation of Synaptic Plasticity Related Proteins in an Animal Model of Ageing." *Antioxidants* 10(8):1235. https://doi.org/10.3390/antiox10081235
55. Vauzour, David, Catarina Rendeiro, Alfonsina D'Amato, Pierre Waffo-Téguo, Tristan Richard, Jean Michel Mérillon, Matthew G. Pontifex, Emily Connell, Michael Müller, Laurie T. Butler, Claire M. Williams, and Jeremy P. E. Spencer. (2021b). "Anthocyanins Promote Learning through Modulation of Synaptic Plasticity Related Proteins in an Animal Model of Ageing." *Antioxidants* 10(8):1235. https://doi.org/10.3390/antiox10081235
56. Xu, Pei, Dejian Pang, Jiayi Zhou, Sihui Li, Dunjin Chen, and Bolan Yu. (2021). "Behavioral Changes and Brain Epigenetic Alterations Induced by Maternal Deficiencies of B Vitamins in a Mouse Model." *Psychopharmacology* 238(4):1213–22. https://doi.org/ 10.1007/s00213-021-05766-2

57. Yook, J. Soo, & Minchul Lee. (2020). "Potential Role of Phytochemicals in Brain Plasticity: Focus on Polyunsaturated Fatty Acids." *Physical Activity and Nutrition* 24(1):14. https://doi.org/10.20463/pan.2020.0003
58. Zaa, César A., Álvaro J. Marcelo, Zhiqiang An, José L. Medina-Franco, and Marco A. Velasco-Velázquez. 2023. "Anthocyanins: Molecular Aspects on Their Neuroprotective Activity". *Biomolecules* 13(11):1598. https://doi.org/10.3390/biom13111598
59. Zarei, Shahab A., Mina Shahriari-Khalaji, Ian Max Andolina, and Gila Behzadi. (2023). "Antinociceptive effects of vitamin B-complex: A behavioral and histochemical study in rats." *IBRO Neuroscience Reports* 15:270–80. https://doi.org/10.1016/j.ibneur.2023.09.005
60. Zulkifly, Mohd Faizal Mohd, Ornela Merkohitaj, Jürgen Brockmöller, and Walter Paulus. (2021). "Confounding effects of caffeine on neuroplasticity induced by transcranial alternating current stimulation and paired associative stimulation." *Clinical Neurophysiology* 132(6):1367–79. https://doi.org/10.1016/j.clinph.2021.01.024

Nutritional Neuroscience: Dietary Influences on Brain Plasticity

Yonca Sevim

1. Introduction

Neuroplasticity allows the brain to adapt and change throughout life. It is vital for recovery, learning, and improving memory (de Oliveira, 2020). Lifestyle factors like exercise, diet, and sleep significantly impact it. Nutrition is key, providing essential nutrients that boost brain function and neuroplasticity (Pickersgill et al., 2022).

Adequate and balanced nutrition is crucial for physical and mental health at all life stages, from prenatal development to old age. Key nutrients include proteins, carbohydrates, iron, copper, zinc, long-chain polyunsaturated fatty acid (PUFA), iodine, and vitamins A, B6, D, and C (Cusick & Georgieff, 2016). Functional foods like omega-3 fatty acids, catechin polyphenols, and curcumin support cognition and protect the brain (McCann & Ames, 2005; Murphy et al., 2014) (*See Chapter 1 Neuroplasticity and Nutrition: Neuro Foods*).

Various diets, such as Calorie Restriction, Intermittent Fasting, Ketogenic, and Mediterranean, promote brain plasticity. This review focuses on how Calorie Restriction, Ketogenic Diets, and Mediterranean Diet affect neuroplasticity.

2. Calorie Restriction

Calorie restriction (CR) means lowering calorie intake without malnutrition. It is the top non-drug, non-gene method to boost health and lifespan in many species, including humans (Almendáriz-Palacios et al., 2020). CR diets include continuous and intermittent energy restrictions, like intermittent fasting (IF) and time-restricted feeding. Figure 1 shows various CR and fasting methods in clinical trials.

Initially, CR diets focused on metabolic health. Recently, research has started to explore their impact on brain health and neuroplasticity, especially regarding neurological diseases.

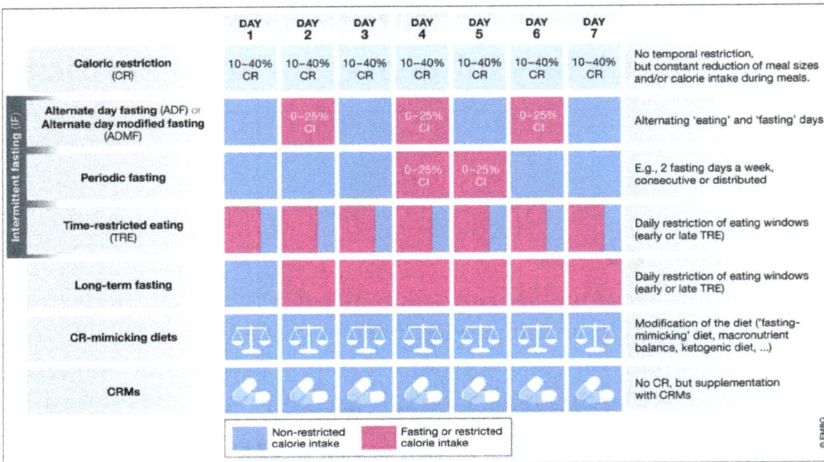

Figure 1: Different concepts of caloric restriction and fasting applications that have transitioned to clinical trials (Hofer et al., 2022). *Blue: non-restricted calorie intake, magenta: fasting or restricted calorie intake, CI: calorie intake, CRM: CR mimetic*

Caloric restriction of 20–40% can extend the lifespan of various species (Fusco & Pani, 2013). Studies from Denmark and Norway showed that involuntary CR for 2–4 years cuts mortality rates (Strom & Jensen, 1951). In Okinawa, centenarians ate 17% fewer calories than typical Japanese adults and 40% fewer than Americans (Suzuki et al., 2001). Moreover, a 30% calorie cut over three months improved verbal memory by 20% in older adults (Witte et al., 2009). Most et al. reviewed and found that human CR triggers metabolic and molecular changes similar to longevity models, thus confirming its health benefits (Most et al., 2017).

CR works in various ways (see Figure 2). During long fasts, the liver runs out of glycogen. Then, the body turns fat into ketones for energy. This switch improves brain function and stress resilience (Mattson et al., 2018). CR boosts DNA repair, cuts oxidative stress, and enhances glucose use. It also strengthens the immune system and reduces age-related gene changes. This is done by raising brain-derived neurotrophic factor (BDNF) levels, which aids hippocampal function (Di Francesco et al., 2018). Similarly, IF protects the brain by improving synaptic strength and increasing neurotrophic factors, especially during injuries (Phillips, 2017).

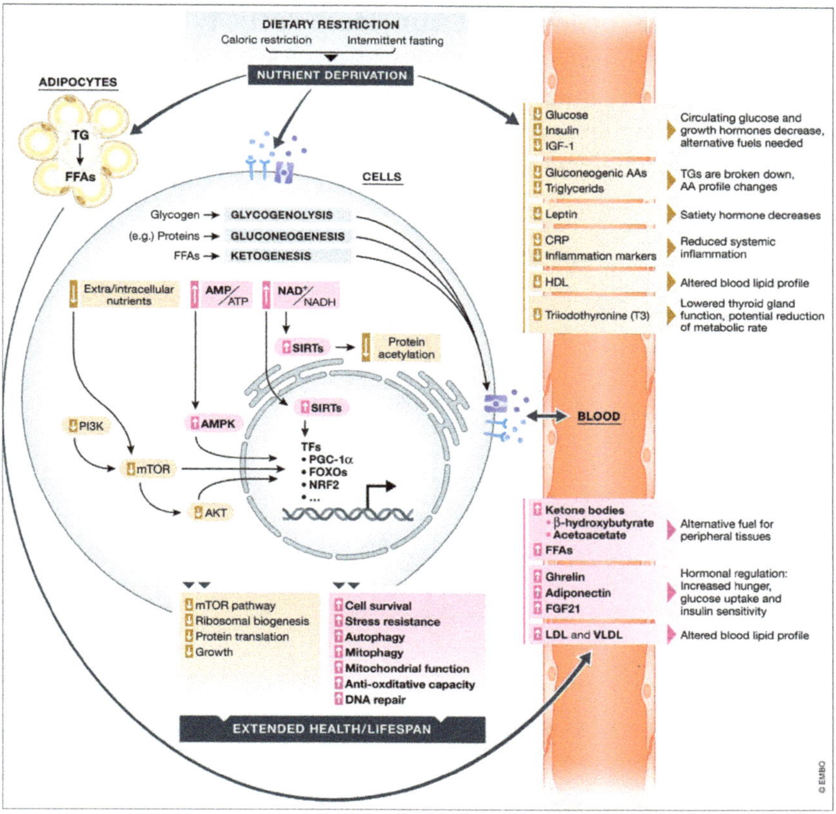

Figure 2: Key molecular events during CR and fasting at the cellular and metabolic levels (Hofer et al., 2022).

AKT: protein kinase B/PKB, AMPK: AMP-activated protein kinase, AMP/ATP: ratio of adenosine monophosphate to adenosine triphosphate, BCAA: branched-chain amino acid, FFA: free fatty acid, FGF21: fibroblast growth factor 21, FOXO: forkhead box O, IGF-1: insulin-like growth factor, (V)LDL : (very) low-density lipoprotein, mTOR: mechanistic target of rapamycin, NAD+/NADH: ratio of oxidized to reduced nicotinamide adenine dinucleotide, NRF2: nuclear factor erythroid 2-related factor 2, PARP1: Poly [ADP-ribose] polymerase 1, PGC-1α: peroxisome proliferation-activated receptor gamma co-activator 1α, TG: triglycerides.

2.1. Mechanisms of Action of CR on Neuroplasticity

CR's effects on neuroplasticity can be explained through four basic mechanisms: Insulin Signaling and Neuroprotection, Autophagy and Cellular Maintenance, Neuroinflammation Modulation, and Synaptic Plasticity and Cognitive Function (Figure 3).

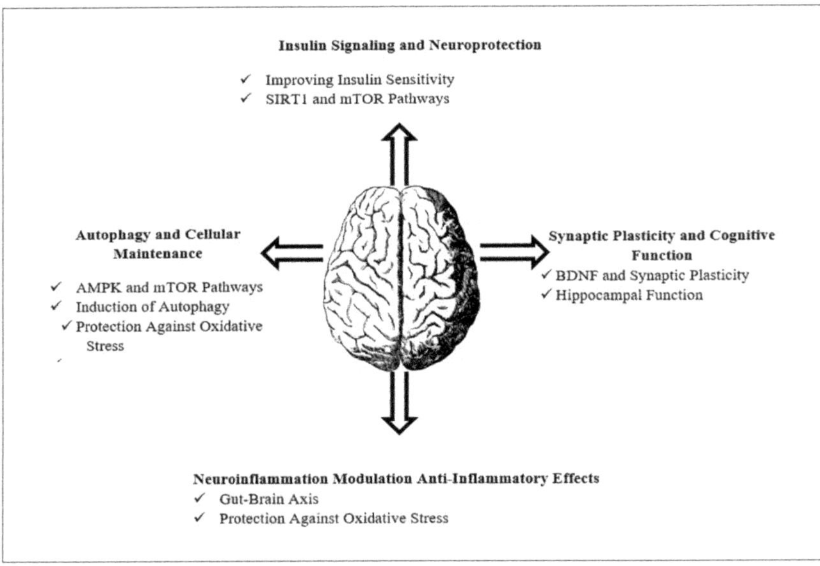

Figure 3: The basic mechanism of CR on neuroplasticity

SIRT1: sirtuin 1, mTOR: mammalian target of rapamycin, AMPK: AMP-activated protein kinase, BDNF: Brain-Derived Neurotrophic Factor

2.1.1. Insulin Signaling and Neuroprotection

Improving Insulin Sensitivity: Insulin is vital for brain development and memory. Impaired insulin signaling can lead to cognitive decline (Bloemer et al., 2018). Insulin crosses into the brain, aiding in protein balance and function. Poor insulin release speeds up brain aging by blocking harmful protein removal (Sun et al., 2020). It also contributes to vascular dementia by affecting blood flow, fat breakdown, and inflammation (Kellar & Craft, 2020). In mice, a low-calorie diet improved insulin sensitivity, reduced stress, and protected memory from high-sugar, high-fat diets. It also reversed some brain damage

(Kim et al., 2016; Ma et al., 2018). CR diets are effective for boosting insulin sensitivity and brain glucose use, vital for energy and cognitive function (Zhang et al., 2023). CR diets improve insulin signaling, which is crucial for brain health, especially in older adults with insulin resistance (Hofer et al., 2022).

SIRT1 and mTOR Pathways: Seven mammalian sirtuins (SIRT) exist, including SIRT1. SIRT1 is an NAD^+-dependent deacetylase. It regulates metabolism and aging. Thus, it protects against brain injury, neurodegenerative diseases, and cerebrovascular damage (Yang et al., 2020). SIRT1 also manages stem cell growth, reduces neuronal damage, and supports vascular cell survival. It helps in conditions like dementia and retinopathy (Maiese, 2018). Furthermore, SIRT1 fights insulin resistance from high-fat diets and boosts insulin sensitivity in the hypothalamus (Schertzer & Lam, 2021). In rodent studies, a 14-day calorie-restricted diet raised SIRT1 levels, enhancing tolerance to cerebral ischemia (Ran et al., 2015). In a study with mice, it has been shown that activation of SIRT1 by CR promotes neuroprotection through the inhibition of the mammalian target of the rapamycin (mTOR) pathway, which is involved in cell growth and survival. This regulation indicates that the SIRT1/mTOR signaling pathway plays a role in the neuroprotective effects of dietary restriction (Ma et al., 2015).

CR activates SIRT1, boosting mitochondrial function and energy in neurons. It also enhances neuroplasticity. Furthermore, SIRT1 aids synaptic plasticity by increasing BDNF expression. BDNF is vital for neuronal growth and synaptic function (Hofer et al., 2022).

2.1.2. Autophagy and Cellular Maintenance

AMPK and mTOR Pathways: Adenosine 5'-Monophosphate (AMP)-Activated Protein Kinase (AMPK) and mTOR are two crucial regulators of cellular metabolism that often work in opposition to maintaining energy balance. AMPK boosts energy production, while mTOR supports growth and protein synthesis (Herzig and Shaw, 2018). Like caloric restriction or fasting, AMPK activates during low energy (Steneberg et al., 2018). It conserves energy by blocking mTOR and promoting fat breakdown and cell recycling. In contrast, mTOR kicks in with plenty of energy, driving growth and replication. When energy is tight, AMPK turns off mTOR by targeting its regulatory proteins, halting protein synthesis (Hofer et al., 2022). This balance is vital for understanding how calorie restriction and low-energy diets affect brain health and adaptability.

During CR, AMPK activates and inhibits mTOR. This shifts cells from growth to maintenance. It preserves cell health and extends lifespan, notably in neurons. AMPK triggers autophagy, recycling damaged components (Green et al., 2022). In the brain, it clears harmful proteins linked to neurodegenerative diseases. This process boosts neuron health and supports better cognitive function, combating age-related decline (Liu et al., 2019; Corti et al., 2020). Additionally, AMPK prevents excessive changes in synapses while allowing necessary ones. This balance is crucial for brain health. Overactive mTOR causes stress and can lead to neurodegeneration. AMPK ensures neurons use energy efficiently, especially during CR (Hofer et al., 2022).

Induction of Autophagy: Fasting and CR boost neuronal autophagy, which is vital for cell upkeep and clearing damaged components (Bagherniya et al., 2018). Autophagy issues trigger protein buildup, mitochondrial problems, and brain inflammation, leading to neurodegeneration (Chandrasekaran et al., 2023). This is crucial in neurons, as damaged proteins can cause diseases (Sun et al., 2020). CR enhances autophagy, protecting brain networks, preventing Alzheimer's and Parkinson's, and aiding synaptic changes.

Protection Against Oxidative Stress: Oxidative stress occurs when there are too many oxidants compared to antioxidants. This often occurs due to excess reactive oxygen species (ROS) or weak antioxidant systems. The brain is especially vulnerable due to its high oxygen use and fat content. Too much ROS can harm neurons and lead to neurodegenerative disorders by causing protein buildup and cell damage (Sahab et al., 2023).

CR diets lower oxidative stress by improving mitochondrial function and boosting antioxidant levels. This reduces ROS. SIRT3 activates manganese superoxide dismutase (SOD2), helping to combat oxidative stress. Meanwhile, nuclear factor erythroid 2-related factor 2 (Nrf2), a key factor in this defense, boosts detoxifying and antioxidant enzymes. Disrupting the Nrf2 signaling pathway raises the brain's risk of damage, protein buildup, and nerve cell death (Zhang et al., 2022). CR diets enhance mitochondrial efficiency and cut down ROS. This supports brain energy needs, which is crucial for preventing damage. They also protect synaptic function and prevent protein aggregation and nerve cell death.

2.1.3. Neuroinflammation Modulation

Anti-Inflammatory Effects: Chronic inflammation is common in many neurological diseases (Furman et al., 2019). CR significantly lowers inflammatory markers like interleukin (IL)-1, IL-8, and tumor necrosis factor α (TNFα) in animals and humans, especially over long periods. This reduction in inflammation, particularly in TNFα and C-reactive protein (CRP), boosts insulin sensitivity and longevity, with the best results seen after 24 months on CR. However, the age at which CR starts and initial inflammation levels can affect its impact (Flanagan et al., 2020). In adult mice, a CR diet sparked autophagy, aiding remyelination and cutting down pro-inflammatory cytokines in blood and the sciatic nerve (Schoffen et al., 2014).

CR has shown that it can reduce obesity-related inflammation. It does this by improving glucose use and lowering inflammatory responses. In cases like ischemic stroke, which often sees high inflammation due to obesity, CR helps. It prevents weight gain and boosts recovery by managing inflammation (Bonaventura et al., 2016). Studies prove CR is safe and helps with weight loss, metabolic health, and emotional well-being in multiple sclerosis patients (Fitzgerald et al., 2018). CR also reduces inflammation by boosting fat breakdown genes and blocking fat creation in fat tissue. Adjusting specific pathways lowers markers like TNFα and nuclear factor kappa-light-chain-enhancer of activated B cells (NF-κB) (Shen et al., 2017). In animal tests, CR cut down obesity and insulin resistance. It also changed gut bacteria and improved glucose use, further reducing inflammation (Li et al., 2017). So, CR aids in managing obesity-induced inflammation by improving metabolism, protecting brain cells, and ensuring a healthy brain environment.

Gut–Brain Axis: CR diets improve gut health, reducing neuroinflammation. This, in turn, benefits brain function. For instance, a 4-week CR diet led to weight loss, better gut integrity, and less inflammation in obese women. However, these benefits vanished once they returned to regular diet (Ott et al., 2017). Moreover, gut bacteria can cause inflammation and insulin resistance. Increasing SIRT1 in the small intestine enhances insulin sensitivity. It does this by tackling insulin resistance linked to high-fat diets. However, such diets disrupt gut bacteria and worsen insulin resistance (Schertzer & Lam, 2021). Changes in bile acids can increase inflammation and affect brain function. Certain bacteria link gut imbalances to neurodegenerative diseases through bile acids (Yeo et al., 2023). Overall, gut metabolites and factors impact the liver, pancreas, and brain, affecting the gut–brain axis.

2.1.4. Synaptic Plasticity and Cognitive Function

BDNF and Synaptic Plasticity: The link between IF and BDNF in boosting cognitive function has garnered much interest. BDNF is crucial for synaptic plasticity, allowing the brain to form new connections. CR boosts BDNF levels by reducing oxidative stress and inflammation. It also activates key pathways related to metabolism and stress (Alkurd et al., 2024). Notch 1 and BDNF interaction is vital for neural development and synaptic function. Interestingly, IF can enhance the growth of new neurons in the hippocampus via the Notch 1 pathway (Baik et al., 2020). These pathways increase BDNF and promote neuron growth, synaptic plasticity, and cell survival.

In various animal studies, CR diets boosted BDNF levels in the brain, particularly in the hippocampus, vital for learning and memory. CR also countered cognitive decline in obese rats by increasing BDNF and its receptor, thanks to its antioxidant properties (Kishi et al., 2015). Additionally, CR diets improved synaptic plasticity and cognitive function by raising BDNF levels (de Carvalho et al., 2021). This is crucial for aging and neurodegenerative diseases, where synaptic plasticity often suffers. These studies underline how diet affects BDNF, thus enhancing cognition and neuroplasticity. However, CR and IF show different effects on BDNF and cognition in healthy people, the overweight, and those with metabolic conditions. Some studies suggest IF might boost BDNF, but results are mixed and debated (Alkurd et al., 2024). The long-term effects of IF on BDNF and cognition in humans remain unclear.

Hippocampal Function: New neurons form throughout life in the mammalian brain. This process, called adult hippocampal neurogenesis, is vital for neuroplasticity, learning, and memory. CR boosts this neurogenesis and memory, linked to the hormone ghrelin from the stomach (Roberts et al., 2023). Recent research shows that acyl-ghrelin raises BDNF mRNA in the hippocampus's granule cells (Hornsby et al., 2020). This increase aligns with acyl-ghrelin activating receptors on hippocampal neurons. It promotes dendritic spine growth, enhances synaptic plasticity, and aids long-term potentiation. All these are critical for learning and memory. Thus, acyl-ghrelin is key for neuroplasticity and cognitive function.

CR diet can boost insulin signaling, protecting memory in mice on high-sugar, high-fat diets (Kim et al., 2016). Moreover, it enhances spatial learning by increasing AMPK and glucose transporter type 4 (GLUT4) levels (Ma et al., 2015). A study on 43 obese middle-aged and older adults explored the effects of intermittent and continuous energy restriction on memory linked to neurogenesis. It found that energy restriction improved memory, especially in those with cognitive decline (Kim et al., 2020). This suggests CR diets could benefit hippocampal function and slow cognitive decline. However, Roberts et al. (2023) noted that different dietary restrictions affect neurogenesis differently. They call for more research to find the best diets for cognitive health in aging.

CR diets can boost brain health and neuroplasticity, especially in neurological diseases. They work by affecting insulin, autophagy, inflammation, and synaptic plasticity. This protection helps against age-related decline and neurodegenerative diseases. More research is needed on long-term effects and best practices. However, current evidence shows that CR diets could improve brain health and cognitive function.

3. Ketogenic Diets

The ketogenic diet has been widely studied for treating neurological disorders. Initially, it was developed for epilepsy. Now, it also shows promise for neurodegenerative diseases, mood disorders, and cognitive enhancement (Barry et al., 2018). The diet benefits brain health, especially for conditions like epilepsy, Alzheimer's, and Parkinson's (Jensen et al., 2020). Figure 4 illustrates the complex mechanisms by which the ketogenic diet may affect neurological diseases.

KD is rich in fats and low in carbohydrates and moderate in protein. It creates ketone bodies like β-hydroxybutyrate (BHB), acetoacetate, and acetone. This process pushes the body into ketosis. In ketosis, the body burns fat for energy instead of carbohydrates. A typical KD includes 70–75% fat, less than 50 grams of carbs, and about 1.0–1.7 grams of protein per kilogram of body weight daily. The body uses fat and ketones for energy during ketosis, avoiding glucose. Fasting or very low-calorie diets can also trigger ketosis, but they do not reach nutritional ketosis. Nutritional ketosis is only found in high-fat diets. KD aims to mimic fasting effects, enhancing fat burning without starvation risks (Dyńka et al., 2022).

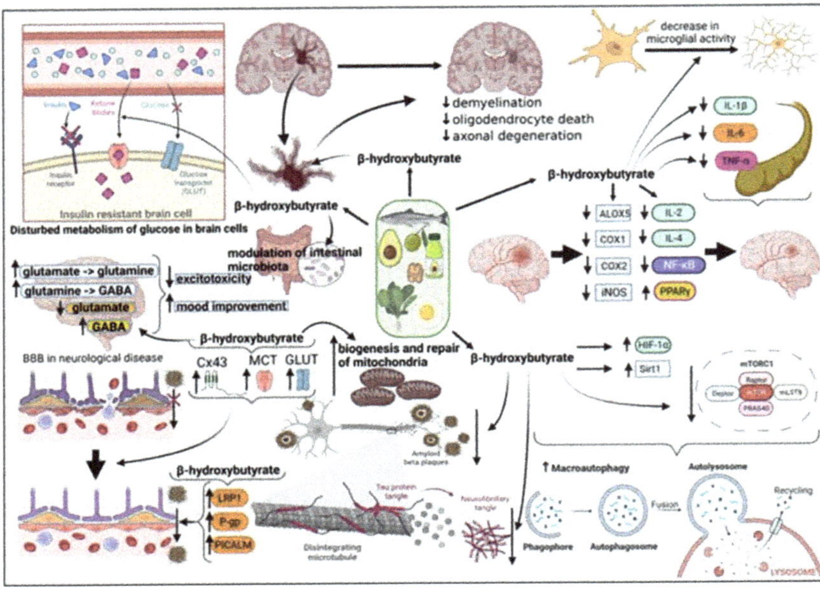

Figure 4: Potential mechanisms of the ketogenic diet effect in neurological diseases (Dyńka et al., 2022).

GABA: Gamma-Aminobutyric Acid, BBB: blood–brain barrier, Cx43: connexin-43, MCT: monocarboxylate transporters, GLUT: glucose transporters, LRP1: LDL receptor-related protein 1, P-gp: glycoprotein P, PICALM: phosphatidylinositol binding clathrin assembly protein, IL-1β: interleukin-1β, IL-6: interleukin-6, TNF-α: tumor necrosis factor α, ALOX5: arachidonate 5-lipoxygenase gene, COX1: cyclooxygenase 1, COX2: cyclooxygenase 2, iNOS: inducible nitric oxide synthase, IL-2: interleukin-2, IL-4: interleukin 4, NF-κB: nuclear factor kappa-light-chain-enhancer of activated B cells, PPARγ: peroxisome proliferator-activated receptor γ, HIF-1α: hypoxia-induced factor 1α, Sirt1: sirtuin 1, mTORC1: mammalian target of rapamycin complex 1, mLST8: mammalian lethal with SEC13 protein 8, PRAS40: proline-rich Akt substrate of 40 kDa, mTOR: mammalian target of rapamycin.

3.1. Mechanisms of Action of KD on Neuroplasticity

KD exerts a range of mechanisms that positively impact neuroplasticity and brain health, including enhanced mitochondrial function, reduced inflammation, increased BDNF expression, promotion of neurogenesis, regulation of glutamate and gamma-aminobutyric acid (GABA) balance, and activation of autophagy, all of which contribute to its neuroprotective effects in various neurological conditions (Figure 5).

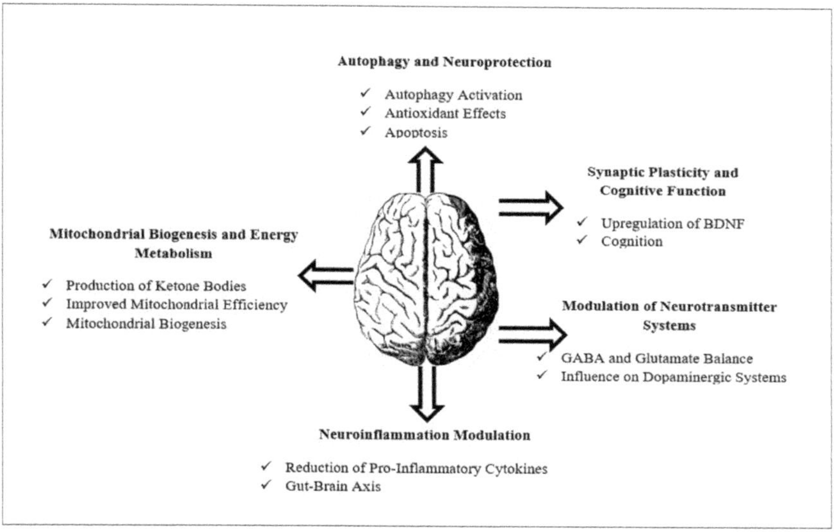

Figure 5: The primary mechanism of KD on neuroplasticity
GABA: Gamma-Aminobutyric Acid, BDNF: Brain-Derived Neurotrophic Factor

3.1.1. Mitochondrial Biogenesis and Energy Metabolism

Production of Ketone Bodies: The ketogenic diet shifts energy from glucose to ketone bodies, like BHB, acetone, and acetoacetate. These are produced from the oxidation of fatty acids. They can cross the blood–brain barrier. This offers an alternative fuel for the brain, beneficial in conditions like Alzheimer's disease (Dyńka et al., 2022).

Improved Mitochondrial Efficiency: Ketone bodies, such as BHB, boost mitochondrial respiration. This increases energy and lowers oxidative stress in neurons. It helps maintain brain function and neuron survival under stress (Svart et al., 2018; Koppel et al., 2023).

Mitochondrial Biogenesis: KD boosts mitochondrial growth by activating the PGC1α-SIRT3-UCP2 pathway. This pathway is vital for improving mitochondrial function and shielding cells from oxidative stress.

Peroxisome proliferator-activated receptor gamma coactivator 1-alpha (PGC1α), a key player, ramps under KD, spurring the creation of proteins needed for energy and replication in mitochondria. This is crucial for energy-hungry cells, like neurons. KD also raises SIRT3 levels, a protein that aids energy production and manages mitochondrial proteins, including those that combat oxidative stress. Furthermore, KD increases uncoupling protein 2 (UCP2) levels, which lowers mitochondrial membrane potential and only ROS abbreviation will be written production. This action cuts down oxidative damage and enhances mitochondrial health. Research shows KD significantly raises UCP2, improving mitochondrial function by curbing excessive ROS. This process ensures a balance, preventing oxidative stress while boosting mitochondrial numbers. Thus, KD enhances mitochondrial health and energy, offering protective benefits for the brain. This has been proven in animal studies and cell experiments, highlighting KD's potential in treating mitochondrial dysfunctions (Hasan-Olive et al., 2019).

Additionally, the KD boosts mitochondrial biogenesis, especially in skeletal muscle, potentially improving muscle function as we age. In studies with mice, KD raised levels of PGC1α, a key regulator for mitochondrial growth, and proteins involved in energy transfer. However, this increase in mitochondrial mass and activity was significant only in skeletal muscle, not in the brain or liver. Elevated protein acetylation levels accompany this increase in skeletal muscle mitochondrial mass, and manipulation of acetylation, for example, through the histone deacetylase (HDAC) inhibitors, similarly boosts mitochondrial mass and fatty acid oxidation. KD might trigger myocyte enhancer factor-2 (MEF2), which raises PGC1α2/3 and boosts mitochondria. Its primary ketone, BHB, could also activate MEF2. Additionally, KD increases AMPK and p38 mitogen-activated protein kinase activity, promoting mitochondrial growth (Pathak & Baar, 2023). However, these results come from animal studies. We still do not know how KD affects humans. Long-term KD might cause heart problems. In rats, KD and extended fasting reduced mitochondrial growth, impaired cellular respiration, and increased heart cell death, leading to cardiac fibrosis (Xu et al., 2021). More studies are needed to see if humans benefit similarly.

3.1.2. Autophagy and Neuroprotection

Autophagy Activation: KD and its key ketone body, BHB, enhance autophagy. This process clears damaged cells and keeps neurons healthy. In low-glucose conditions, KD increases autophagy. It helps remove toxic proteins, like amyloid-β and tau, linked to Alzheimer's and other neurodegenerative diseases. Mechanistically, BHB activates the SIRT1-AMPK pathway, promoting autophagy by phosphorylating key proteins like unc-51-like autophagy-activating kinase (ULK1) while inhibiting mTORC1, a known autophagy suppressor. Additionally, KD upregulates the transcription factor EB (TFEB), a master regulator of lysosomal biogenesis, increasing lysosomal function and promoting the clearance of damaged organelles and protein aggregates. These effects support the neuroprotective role of KD. They reduce neuronal loss and help brain health, especially after ischemic stroke and hypoglycemia (Jang et al., 2023).

Antioxidant Effects: Ketone bodies are well-documented for their antioxidant properties and ability to reduce oxidative stress, particularly in protecting the brain (Puchalska & Crawford, 2017). They help lower oxidative stress in the brain. For example, BHB reduces ROS and boosts antioxidant enzymes like superoxide dismutase (SOD). Many studies associate BHB with increased expression of antioxidant enzymes, such SOD, and decreased oxidative stress, as it inhibits ROS and superoxide generation, prevents lipid and protein oxidation, increases antioxidant protein levels, and enhances mitochondrial respiration and ATP synthesis (Puchalska & Crawford, 2017). This reduction in oxidative stress protects neurons from damage and supports overall brain health.

Apoptosis: Oxidative stress causes neuron loss via apoptosis. Mitochondria drive this process. KD reduces oxidative stress and cell and animal apoptosis. It boosts the coenzyme Q/QH2 ratio, lowering mitochondrial ROS. Ketone bodies also help. They reduce mitochondrial ROS, prevent the mitochondrial permeability transition pore (mPTP) opening, and stop cytochrome C release. Moreover, KD may enhance mitochondrial function, protecting against conditions like Alzheimer's and Parkinson's (Jang et al., 2023).

3.1.3. Neuroinflammation Modulation

Reduction of Pro-Inflammatory Cytokines: KD influences central and peripheral inflammation (Dyńka et al., 2022). It also reduces microglial activation and lowers pro-inflammatory markers like IL-1β, IL-6, and TNFα in the hippocampus. This is achieved by activating peroxisome proliferator-activated receptor γ (PPARγ), downregulating inflammatory signaling, including pathways like cyclooxygenase 2 (COX2) and NF-κB. This reduction in inflammation helps to protect neurons. It guards against neurodegenerative diseases (Jeong et al., 2011).

Gut–Brain Axis: The gut microbiota is critical in regulating host metabolism, cognition, and neuroinflammation by producing metabolites like short-chain fatty acids (SCFAs), lactate, and neurotransmitters such as GABA, dopamine, and serotonin. KD improves gut dysbiosis by increasing the number of beneficial bacteria and reducing those that cause inflammation while restoring the balance between Firmicutes and Bacteroidetes. The metabolites produced by bacteria, such as SCFAs, neurotransmitters, hydrogen sulfide, and lipopolysaccharide, affect the permeability of the gut–brain barrier and the production of pro-inflammatory cytokines, contributing to the diet's anti-neuroinflammatory effects. However, some studies suggest that KD may negatively impact gut microbiota diversity due to its low carbohydrate content, as many microbes rely on fiber for energy. Moreover, the neuroprotective effects of the KD through the gut microbiome may vary depending on individual factors such as sex, age, and race, indicating the need for further research to understand these mechanisms fully (Jang et al., 2023).

3.1.4. Synaptic Plasticity and Cognitive Function

Upregulation of BDNF: KD increases levels of BDNF. BDNF is vital for myelin repair, learning, and memory. KD enhances BDNF by producing ROS, activating NF-κB, and boosting histone acetyltransferase activity (Marosi et al., 2016).

Cognition: KD may improve cognition by lowering inflammation and stress and enhancing mitochondrial function. Most evidence comes from animal studies on epilepsy, traumatic brain injury (TBI), Alzheimer's, and aging. KD notably enhances cognitive function in aging models. It improves spatial learning and memory in older mice. This is linked to better mitochondrial function in the hippocampus and prefrontal cortex. In TBI models, KD benefits cognition and recovery. Mice on KD show better memory and less brain damage. Younger rats benefit more from KD after TBI than older ones. In Alzheimer's models, results vary. KD reduces amyloid-beta plaques and inflammation but often does not improve cognition. Some studies show memory improvements in female mice with Alzheimer's, while others do not. Findings in healthy animals suggest that KD's effects on cognition are age and species-dependent. Overall, KD shows promise for cognitive improvement in epilepsy, TBI, and aging. Its effects on Alzheimer's and in healthy animals are unclear. Human studies are limited. More research is needed to confirm human-animal model findings (Grabowska et al., 2024).

3.1.5. Modulation of Neurotransmitter Systems

Regulation of GABA and Glutamate Balance: KD protects the brain by lowering excitotoxicity. It boosts GABA, an inhibitory neurotransmitter, and improves the GABA-to-glutamate ratio. This balance stabilizes brain activity and aids cognitive function (Field et al., 2022).

Influence on Dopaminergic Systems: KD also reduces motor dysfunction and dopaminergic neuron loss in models of Alzheimer's and Parkinson's diseases. It does this by curbing microglial activation and pro-inflammatory cytokines (Jang et al., 2023). Furthermore, KD enhances dopamine and lessens damage to dopaminergic neurons. This suggests potential benefits for mood disorders and Parkinson's disease (Zhang et al., 2022).

While the KD has several positive effects, as previously discussed, there are also important considerations. Diets high in fat have been linked to disruptions in neurogenesis. Consuming such diets over a long period can lead to mitochondrial oxidative stress in the brain. For example, a study found that rats who fed a ketogenic diet consisting of 42% fat by energy and reduced carbohydrates experienced impaired neurogenesis in the hippocampus (Lindqvist et al., 2006). However, other research showed that a similar high-fat diet did not inhibit neurogenesis (Strandberg et al., 2008). In a recent study involving a diet composed of 4.9% carbohydrates, 48.7% protein, and 46.35% fat, no significant changes in neurogenesis within the dentate gyrus were observed (Irfannuddin et al., 2021). Therefore, while the ketogenic diet may offer benefits, its potential adverse effects on brain health, particularly regarding neurogenesis, should not be overlooked.

KD is popular and beneficial, but not for everyone. It can cause constipation, bad breath, and electrolyte imbalances. These problems can be serious without proper management. There is also concern that it might raise heart disease risk. So, careful evaluation is key to understanding its long-term effects (Meng et al., 2024).

CR and KD show a complex link between nutrition and brain health. They boost neuroplasticity, increase mitochondrial growth, and reduce oxidative stress, leading to better cognitive function and protection against brain damage. Yet, these diets aren't for everyone and can be risky. More research is needed to understand their long-term effects on brain health and cognition, ensuring people can enjoy the benefits without facing risks.

4. The Mediterranean Diet

The Mediterranean diet (MD) is characterized by a high intake of various plant-based foods, including fruits, vegetables, legumes, whole grains, nuts, seeds, and olive oil as a primary fat source. It also has high fish consumption and allows for moderate red wine intake during meals. In contrast, foods like red meat, high-fat dairy, and processed items are only occasionally included. This diet is abundant in bioactive nutrients, such as vitamins, minerals, polyphenols, fiber, nitrates, and both mono- and polyunsaturated fatty acids—all of which, alone or synergistically, have been linked to positive health effects (Shannon et al., 2021). MD has been widely studied for its positive effects on overall health (Guasch-Ferré & Willett, 2021). Numerous randomized clinical trials indicate that the MD substantially reduces oxidative stress, inflammation, and the progression of conditions such as cardiovascular disease, type 2 diabetes, and other metabolic disorders—factors closely associated with the development of neurodegenerative diseases. Prospective studies further show that adherence to the MD is linked to lower overall mortality, particularly from cardiovascular causes, contributing to increased longevity. This dietary pattern has also been associated with a reduced risk of age-related cognitive decline and a lower incidence of neurodegenerative diseases, especially Alzheimer's disease (Guasch-Ferré et al., 2021). MD has been associated with various neuroprotective effects, primarily attributed to its anti-oxidant, anti-inflammatory, and cardio-metabolic benefits. Critical elements of MD, such as essential fatty acids, polyphenols, and vitamins, have been shown to decrease oxidative stress. Although observational studies suggest that the diet benefits brain health, findings from clinical trials have been somewhat inconsistent. While further longitudinal studies and larger-scale clinical trials are needed, MD remains a promising strategy for supporting cognitive function, mainly due to its well-documented cardiovascular and metabolic advantages (Siervo et al., 2021). Recent research has increasingly focused on its impact on brain health and neuroplasticity.

MD aligns remarkably well with the nine hallmark components of aging outlined by López-Otín et al., which encompass genomic instability, telomere shortening, epigenetic changes, loss of protein homeostasis, deregulated nutrient sensing, mitochondrial dysfunction, cellular senescence, stem cell depletion, and disrupted intercellular communication (López-Otín et al., 2013; Shannon et al., 2021) (Figure 6). Given its positive impact on these aging processes, it is not surprising that MD is also associated with substantial benefits for brain health. This comprehensive influence on cellular aging mechanisms supports MD's role in promoting longevity and resilience against age-related cognitive decline.

Figure 6: Mediterranean diet and the hallmarks of aging. MD, both in its entirety and through individual components such as olive oil, red wine, fish, vegetables, fruits, nuts, folate, tomatoes, and polyphenols, may beneficially influence each of the nine hallmarks of aging. The mechanisms involved vary by component; for example, this diet's high intake of olive oil, nuts, tomatoes, and red wine may reduce genomic instability (upper right corner) by lowering oxidative stress and mitigating DNA damage. Each mechanism is illustrated to show how specific aspects of MD positively interact with these hallmarks (Shannon et al., 2021).

4.1. Mechanisms of Action of MD on Neuroplasticity

MD effects on neuroplasticity can be explained through four primary mechanisms: insulin signaling and neuroprotection, oxidative stress reduction, neuroinflammation modulation, and synaptic plasticity and cognitive function (Figure 7).

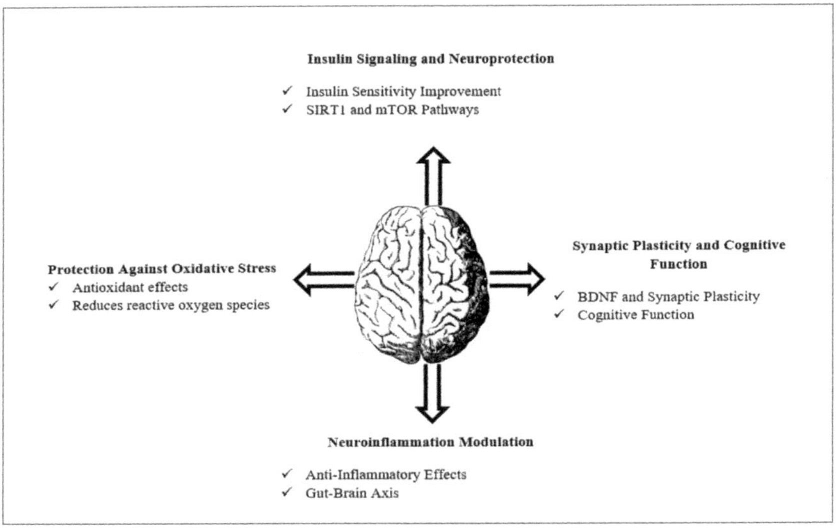

Figure 7: The primary mechanism of MD on neuroplasticity
SIRT1: sirtuin 1, mTOR: mammalian target of rapamycin, BDNF: Brain-Derived Neurotrophic Factor

4.1.1. Insulin Signaling and Neuroprotection

Insulin Sensitivity Improvement: MD may reduce insulin resistance (IR) through its balanced profile of healthy fats, fiber-rich carbohydrates, and polyphenol-packed plant foods, collectively improving insulin sensitivity. Critical components like monounsaturated fats from olive oil and polyphenols from fruits and vegetables positively influence glucose metabolism, reduce systemic inflammation, and modulate lipid profiles, all decreasing IR (Mirabelli et al., 2020). Polyphenols, particularly, have been shown to activate SIRT1. Activation of SIRT1 enhances mitochondrial function, reduces oxidative stress, and supports glucose homeostasis, thereby mitigating IR

(Chung et al., 2010). SIRT1 also has neuroprotective effects; it helps maintain neuronal integrity and modulates pathways involved in amyloid processing, reducing neuroinflammation and oxidative damage in brain cells. Thus, by activating SIRT1, MD addresses IR and promotes brain health, providing a dual mechanism that may lower the risk of neurodegenerative diseases.

Studies indicate that MD may reduce the accumulation of β-amyloid and improve cerebral glucose utilization, potentially delaying Alzheimer's onset by 1.5 to 3.5 years in cognitively normal, middle-aged individuals (Berti et al., 2018). Additionally, higher MD adherence has been linked with greater brain volume, improved cerebral connectivity, and fewer white matter lesions in older adults (Mirabelli et al., 2020). Anthocyanins, which accumulate in brain tissue, have shown a potential to improve cognitive functions by reducing β-amyloid processing and neuroinflammation through activating the PPARγ pathway. This activation aids in managing lipid metabolism and reducing IR in both peripheral tissues and the brain. As a result, anthocyanins may help prevent cognitive decline linked to insulin dysfunction, mainly through insulin sensitization that supports cerebral glucose uptake (Vepsäläinen et al., 2013). Further supporting MD's benefits, medium-chain fatty acids from coconut oil may enhance cognitive performance in elderly patients with dementia by elevating circulating ketone bodies, providing an alternative brain energy source to glucose (de la Rubia Ortí et al., 2018).

SIRT1 and mTOR Pathways: Polyphenols like resveratrol, curcumin, quercetin, and catechins exhibit antioxidant and anti-inflammatory effects by modulating pathways such as NF-κB and the Mitogen-Activated Protein Kinases (MAPKs). Additionally, these polyphenols can activate SIRT1, either directly or indirectly, across various models. SIRT1 activation by polyphenols supports numerous physiological processes, including calorie restriction, oxidative stress regulation, inflammation reduction, cellular aging, and autophagy. This activation also influences apoptosis, metabolism, adipogenesis, circadian rhythms, muscle function, mitochondrial biogenesis, and vascular health. Thus, dietary polyphenols contribute to a broad spectrum of cellular functions that promote overall health (Chung et al., 2010).

MD with a high ω-3/omega 6 (ω-6) ratio was associated with reduced insulin levels, body weight, atherosclerosis indicators, and liver fat accumulation. These outcomes were linked to lowered mTOR expression, alongside an increase in mitochondrial electron transport chain and Krebs cycle pathway activity, compared to diets with lower ω-3/ω-6 ratios. Consequently, a high ω-3/ω-6 ratio diet promoted improved mitochondrial complex function and reduced levels of fumaric acid and oxidative stress (Khalil et al., 2022).

4.1.2. Oxidative Stress Reduction

Protection Against Oxidative Stress: MD includes a variety of nutrients with antioxidant effects, such as essential fatty acids from oily fish, polyphenols from olive oil and nuts, berries, and grapes (esp. Resveratrol), and antioxidants found in brightly colored fruits and vegetables like berries, tomatoes, and carrots. These include vitamins like C and E and minerals like selenium and zinc. These compounds neutralize free radicals, reducing oxidative damage to neuronal membranes and proteins, a critical factor in neurodegeneration (Górska et al., 2021).

Reduces Reactive Oxygen Species: MD has been shown to reduce the production of ROS, which are highly reactive molecules that contribute to oxidative stress and neurodegeneration. The MD enhances the body's natural antioxidant defenses, rich in antioxidants such as polyphenols from olive oil, flavonoids from fruits and vegetables, and ω-3 fatty acids from fish. These nutrients help neutralize ROS and reduce oxidative damage to cells, including neurons, thereby protecting brain function. ROS causes damage to cells, particularly in the brain, which is highly metabolically active. This oxidative damage is a significant contributor to neurodegenerative diseases such as Alzheimer's and Parkinson's (Sahab et al., 2023). Additionally, MD is rich in inorganic nitrates, mainly from green leafy and root vegetables. Inorganic nitrates have been linked to reduced oxidative stress by enhancing nitric oxide signaling, which improves mitochondrial efficiency and reduces ROS production by inhibiting NADPH oxidase (Sui et al., 2019).

4.1.3. Neuroinflammation Modulation

Anti-Inflammatory Effects: Chronic inflammation is a well-known contributor to neurodegenerative diseases. Several clinical trials suggest that MD provides neuroprotection against cognitive decline primarily through its antioxidant and anti-inflammatory properties (Baumgart et al., 2015). Compounds such as β-carotene, polyphenols, vitamin C, and vitamin E, which are abundant in MD, support the body's antioxidant systems and reduce neuroinflammation. Flavonoids and polyphenols help inhibit transcription factors like NF-κB and reduce the expression of pro-inflammatory cytokines like IL-1β, TNF-α, and more. These effects are partly mediated through the activation of PPAR-γ, which stimulates anti-inflammatory cytokine production and suppresses NF-κB activity (Schwingshackl et al., 2020). Studies show that diets enriched with extra virgin olive oil (EVOO) downregulate genes linked to inflammatory pathways, and resveratrol (a polyphenol in red wine, grapes, and peanuts) demonstrates anti-inflammatory effects by inhibiting NF-κB in vitro. Additionally, **ω-3** PUFAs in fish oil reduce systemic and neural inflammation, contrasting with the inflammation-promoting effects of saturated fatty acids (Armeli et al., 2021).

In the PREDIMED (Prevención et al.) trial, older adults with vascular risk consuming an MD enriched with EVOO or nuts exhibited improved cognitive performance compared to those on a low-fat diet (Martínez-Lapiscina et al., 2013). Further randomized trials have also shown the cognitive benefits of MD in high-risk elderly populations. For example, Parkinson's disease patients who followed a personalized MD for 10 weeks achieved better cognitive scores (Paknahad et al., 2020). Trials in Japan revealed that diets rich in fruits and vegetables reduce the risk of developing Parkinson's disease (Okubo et al., 2012) and amyotrophic lateral sclerosis (Okamoto et al., 2009). Moreover, a recent research study indicated that MD has protective effects against both peripheral and central inflammation in primates (*Macaca fascicularis*), resulting in preserved brain structure and improved socioemotional behavior (Frye et al., 2024). In contrast to the Western diet, which promotes Alzheimer 's-related neuroinflammation, MD prevents such inflammation and is associated with early increases in gray matter volume in female nonhuman primates (Shively et al., 2023).

Gut–Brain Axis: MD, rich in fiber, polyphenols, unsaturated fats, and **ω-3** fatty acids, supports a balanced and healthy gut microbiome. Dietary fiber from whole grains, fruits, and vegetables serves as a prebiotic, nourishing beneficial bacteria such as *Lactobacillus* and *Bifidobacterium*, which are instrumental in producing short-chain fatty acids (SCFAs) like butyrate (Vulevic et al., 2013). Butyrate and other SCFAs help maintain the integrity of the blood–brain barrier and regulate neuroinflammatory responses (Jiang et al., 2021). Furthermore, SCFAs are linked to increased BDNF production, a crucial neuroplasticity molecule (Church et al., 2023). In addition, the MD's key components—dietary fibers, EVOO, and PUFAs (notably **ω-3**)—encourage the growth of SCFA-producing bacteria, such as *Clostridium leptum* and *Eubacterium rectale*, as well as other beneficial species like *Bifidobacteria*, *Bacteroides*, and *Faecalibacterium prausnitzii*. This dietary pattern simultaneously inhibits the growth of less favorable species, including *Firmicutes* and *Blautia*. These shifts in the gut microbiota are associated with favorable effects on inflammation and oxidative stress, reduced cancer risk, and improved metabolic health, illustrating how the MD supports a microbiome that promotes overall physical and cognitive well-being. An overview of the impact of Mediterranean diet components on gut microbiota and the related health benefits can be seen in Table 1 (Barber et al., 2023).

Table 1: Overview of the impact of Mediterranean diet components on gut microbiota and the related health benefits.

Dietary Component of the MD	Dietary Origins within the MD	Effects on Gut Microbiota	Associated Health Benefits
Dietary fibers	Plant-based foods (vegetables, fruit, cereals)	Improved diversity; Increased Bifidobacteria and Bacteroides species, and SCFA-producing bacteria (*Clostridium leptum* and *Eubacterium rectale*)	Improved cardio-metabolic health, insulin sensitivity, and risk of developing colorectal carcinoma
Polyphenols	Extra virgin olive oil	Changes in lactic acid bacteria; Reduced *H. pylori*	Improved inflammatory, oxidative, endothelial, and general metabolic health status; Prevention of gastric ulcer
PUFAs (including ω-3)	Oily fish Seafood Nuts	Reduced Firmicutes and Blautia species	Improved inflammatory and immune status; Improved intestinal epithelial barrier
Magnesium and Calcium	Fresh vegetables and fruit	Changes in cecal SCFAs; Favored growth of lactobacilli; Regulation of intestinal tight junction gene expression	Improved gastrointestinal and psychiatric disorders; Improved growth performance

(Continued)

Table 1: (Continued)

Dietary Component of the MD	Dietary Origins within the MD	Effects on Gut Microbiota	Associated Health Benefits
Selenium, Phytate, Zinc and Vitamin B6	Fresh vegetables and fruit	Regulation of intestinal brush border functionality and composition and diversity of gut microbiota	Possible impact on cardio-metabolic risk; Iron status
Relative lack of RCs, SFAs and TFAs	Lack of processed foods that typify a Western diet	Reduced propensity for diminishment of diversity and pro-inflammatory nature of gut microbiota	The reduced propensity for the development of obesity, insulin resistance, endotoxemia, leaky gut, and metabolic dysfunction

Barber et al., 2023 (*H. pylori: Helicobacter pylori; MD: Mediterranean Diet; ω-3: Omega 3; PUFAs: Polyunsaturated Fatty Acids; RCs: Refined Carbohydrates; SCFA: Short-Chain Fatty Acid; SFAs: Saturated Fatty Acids; TFAs: Trans Fatty Acids*).

4.1.4. Synaptic Plasticity and Cognitive Function

BDNF and Synaptic Plasticity: MD is associated with enhanced BDNF levels, which are crucial to neuroplasticity and cognitive function. In animal studies, dietary components of the MD, like ω-3 fatty acids, vitamin E, and flavonoids, have shown positive effects on BDNF levels in the brain. For instance, ω-3 fatty acids, particularly alpha-linolenic acid and docosahexaenoic acid, as well as antioxidants such as vitamin E, have been linked to increased BDNF expression in various animal models, including healthy, parkinsonian, epileptic, and trauma-injured rats (Cysneiros et al., 2010; Hou et al., 2010). This suggests a protective effect on the brain by promoting neuronal health and synaptic plasticity. On the contrary, diets high in saturated fats or total fat intake tend to lower BDNF levels, negatively affecting neuronal plasticity and cognitive function, as observed in similar animal studies (Sánchez-Villegas et al., 2011). These findings highlight the MD's favorable impact due to its anti-inflammatory and antioxidant properties. Nutrients in the MD, such as polyphenols, flavonoids, and ω-3 fatty acids, counteract oxidative stress, which otherwise reduces BDNF expression by impairing cellular processes. Human studies on MD's effects on plasma BDNF are limited. However, findings from the PREDIMED study—a large-scale randomized trial—indicated that MD supplemented with nuts or EVOO positively influenced BDNF levels in depressed patients. However, after a three-year intervention, no significant difference in plasma BDNF was observed across the general study population. This suggests that the MD's beneficial effects on BDNF might be more pronounced in individuals with specific conditions, such as depression, potentially through improvements in endothelial function and a reduction in inflammatory cytokines like IL-6 and TNF-α, which are known inhibitors of BDNF expression (Sánchez-Villegas et al., 2011). While more human studies are needed to evaluate BDNF responses in broader populations, current animal and preliminary human evidence suggest that adherence to MD can enhance BDNF production and support cognitive health through its nutrient-dense, anti-inflammatory, and antioxidant profile.

Cognitive Function: MD has been widely studied for its effects on cognitive function, with evidence drawn primarily from observational studies and some experimental trials. Greater adherence to the MD has shown promising links to improved performance in various cognitive domains—such as memory, executive function, attention, and global cognition—and slower cognitive decline over time (Siervo et al., 2021). In early studies, like the 3-City study, higher MD adherence correlated with fewer errors on cognitive tests and greater resilience against cognitive decline over five years (Féart et al., 2009). Follow-up imaging studies in a subgroup of the same cohort revealed that higher MD adherence was associated with preserved white matter microstructure and better structural connectivity, indicating potential protective effects on brain structure (Pelletier et al., 2015). Similarly, other research has observed that MD adherence may help maintain brain volume, cortical thickness, and overall brain health, with some studies linking it to lower Alzheimer's biomarkers (Siervo et al., 2021). Clinical findings suggest that anthocyanin supplementation can enhance memory and verbal fluency in elderly individuals with mild cognitive impairment, highlighting their promise for cognitive health (Kent et al., 2017).

Notably, research conducted in Mediterranean countries consistently reports more substantial protective effects of the MD on cognitive health. For instance, in the InChianti study, individuals with higher MD adherence experienced a significantly reduced decline in cognitive scores over a decade (Tanaka et al., 2018). The PREDIMED trial further demonstrated that MD, particularly when supplemented with EVOO or nuts, was linked to better cognitive scores over 6.5 years (Valls-Pedret et al., 2015). However, results from studies conducted in non-Mediterranean countries have been more varied, with some finding positive associations between MD adherence and reduced cognitive decline, while others report no significant link. These mixed results may stem from differing interpretations of the MD and variations in cognitive assessment methods. Despite these inconsistencies, the overall trend suggests that the MD could support cognitive health, especially in populations with a traditional Mediterranean dietary pattern (Siervo et al., 2021).

In summary, MD offers a compelling strategy for supporting neuroplasticity and cognitive health, primarily through its rich content of antioxidants, anti-inflammatory compounds, and essential nutrients. By promoting insulin sensitivity, reducing oxidative stress, and modulating neuroinflammation, MD appears to mitigate critical factors linked to neurodegeneration. Its beneficial impact on brain structure and function—demonstrated by increased brain volume, improved cerebral connectivity, and reduced amyloid accumulation—suggests potential protection against age-related cognitive decline and neurodegenerative diseases. Though further longitudinal and clinical studies are needed to solidify these findings, the MD shows promise as an accessible, sustainable dietary approach to bolster brain health and resilience across aging populations.

5. Conclusion

Dietary interventions such as MD, CR, and KD all present unique neuroprotective mechanisms, yet they converge on several critical pathways beneficial for neuroplasticity. MD, emphasizing antioxidants, polyphenols, and omega-3 fatty acids, excels in reducing oxidative stress, neuroinflammation, and insulin resistance—factors crucial for supporting long-term brain health. Meanwhile, CR's primary influence is through metabolic adaptations that boost autophagy, reduce oxidative damage, and improve mitochondrial function, all essential for maintaining neuronal integrity and enhancing cognitive resilience. On the other hand, KD is distinguished by its ketone production, which provides an alternative energy source to glucose, enhances synaptic plasticity, and reduces neuroinflammation, making it especially beneficial for certain neurodegenerative conditions.

While all three dietary approaches offer neuroprotective benefits, MD stands out due to its extensive research base and adaptability across diverse populations. CR and KD, though powerful, require more rigorous adherence and can be challenging for long-term implementation, particularly in broader populations. MD's versatility and accessibility make it a more widely applicable option, and it has shown consistent associations with improved cognitive health in both Mediterranean and non-Mediterranean cohorts. Therefore, MD may be the most practical choice for supporting neuroplasticity at the population level. However, targeted uses of CR and KD could be valuable in specific clinical contexts, mainly where metabolic and neurodegenerative risks are high.

6. Takeaways

1. Both CR and KD boost mitochondrial function, enhancing energy production and reducing oxidative stress in neurons.
2. CR, KD, and MD activate pathways such as AMPK and SIRT1, which promote autophagy and cellular maintenance. These pathways clear damaged proteins like amyloid-beta and tau, helping to prevent neurodegenerative diseases.
3. Each diet increases BDNF levels, which are crucial for synaptic plasticity, memory formation, and cognitive health.
4. These diets enhance synaptic resilience, improving cognitive performance, especially in aging and neurodegenerative conditions.
5. CR and KD lower levels of pro-inflammatory cytokines, preserving neuronal health and protecting against neurodegeneration.
6. MD's polyphenols and omega-3 fatty acids lower inflammatory markers, potentially decreasing neuroinflammation and associated neurodegeneration.
7. Antioxidants in MD, including vitamins, polyphenols, and minerals, help neutralize reactive oxygen species, protecting neurons from oxidative stress.
8. All three diets positively affect the gut microbiota, which reduces neuroinflammation and supports brain function.

9. CR and MDs improve insulin signaling in the brain, supporting glucose metabolism and cognitive function.
10. KD improves the balance between excitatory (glutamate) and inhibitory (GABA) neurotransmitters, reducing excitotoxicity.
11. CR enhances neurogenesis in the hippocampus by increasing BDNF and reducing oxidative stress, supporting the formation of new neurons crucial for learning and memory.
12. KD may not be suitable for everyone due to potential risks like heart disease, electrolyte imbalances, and impaired neurogenesis in some cases.

Abbreviations

AKT: Protein Kinase B/PKB
ALOX5: Arachidonate 5-Lipoxygenase Gene
AMP/ATP: Ratio of Adenosine Monophosphate to Adenosine Triphosphate
AMPK: Adenosine 5'-Monophosphate (AMP)-Activated Protein Kinase
BBB: Blood–Brain Barrier
BCAA: Branched-Chain Amino Acid
BDNF: Brain-Derived Neurotrophic Factor
BHB: β-Hydroxybutyrate
Blue: Non-Restricted Calorie Intake,
CI: Calorie Intake
COX1: Cyclooxygenase 1
COX2: Cyclooxygenase 2
CR: Calorie Restriction
CRM: CR Mimetic
CRP: C-Reactive Protein
Cx43: Connexin-43
EVOO: Extra Virgin Olive Oil
FFA: Free Fatty Acid
FGF21: Fibroblast Growth Factor 21
FOXO: Forkhead Box O
GABA: Gamma-Aminobutyric Acid
GLUT: Glucose Transporters
GLUT4: Glucose Transporter Type 4
H. pylori: Helicobacter Pylori

HDAC: The Histone Deacetylase
HIF-1α: Hypoxia-İnduced Factor 1α
IF: Intermittent Fasting
IGF-1: Insulin-Like Growth Factor
IL: Interleukin
IL-1β: Interleukin-1β
IL-2: Interleukin-2
IL-4: Interleukin-4
IL-6: Interleukin-6
IR: Insulin Resistance
iNOS: Inducible Nitric Oxide Synthase
KD: Ketogenic Diet
LRP1: LDL Receptor-Related Protein 1
Magenta: Fasting or Restricted Calorie Intake
MAPKs: the Mitogen-Activated Protein Kinases
MCT: Monocarboxylate Transporters
MEF2: Myocyte Enhancer Factor-2
mLST8: Mammalian Lethal With SEC13 Protein 8
mPTP: The Mitochondrial Permeability Transition Pore
mTOR: The Mammalian Target Of Rapamycin
mTORC1: mTOR Complex 1
NAD+/NADH: Ratio of Oxidized to Reduced Nicotinamide Adenine Dinucleotide
NADH: Nicotinamide Adenine Dinucleotide (NAD) + hydrogen (H)
NF-κB: Nuclear Factor Kappa-Light-Chain-Enhancer of Activated B Cells
Nrf2: Nuclear Factor Erythroid 2-Related Factor 2
PARP1: Poly [ADP-Ribose] Polymerase 1
PGC1α: Peroxisome Proliferator-Activated Receptor Gamma Coactivator 1-Alpha
P-gp: Glycoprotein P
PICALM: Phosphatidylinositol Binding Clathrin Assembly Protein
PPARγ: Peroxisome Proliferator-Activated Receptor γ
PRAS40: proline-rich Akt substrate of 40 kDa
PREDIMED: Prevención con Dieta Mediterránea
PUFAs: Polyunsaturated Fatty Acids

RCs: Refined Carbohydrates
ROS: Reactive Oxygen Species
SCFAs: Short-Chain Fatty Acids
SCFAs: Short-chain Chain Fatty Acids
SFAs: Saturated Fatty Acids
SIRT1: The Seven Mammalian Sirtuins 1
SOD2: Manganese Superoxide Dismutase
TBI: Traumatic Brain Injury
TFAs: Trans Fatty Acids
TFEB: The Transcription Factor EB
TG: Triglycerides
TNFα: Tumor Necrosis Factor α
UCP2: Uncoupling Protein 2
ULK1: Unc-51-Like Autophagy-Activating Kinase
VLDL: Very Low-Density Lipoprotein
ω-3: Omega 3
ω-6: Omega 6

Bibliography

1. Alkurd, R., Mahrous, L., Zeb, F., Khan, M. A., Alhaj, H., Khraiwesh, H. M., & Faris, M. E. (2024). Effect of Calorie Restriction and Intermittent Fasting Regimens on Brain-Derived Neurotrophic Factor Levels and Cognitive Function in Humans: A Systematic Review. *Medicina (Kaunas, Lithuania)*, *60*(1), 191. https://doi.org/10.3390/medicina60010191
2. Almendáriz-Palacios, C., Mousseau, D. D., Eskiw, C. H., & Gillespie, Z. E. (2020). Still Living Better through Chemistry: An Update on Caloric Restriction and Caloric Restriction Mimetics as Tools to Promote Health and Lifespan. *International journal of molecular sciences*, *21*(23), 9220. https://doi.org/10.3390/ijms21239220
3. Armeli, F., Bonucci, A., Maggi, E., Pinto, A., & Businaro, R. (2021). Mediterranean Diet and Neurodegenerative Diseases: The Neglected Role of Nutrition in Modulating the Endocannabinoid System. *Biomolecules*, *11*(6), 790. https://doi.org/10.3390/biom11060790
4. Bagherniya, M., Butler, A. E., Barreto, G. E., & Sahebkar, A. (2018). The effect of fasting or calorie restriction on autophagy induction:

A review of the literature. Ageing research reviews, 47, 183–197. https://doi.org/10.1016/j.arr.2018.08.004

5. Baik, S. H., Rajeev, V., Fann, D. Y., Jo, D. G., & Arumugam, T. V. (2020). Intermittent fasting increases adult hippocampal neurogenesis. Brain and behavior, 10(1), e01444. https://doi.org/10.1002/brb3.1444
6. Barber, T. M., Kabisch, S., Pfeiffer, A. F. H., & Weickert, M. O. (2023). The Effects of the Mediterranean Diet on Health and Gut Microbiota. *Nutrients*, *15*(9), 2150. https://doi.org/10.3390/nu15092150
7. Barry, D., Ellul, S., Watters, L., Lee, D., Haluska, R., Jr., & White, R. (2018). The ketogenic diet in disease and development. International journal of developmental neuroscience: the official journal of the International Society for Developmental Neuroscience, 68, 53–58. https://doi.org/10.1016/j.ijdevneu.2018.04.005
8. Baumgart, M., Snyder, H. M., Carrillo, M. C., Fazio, S., Kim, H., & Johns, H. (2015). Summary of the evidence on modifiable risk factors for cognitive decline and dementia: A population-based perspective. *Alzheimer's & dementia: The Journal of the Alzheimer's Association*, *11*(6), 718–726. https://doi.org/10.1016/j.jalz.2015.05.016
9. Berti, V., Walters, M., Sterling, J., Quinn, C. G., Logue, M., Andrews, R., Matthews, D. C., Osorio, R. S., Pupi, A., Vallabhajosula, S., Isaacson, R. S., de Leon, M. J., & Mosconi, L. (2018). Mediterranean diet and 3-year Alzheimer brain biomarker changes in middle-aged adults. Neurology, 90(20), e1789–e1798. https://doi.org/10.1212/WNL.0000000000005527
10. Bloemer, J., Pinky, P. D., Govindarajulu, M., Hong, H., Judd, R., Amin, R. H., Moore, T., Dhanasekaran, M., Reed, M. N., & Suppiramaniam, V. (2018). Role of Adiponectin in Central Nervous System Disorders. Neural plasticity, 2018, 4593530. https://doi.org/10.1155/2018/4593530
11. Bonaventura, A., Liberale, L., Vecchié, A., Casula, M., Carbone, F., Dallegri, F., & Montecucco, F. (2016). Update on Inflammatory Biomarkers and Treatments in Ischemic Stroke. International journal of molecular sciences, 17(12), 1967. https://doi.org/10.3390/ijms17121967
12. Chandrasekaran, V., Hediyal, T. A., Anand, N., Kendaganna, P. H., Gorantla, V. R., Mahalakshmi, A. M., Ghanekar, R. K., Yang, J., Sakharkar, M. K., & Chidambaram, S. B. (2023). Polyphenols, Autophagy and Neurodegenerative Diseases: A Review. Biomolecules, 13(8), 1196. https://doi.org/10.3390/biom13081196

13. Chung, S., Yao, H., Caito, S., Hwang, J. W., Arunachalam, G., & Rahman, I. (2010). Regulation of SIRT1 in cellular functions: role of polyphenols. Archives of biochemistry and biophysics, 501(1), 79–90. https://doi.org/10.1016/j.abb.2010.05.003
14. Church, J. S., Bannish, J. A. M., Adrian, L. A., Rojas Martinez, K., Henshaw, A., & Schwartzer, J. J. (2023). Serum short-chain fatty acids mediate hippocampal BDNF and correlate with decreasing neuroinflammation following a high pectin fiber diet in mice. Frontiers in neuroscience, 17, 1134080. https://doi.org/10.3389/fnins.2023.1134080
15. Corti, O., Blomgren, K., Poletti, A., & Beart, P. M. (2020). Autophagy in neurodegeneration: New insights underpinning therapy for neurological diseases. Journal of Neurochemistry, 154(4), 354–371. https://doi.org/10.1111/jnc.15002
16. Cusick, S. E., & Georgieff, M. K. (2016). The Role of Nutrition in Brain Development: The Golden Opportunity of the "First 1000 Days". *The Journal of Pediatrics*, *175*, 16–21. https://doi.org/10.1016/j.jpeds.2016.05.013
17. Cysneiros, R. M., Ferrari, D., Arida, R. M., Terra, V. C., de Almeida, A. C., Cavalheiro, E. A., & Scorza, F. A. (2010). Qualitative analysis of hippocampal plastic changes in rats with epilepsy supplemented with oral omega-3 fatty acids. Epilepsy & behavior: E&B, 17(1), 33–38. https://doi.org/10.1016/j.yebeh.2009.11.006
18. de Carvalho, T. S., Sanchez-Mendoza, E. H., Schultz Moreira, A. R., Nascentes Melo, L. M., Wang, C., Sardari, M., Hagemann, N., Doeppner, T. R., Kleinschnitz, C., & Hermann, D. M. (2021). Hypocaloric Diet Initiated Post-Ischemia Provides Long-Term Neuroprotection and Promotes Peri-Infarct Brain Remodeling by Regulating Metabolic and Survival-Promoting Proteins. Molecular neurobiology, 58(4), 1491–1503. https://doi.org/10.1007/s12035-020-02207-7
19. de la Rubia Ortí, J. E., García-Pardo, M. P., Drehmer, E., Sancho Cantus, D., Julián Rochina, M., Aguilar, M. A., & Hu Yang, I. (2018). Improvement of Main Cognitive Functions in Patients with Alzheimer's Disease after Treatment with Coconut Oil Enriched Mediterranean Diet: A Pilot Study. Journal of Alzheimer's disease: JAD, 65(2), 577–587. https://doi.org/10.3233/JAD-180184
20. de Oliveira R. M. W. (2020). Neuroplasticity. *Journal of chemical neuroanatomy*, *108*, 101822. https://doi.org/10.1016/j.jchemneu.2020.101822

21. Di Francesco, A., Di Germanio, C., Bernier, M., & de Cabo, R. (2018). A time to fast. *Science (New York, N.Y.)*, *362*(6416), 770–775. https://doi.org/10.1126/science.aau2095
22. Dyńka, D., Kowalcze, K., & Paziewska, A. (2022). The Role of Ketogenic Diet in the Treatment of Neurological Diseases. Nutrients, 14(23), 5003. https://doi.org/10.3390/nu14235003
23. Féart, C., Samieri, C., Rondeau, V., Amieva, H., Portet, F., Dartigues, J. F., Scarmeas, N., & Barberger-Gateau, P. (2009). Adherence to a Mediterranean diet, cognitive decline, and risk of dementia. JAMA, 302(6), 638–648. https://doi.org/10.1001/jama.2009.1146
24. Field, R., Field, T., Pourkazemi, F., & Rooney, K. (2022). Ketogenic diets and the nervous system: a scoping review of neurological outcomes from nutritional ketosis in animal studies. Nutrition research reviews, 35(2), 268–281. https://doi.org/10.1017/S0954422421000214
25. Fitzgerald, K. C., Vizthum, D., Henry-Barron, B., Schweitzer, A., Cassard, S. D., Kossoff, E., Hartman, A. L., Kapogiannis, D., Sullivan, P., Baer, D. J., Mattson, M. P., Appel, L. J., & Mowry, E. M. (2018). Effect of intermittent vs. daily calorie restriction on changes in weight and patient-reported outcomes in people with multiple sclerosis. Multiple sclerosis and related disorders, 23, 33–39. https://doi.org/10.1016/j.msard.2018.05.002
26. Flanagan, E. W., Most, J., Mey, J. T., & Redman, L. M. (2020). Calorie Restriction and Aging in Humans. Annual review of nutrition, 40, 105–133. https://doi.org/10.1146/annurev-nutr-122319-034601
27. Frye, B. M., Negrey, J. D., Johnson, C. S. C., Kim, J., Barcus, R. A., Lockhart, S. N., Whitlow, C. T., Chiou, K. L., Snyder-Mackler, N., Montine, T. J., Craft, S., Shively, C. A., & Register, T. C. (2024). Mediterranean diet protects against a neuroinflammatory cortical transcriptome: Associations with brain volumetrics, peripheral inflammation, social isolation, and anxiety in nonhuman primates (Macaca fascicularis). *Brain, behavior, and immunity*, *119*, 681–692. https://doi.org/10.1016/j.bbi.2024.04.016
28. Furman, D., Campisi, J., Verdin, E., Carrera-Bastos, P., Targ, S., Franceschi, C., Ferrucci, L., Gilroy, D. W., Fasano, A., Miller, G. W., Miller, A. H., Mantovani, A., Weyand, C. M., Barzilai, N., Goronzy, J. J., Rando, T. A., Effros, R. B., Lucia, A., Kleinstreuer, N., & Slavich, G. M. (2019). Chronic inflammation in the etiology of disease across the life span. Nature medicine, 25(12), 1822–1832. https://doi.org/10.1038/s41591-019-0675-0

29. Fusco, S., & Pani, G. (2013). Brain response to calorie restriction. *Cellular and molecular life sciences: CMLS*, *70*(17), 3157–3170. https://doi.org/10.1007/s00018-012-1223-y
30. Górska, P., Górna, I., & Przysławski, J. (2021). Mediterranean diet and oxidative stress. Nutrition & Food Science. 51(4): 677-689. https://doi.org/10.1108/NFS-07-2020-0264
31. Grabowska, K., Grabowski, M., Przybyła, M., Pondel, N., Barski, J. J., Nowacka-Chmielewska, M., & Liśkiewicz, D. (2024). Ketogenic diet and behavior: insights from experimental studies. Frontiers in nutrition, 11, 1322509. https://doi.org/10.3389/fnut.2024.1322509
32. Green, C. L., Lamming, D. W., & Fontana, L. (2022). Molecular mechanisms of dietary restriction promoting health and longevity. Nature reviews. Molecular cell biology, 23(1), 56–73. https://doi.org/10.1038/s41580-021-00411-4
33. Guasch-Ferré, M., & Willett, W. C. (2021). The Mediterranean diet and health: a comprehensive overview. Journal of Internal Medicine, 290(3), 549–566. https://doi.org/10.1111/joim.13333
34. Hasan-Olive, M. M., Lauritzen, K. H., Ali, M., Rasmussen, L. J., Storm-Mathisen, J., & Bergersen, L. H. (2019). A Ketogenic Diet Improves Mitochondrial Biogenesis and Bioenergetics via the PGC1α-SIRT3-UCP2 Axis. Neurochemical research, 44(1), 22–37. https://doi.org/10.1007/s11064-018-2588-6
35. Herzig, S., & Shaw, R. J. (2018). AMPK: guardian of metabolism and mitochondrial homeostasis. Nature reviews. Molecular cell biology, 19(2), 121–135. https://doi.org/10.1038/nrm.2017.95
36. Hofer, S. J., Carmona-Gutierrez, D., Mueller, M. I., & Madeo, F. (2022). The ups and downs of caloric restriction and fasting: from molecular effects to clinical application. *EMBO molecular medicine*, *14*(1), e14418. https://doi.org/10.15252/emmm.202114418
37. Hornsby, A. K. E., Buntwal, L., Carisi, M. C., Santos, V. V., Johnston, F., Roberts, L. D., Sassi, M., Mequinion, M., Stark, R., Reichenbach, A., Lockie, S. H., Siervo, M., Howell, O., Morgan, A. H., Wells, T., Andrews, Z. B., Burn, D. J., & Davies, J. S. (2020). Unacylated-Ghrelin Impairs Hippocampal Neurogenesis and Memory in Mice and Is Altered in Parkinson's Dementia in Humans. Cell reports. Medicine, 1(7), 100120. https://doi.org/10.1016/j.xcrm.2020.100120
38. Hou, Y., Aboukhatwa, M. A., Lei, D. L., Manaye, K., Khan, I., & Luo, Y. (2010). Anti-depressant natural flavonols modulate BDNF and beta amyloid in neurons and hippocampus of double TgAD

mice. Neuropharmacology, 58(6), 911–920. https://doi.org/10.1016/j.neuropharm.2009.11.002
39. Irfannuddin, I., Sarahdeaz, S. F. P., Murti, K., Santoso, B., & Koibuchi, N. (2021). The effect of ketogenic diets on neurogenesis and apoptosis in the dentate gyrus of the male rat hippocampus. The Journal of Physiological Sciences: JPS, 71(1), 3. https://doi.org/10.1186/s12576-020-00786-7
40. Jang, J., Kim, S. R., Lee, J. E., Lee, S., Son, H. J., Choe, W., Yoon, K. S., Kim, S. S., Yeo, E. J., & Kang, I. (2023). Molecular Mechanisms of Neuroprotection by Ketone Bodies and Ketogenic Diet in Cerebral Ischemia and Neurodegenerative Diseases. International journal of molecular sciences, 25(1), 124. https://doi.org/10.3390/ijms25010124
41. Jensen, N. J., Wodschow, H. Z., Nilsson, M., & Rungby, J. (2020). Effects of Ketone Bodies on Brain Metabolism and Function in Neurodegenerative Diseases. International journal of molecular sciences, 21(22), 8767. https://doi.org/10.3390/ijms21228767
42. Jeong, E. A., Jeon, B. T., Shin, H. J., Kim, N., Lee, D. H., Kim, H. J., Kang, S. S., Cho, G. J., Choi, W. S., & Roh, G. S. (2011). Ketogenic diet-induced peroxisome proliferator-activated receptor-γ activation decreases neuroinflammation in the mouse hippocampus after kainic acid-induced seizures. Experimental neurology, 232(2), 195–202. https://doi.org/10.1016/j.expneurol.2011.09.001
43. Jiang, Y., Li, K., Li, X., Xu, L., & Yang, Z. (2021). Sodium butyrate ameliorates the impairment of synaptic plasticity by inhibiting the neuroinflammation in 5XFAD mice. *Chemico-biological interactions*, *341*, 109452. https://doi.org/10.1016/j.cbi.2021.109452
44. Kellar, D., & Craft, S. (2020). Brain insulin resistance in Alzheimer's disease and related disorders: mechanisms and therapeutic approaches. *The Lancet. Neurology*, *19*(9), 758–766. https://doi.org/10.1016/S1474-4422(20)30231-3
45. Kent, K., Charlton, K., Roodenrys, S., Batterham, M., Potter, J., Traynor, V., Gilbert, H., Morgan, O., & Richards, R. (2017). Consumption of anthocyanin-rich cherry juice for 12 weeks improves memory and cognition in older adults with mild-to-moderate dementia. European journal of nutrition, 56(1), 333–341. https://doi.org/10.1007/s00394-015-1083-y
46. Khalil, M., Shanmugam, H., Abdallah, H., John Britto, J. S., Galerati, I., Gómez-Ambrosi, J., Frühbeck, G., & Portincasa, P. (2022). The

Potential of the Mediterranean Diet to Improve Mitochondrial Function in Experimental Models of Obesity and Metabolic Syndrome. *Nutrients, 14*(15), 3112. https://doi.org/10.3390/nu14153112

47. Kim, C., Pinto, A. M., Bordoli, C., Buckner, L. P., Kaplan, P. C., Del Arenal, I. M., Jeffcock, E. J., Hall, W. L., & Thuret, S. (2020). Energy Restriction Enhances Adult Hippocampal Neurogenesis-Associated Memory after Four Weeks in an Adult Human Population with Central Obesity; a Randomized Controlled Trial. *Nutrients, 12*(3), 638. https://doi.org/10.3390/nu12030638
48. Kim, H., Kang, H., Heo, R. W., Jeon, B. T., Yi, C. O., Shin, H. J., Kim, J., Jeong, S. Y., Kwak, W., Kim, W. H., Kang, S. S., & Roh, G. S. (2016). Caloric restriction improves diabetes-induced cognitive deficits by attenuating neurogranin-associated calcium signaling in high-fat diet-fed mice. *Journal of cerebral blood flow and metabolism: official journal of the International Society of Cerebral Blood Flow and Metabolism, 36*(6), 1098–1110. https://doi.org/10.1177/0271678X15606724
49. Kishi, T., Hirooka, Y., Nagayama, T., Isegawa, K., Katsuki, M., Takesue, K., & Sunagawa, K. (2015). Calorie restriction improves cognitive decline via up-regulation of brain-derived neurotrophic factor: tropomyosin-related kinase B in hippocampus ofobesity-induced hypertensive rats. International Heart Journal, 56(1), 110–115. https://doi.org/10.1536/ihj.14-168
50. Koppel, S. J., Wilkins, H. M., Weidling, I. W., Wang, X., Menta, B. W., & Swerdlow, R. H. (2023). β-Hydroxybutyrate preferentially enhances neuron over astrocyte respiration while signaling cellular quiescence. Mitochondrion, 68, 125–137. https://doi.org/10.1016/j.mito.2022.12.004
51. Li, G., Xie, C., Lu, S., Nichols, R. G., Tian, Y., Li, L., Patel, D., Ma, Y., Brocker, C. N., Yan, T., Krausz, K. W., Xiang, R., Gavrilova, O., Patterson, A. D., & Gonzalez, F. J. (2017). Intermittent Fasting Promotes White Adipose Browning and Decreases Obesity by Shaping the Gut Microbiota. Cell metabolism, 26(5), 801. https://doi.org/10.1016/j.cmet.2017.10.007
52. Lindqvist, A., Mohapel, P., Bouter, B., Frielingsdorf, H., Pizzo, D., Brundin, P., & Erlanson-Albertsson, C. (2006). A high-fat diet impairs hippocampal neurogenesis in male rats. European journal of neurology, 13(12), 1385–1388. https://doi.org/10.1111/j.1468-1331.2006.01500.x

53. Liu, Y., Cheng, A., Li, Y. J., Yang, Y., Kishimoto, Y., Zhang, S., Wang, Y., Wan, R., Raefsky, S. M., Lu, D., Saito, T., Saido, T., Zhu, J., Wu, L. J., & Mattson, M. P. (2019). SIRT3 mediates hippocampal synaptic adaptations to intermittent fasting and ameliorates deficits in APP mutant mice. Nature communications, 10(1), 1886. https://doi.org/10.1038/s41467-019-09897-1
54. López-Otín, C., Blasco, M. A., Partridge, L., Serrano, M., & Kroemer, G. (2013). The hallmarks of aging. Cell, 153(6), 1194–1217. https://doi.org/10.1016/j.cell.2013.05.039
55. Ma, L., Dong, W., Wang, R., Li, Y., Xu, B., Zhang, J., Zhao, Z., & Wang, Y. (2015). Effect of caloric restriction on the SIRT1/mTOR signaling pathways in senile mice. *Brain research bulletin*, *116*, 67–72. https://doi.org/10.1016/j.brainresbull.2015.06.004
56. Ma, L., Wang, R., Dong, W., & Zhao, Z. (2018). Caloric restriction can improve learning and memory in C57/BL mice, probably via regulation of the AMPK signaling pathway. *Experimental gerontology*, *102*, 28–35. https://doi.org/10.1016/j.exger.2017.11.013
57. Maiese K. (2018). The mechanistic target of rapamycin (mTOR) and the silent mating-type information regulation 2 homolog 1 (SIRT1): oversight for neurodegenerative disorders. *Biochemical Society transactions*, *46*(2), 351–360. https://doi.org/10.1042/BST20170121
58. Marosi, K., Kim, S. W., Moehl, K., Scheibye-Knudsen, M., Cheng, A., Cutler, R., Camandola, S., & Mattson, M. P. (2016). 3-Hydroxybutyrate regulates energy metabolism and induces BDNF expression in cerebral cortical neurons. Journal of neurochemistry, 139(5), 769–781. https://doi.org/10.1111/jnc.13868
59. Martínez-Lapiscina, E. H., Clavero, P., Toledo, E., Estruch, R., Salas-Salvadó, J., San Julián, B., Sanchez-Tainta, A., Ros, E., Valls-Pedret, C., & Martinez-Gonzalez, M. Á. (2013). Mediterranean diet improves cognition: the PREDIMED-NAVARRA randomized trial. Journal of neurology, neurosurgery, and psychiatry, 84(12), 1318–1325. https://doi.org/10.1136/jnnp-2012-304792
60. Mattson, M. P., Moehl, K., Ghena, N., Schmaedick, M., & Cheng, A. (2018). Intermittent metabolic switching, neuroplasticity and brain health. *Nature reviews. Neuroscience*, *19*(2), 63–80. https://doi.org/10.1038/nrn.2017.156
61. McCann, J. C., & Ames, B. N. (2005). Is docosahexaenoic acid, an n-3 long-chain polyunsaturated fatty acid, required for development of normal brain function? An overview of evidence from cognitive

and behavioral tests in humans and animals. *The American journal of clinical nutrition*, *82*(2), 281–295. https://doi.org/10.1093/ajcn.82.2.281

62. Meng, Y., Sun, J., & Zhang, G. (2024). Take the bull by the horns and tackle the potential downsides of the ketogenic diet. Nutrition (Burbank, Los Angeles County, Calif.), 125, 112480. https://doi.org/10.1016/j.nut.2024.112480
63. Mirabelli, M., Chiefari, E., Arcidiacono, B., Corigliano, D. M., Brunetti, F. S., Maggisano, V., Russo, D., Foti, D. P., & Brunetti, A. (2020). Mediterranean Diet Nutrients to Turn the Tide against Insulin Resistance and Related Diseases. Nutrients, 12(4), 1066. https://doi.org/10.3390/nu12041066
64. Most, J., Tosti, V., Redman, L. M., & Fontana, L. (2017). Calorie restriction in humans: An update. *Ageing research reviews*, *39*, 36–45. https://doi.org/10.1016/j.arr.2016.08.005
65. Murphy, T., Dias, G. P., & Thuret, S. (2014). Effects of diet on brain plasticity in animal and human studies: mind the gap. *Neural plasticity*, *2014*, 563160. https://doi.org/10.1155/2014/563160
66. Okamoto, K., Kihira, T., Kobashi, G., Washio, M., Sasaki, S., Yokoyama, T., Miyake, Y., Sakamoto, N., Inaba, Y., & Nagai, M. (2009). Fruit and vegetable intake and risk of amyotrophic lateral sclerosis in Japan. *Neuroepidemiology*, *32*(4), 251–256. https://doi.org/10.1159/000201563
67. Okubo, H., Miyake, Y., Sasaki, S., Murakami, K., Tanaka, K., Fukushima, W., Kiyohara, C., Tsuboi, Y., Yamada, T., Oeda, T., Shimada, H., Kawamura, N., Sakae, N., Fukuyama, H., Hirota, Y., Nagai, M., & Fukuoka Kinki Parkinson's Disease Study Group (2012). Dietary patterns and risk of Parkinson's disease: a case-control study in Japan. European journal of neurology, 19(5), 681–688. https://doi.org/10.1111/j.1468-1331.2011.03600.x
68. Ott, B., Skurk, T., Hastreiter, L., Lagkouvardos, I., Fischer, S., Büttner, J., Kellerer, T., Clavel, T., Rychlik, M., Haller, D., & Hauner, H. (2017). Effect of caloric restriction on gut permeability, inflammation markers, and fecal microbiota in obese women. Scientific reports, 7(1), 11955. https://doi.org/10.1038/s41598-017-12109-9
69. Paknahad, Z., Sheklabadi, E., Derakhshan, Y., Bagherniya, M., & Chitsaz, A. (2020). The effect of the Mediterranean diet on cognitive function in patients with Parkinson's disease: A randomized clinical controlled trial. *Complementary therapies in medicine*, *50*, 102366. https://doi.org/10.1016/j.ctim.2020.102366

70. Pathak, S. J., & Baar, K. (2023). Ketogenic Diets and Mitochondrial Function: Benefits for Aging But Not for Athletes. Exercise and Sport Sciences Reviews, 51(1), 27–33. https://doi.org/10.1249/JES.0000000000000307
71. Pelletier, A., Barul, C., Féart, C., Helmer, C., Bernard, C., Periot, O., Dilharreguy, B., Dartigues, J. F., Allard, M., Barberger-Gateau, P., Catheline, G., & Samieri, C. (2015). Mediterranean diet and preserved brain structural connectivity in older subjects. Alzheimer's & dementia: The Journal of the Alzheimer's Association, 11(9), 1023–1031. https://doi.org/10.1016/j.jalz.2015.06.1888
72. Phillips C. Lifestyle Modulators of Neuroplasticity: How Physical Activity, Mental Engagement, and Diet Promote Cognitive Health during Aging. Neural Plast. 2017;2017:3589271. https://doi.org/10.1155/2017/3589271
73. Pickersgill, J. W., Turco, C. V., Ramdeo, K., Rehsi, R. S., Foglia, S. D., & Nelson, A. J. (2022). The Combined Influences of Exercise, Diet and Sleep on Neuroplasticity. *Frontiers in psychology*, *13*, 831819. https://doi.org/10.3389/fpsyg.2022.831819
74. Puchalska, P., & Crawford, P. A. (2017). Multi-dimensional Roles of Ketone Bodies in Fuel Metabolism, Signaling, and Therapeutics. *Cell metabolism*, *25*(2), 262–284. https://doi.org/10.1016/j.cmet.2016.12.022
75. Ran, M., Li, Z., Yang, L., Tong, L., Zhang, L., & Dong, H. (2015). Calorie restriction attenuates cerebral ischemic injury via increasing SIRT1 synthesis in the rat. Brain research, 1610, 61–68. https://doi.org/10.1016/j.brainres.2015.03.043
76. Roberts, L. D., Hornsby, A. K., Thomas, A., Sassi, M., Kinzett, A., Hsiao, N., David, B. R., Good, M., Wells, T., & Davies, J. S. (2023). The 5:2 diet does not increase adult hippocampal neurogenesis or enhance spatial memory in mice. EMBO reports, 24(12), e57269. https://doi.org/10.15252/embr.202357269
77. Sahab Negah, S., & Forouzanfar, F. (2023). Oxidative stress is a new avenue for treatment of neuropsychiatric disorders: hype of hope? Current molecular medicine https://doi.org/10.2174/1566524023666230904150907
78. Sánchez-Villegas, A., Galbete, C., Martinez-González, M. A., Martinez, J. A., Razquin, C., Salas-Salvadó, J., Estruch, R., Buil-Cosiales, P., & Martí, A. (2011). The effect of the Mediterranean diet on plasma brain-derived neurotrophic factor (BDNF) levels: the PREDIMED-NAVARRA randomized trial. Nutritional neuroscience, 14(5), 195–201. https://doi.org/10.1179/1476830511Y.0000000011

79. Schertzer, J. D., & Lam, T. K. T. (2021). Peripheral and central regulation of insulin by the intestine and microbiome. American journal of physiology. Endocrinology and metabolism, 320(2), E234–E239. https://doi.org/10.1152/ajpendo.00547.2020
80. Schoffen, J. P., Santi Rampazzo, A. P., Cirilo, C. P., Zapater, M. C., Vicentini, F. A., Comar, J. F., Bracht, A., & Natali, M. R. (2014). Food restriction enhances oxidative status in aging rats with neuroprotective effects on myenteric neuron populations in the proximal colon. Experimental gerontology, 51, 54–64. https://doi.org/10.1016/j.exger.2014.01.001
81. Schwingshackl, L., Morze, J., & Hoffmann, G. (2020). Mediterranean diet and health status: Active ingredients and pharmacological mechanisms. *British journal of pharmacology*, *177*(6), 1241–1257. https://doi.org/10.1111/bph.14778
82. Shannon, O. M., Ashor, A. W., Scialo, F., Saretzki, G., Martin-Ruiz, C., Lara, J., Matu, J., Griffiths, A., Robinson, N., Lillà, L., Stevenson, E., Stephan, B. C. M., Minihane, A. M., Siervo, M., & Mathers, J. C. (2021). Mediterranean diet and the hallmarks of aging. European journal of clinical nutrition, 75(8), 1176–1192. https://doi.org/10.1038/s41430-020-00841-x
83. Shen, Y., Kapfhamer, D., Minnella, A. M., Kim, J. E., Won, S. J., Chen, Y., Huang, Y., Low, L. H., Massa, S. M., & Swanson, R. A. (2017). Bioenergetic state regulates innate inflammatory responses through the transcriptional co-repressor CtBP. Nature communications, 8(1), 624. https://doi.org/10.1038/s41467-017-00707-0
84. Shively, C. A., Negrey, J. D., Frye, B. M., Johnson, C. S. C., Kim, J., Barcus, R. A., & et al. (2023). Mediterranean diet protects against neuroinflammation and degeneration in female nonhuman primates. The journal of the Alzheimer's Association. 19:S10 Alzheimer's Imaging Consortium. e081770 https://doi.org/10.1002/alz.081770
85. Siervo, M., Shannon, O. M., Llewellyn, D. J., Stephan, B. C., & Fontana, L. (2021). Mediterranean diet and cognitive function: From methodology to mechanisms of action. Free radical biology & medicine, 176, 105–117. https://doi.org/10.1016/j.freeradbiomed.2021.09.018
86. Steneberg, P., Lindahl, E., Dahl, U., Lidh, E., Straseviciene, J., Backlund, F., Kjellkvist, E., Berggren, E., Lundberg, I., Bergqvist, I., Ericsson, M., Eriksson, B., Linde, K., Westman, J., Edlund, T., & Edlund, H. (2018). PAN-AMPK activator O304 improves glucose homeostasis

and microvascular perfusion in mice and type 2 diabetes patients. JCI insight, 3(12), e99114. https://doi.org/10.1172/jci.insight.99114

87. Strandberg, J., Kondziella, D., Thorlin, T., & Asztely, F. (2008). Ketogenic diet does not disturb neurogenesis in the dentate gyrus in rats. Neuroreport, 19(12), 1235–1237. https://doi.org/10.1097/WNR.0b013e32830a7109
88. Strom, A., & Jensen, R. A. (1951). Mortality from circulatory diseases in Norway 1940-1945. *Lancet (London, England)*, *1*(6647), 126–129. https://doi.org/10.1016/s0140-6736(51)91210-x
89. Sui, Y., Tian, R., & Lu, N. (2019). NADPH oxidase is a primary target for antioxidant effects by inorganic nitrite in lipopolysaccharide-induced oxidative stress in mice and macrophage cells. *Nitric oxide: biology and chemistry*, *89*, 46–53. https://doi.org/10.1016/j.niox.2019.05.002
90. Sun, Y., Ma, C., Sun, H., Wang, H., Peng, W., Zhou, Z., Wang, H., Pi, C., Shi, Y., & He, X. (2020). Metabolism: A Novel Shared Link between Diabetes Mellitus and Alzheimer's Disease. Journal of diabetes research, 2020, 4981814. https://doi.org/10.1155/2020/4981814
91. Suzuki, H., Morikawa, Y., & Takahashi, H. (2001). Effect of DHA oil supplementation on intelligence and visual acuity in the elderly. *World review of nutrition and dietetics*, *88*, 68–71. https://doi.org/10.1159/000059767
92. Svart, M., Gormsen, L. C., Hansen, J., Zeidler, D., Gejl, M., Vang, K., Aanerud, J., & Moeller, N. (2018). Regional cerebral effects of ketone body infusion with 3-hydroxybutyrate in humans: Reduced glucose uptake, unchanged oxygen consumption and increased blood flow by positron emission tomography. A randomized, controlled trial. PloS one, 13(2), e0190556. https://doi.org/10.1371/journal.pone.0190556
93. Tanaka, T., Talegawkar, S. A., Jin, Y., Colpo, M., Ferrucci, L., & Bandinelli, S. (2018). Adherence to a Mediterranean Diet Protects from Cognitive Decline in the Invecchiare in Chianti Study of Aging. Nutrients, 10(12), 2007. https://doi.org/10.3390/nu10122007
94. Valls-Pedret, C., Sala-Vila, A., Serra-Mir, M., Corella, D., de la Torre, R., Martínez-González, M. Á., Martínez-Lapiscina, E. H., Fitó, M., Pérez-Heras, A., Salas-Salvadó, J., Estruch, R., & Ros, E. (2015). Mediterranean Diet and Age-Related Cognitive Decline: A Randomized Clinical Trial. JAMA internal medicine, 175(7), 1094–1103. https://doi.org/10.1001/jamainternmed.2015.1668

95. Vepsäläinen, S., Koivisto, H., Pekkarinen, E., Mäkinen, P., Dobson, G., McDougall, G. J., Stewart, D., Haapasalo, A., Karjalainen, R. O., Tanila, H., & Hiltunen, M. (2013). Anthocyanin-enriched bilberry and blackcurrant extracts modulate amyloid precursor protein processing and alleviate behavioral abnormalities in the APP/PS1 mouse model of Alzheimer's disease. The Journal of Nutritional Biochemistry, 24(1), 360–370. https://doi.org/10.1016/j.jnutbio.2012.07.006
96. Vulevic, J., Juric, A., Tzortzis, G., & Gibson, G. R. (2013). A mixture of trans-galactooligosaccharides reduces markers of metabolic syndrome and modulates overweight adults' fecal microbiota and immune function. The Journal of Nutrition, 143(3), 324–331. https://doi.org/10.3945/jn.112.166132
97. Witte, A. V., Fobker, M., Gellner, R., Knecht, S., & Flöel, A. (2009). Caloric restriction improves memory in elderly humans. *Proceedings of the National Academy of Sciences of the United States of America*, *106*(4), 1255–1260. <-https://doi.org/10.1073/pnas.0808587106
98. Xu, S., Tao, H., Cao, W., Cao, L., Lin, Y., Zhao, S., Xu, W., Cao, J., & Zhao, J. (2021). Ketogenic diets inhibit mitochondrial biogenesis and induce cardiac fibrosis. Sig Transduct Target Ther 6, 54. https://doi.org/10.1038/s41392-020-00411-4
99. Yang, Q., Zhou, Y., Sun, Y., Luo, Y., Shen, Y., & Shao, A. (2020). Will Sirtuins Be Promising Therapeutic Targets for TBI and Associated Neurodegenerative Diseases? *Frontiers in neuroscience*, *14*, 791. https://doi.org/10.3389/fnins.2020.00791
100. Yeo, X. Y., Tan, L. Y., Chae, W. R., Lee, D. Y., Lee, Y. A., Wuestefeld, T., & Jung, S. (2023). Liver's influence on the brain through the action of bile acids. Frontiers in neuroscience, 17, 1123967. https://doi.org/10.3389/fnins.2023.1123967
101. Zhang, B., Huang, R., Xu, S., He, Y., Li, J., Wan, L., Xiao, B., & Luo, Z. (2023). Dietary energy restriction in neurological diseases: what's new? *European journal of nutrition*, *62*(2), 573–588. https://doi.org/10.1007/s00394-022-03036-1

Exercise-Induced Neuroplasticity: Mechanisms and Benefits

İdil Özkoç

1. Introduction

Environmental factors can change organs and tissues throughout life (Westneat et al., 2019). Such changes are best in adaptable structures like the brain, which evolves during growth and aging. Neuroplasticity is the central nervous system's ability to grow and adapt due to psychological and environmental factors (De Sousa Fernandes et al., 2020). This process is driven by increased neurotrophin levels, which affect cell signaling development (Kempermann et al., 2018).

Exercise boosts neuroplasticity (NP), improving symptoms and outcomes in various conditions. This includes tremors in Parkinson's disease, gait speed, working memory, and behavior in multiple sclerosis, Parkinson's disease, and stroke (Silva-Batista et al., 2020; Tymofiyeva & Gaschler, 2021). This is true for healthy young adults, older adults, and those with neurological conditions.

2. Exercise and Brain

Exercise boosts brain health by improving cellular balance, learning, memory, and thinking. Various studies have shown its benefits, but the exact pathways affected by different exercise types (like type, frequency, volume, and duration) remain unclear (Vilela et al., 2017). However, increased cardiorespiratory fitness is linked to changes in the hippocampus (Kleemeyer et al., 2016; Maass et al., 2015).

Exercise also enhances brain plasticity. It increases neuron activity, boosts synaptic strength, and aids in the growth of dendrites and myelination of axons. These changes improve recovery in thinking and movement, making exercise a top rehabilitation method (K. Chen et al., 2019).

Exercise affects neuroplasticity based on its length and intensity. Tailored regimens can boost cognitive function after a stroke. Aerobic exercise enhances the hippocampus and improves verbal, learning, and spatial memory (Erickson et al., 2009; Ten Brinke et al., 2015). Meanwhile, resistance training aids older adults with executive skills (Liu-Ambrose et al., 2012).

Recent studies show that exercise boosts cognitive function in healthy adults. It improves motor learning, cognitive flexibility, learning, and executive control (Masley et al., 2009; Pontifex et al., 2009; Winter et al., 2007).

2.1. Hippocampal Volume

The hippocampus is key for neuroplasticity. It is highly affected by physical activity, especially aerobic exercise (Cotman et al., 2007). Voluntary exercise boosts neurogenesis in the hippocampus (Van Praag et al., 2005) and prevents age-related declines in cell growth. Trophic factors, including brain-derived neurotrophic factor (BDNF) and insulin-like growth factor 1 (IGF-1), may mediate changes in hippocampus shape (Ferris et al., 2007; Vaynman et al., 2004). Exercise causes the hippocampal Akt and CREB signal pathways to become active (Cechella et al., 2014).

2.2. Dendrit

Dendritic shapes vary widely among neurons and are vital for the nervous system (Bozelos et al., 2016). Dendrites and axons are key for exercise-related brain changes (Nesin et al., 2019). Exercise can enhance dendritic complexity in several hippocampal areas (Solstad et al., 2006). It also prevents loss of spine density and boosts brain cell growth. In stressed rats, running reduces depression and improves memory by increasing dendritic length, promoting brain cell growth, and restoring spine density (Yau et al., 2011). Exercise also fixes dendritic issues in the prefrontal cortex (Ghalandari-Shamami et al., 2019).

2.3. Synaps

After a brain injury, synaptic plasticity aids neuronal repair. Cerebral ischemia might disrupt synapse structure, leading to dysfunction (Yepes, 2020). This plasticity includes short-term and long-term changes (Citri & Malenka, 2008).

Exercise enhances long-term potentiation by strengthening connections between the dentate gyrus and the entorhinal cortex (Liang et al., 2021). It also promotes dendritic growth, reduces neuron and dendrite loss in the hippocampus, and boosts cognitive function (Y. Lin et al., 2015). Notably, induced by electrical stimulation, involuntary exercise offers more excellent cognitive protection than forced or voluntary activity (Liang et al., 2021).

Reports show that synapses near the damaged area have more vesicles and are less damaged (Xing & Bai, 2020). Furthermore, rehabilitation training can make new synapses in rats more stable after a brain infarction (Clark et al., 2019).

2.4. Neurotrophins

Neurotrophins are vital proteins for our nervous systems. They support survival, growth, and function (Kozorovitskiy & Gould, 2010; Yamaguchi et al., 2016). Key neurotrophins include GDNF, nerve growth factor (NGF), neurotrophin 3 (NT3), neurotrophin 4 (NT4) and BDNF. These are crucial for neuroplasticity. Research shows that low levels of these proteins, especially BDNF, harm cognitive abilities, including learning, memory, and focus (De Sousa Fernandes et al., 2020).

Tropomyosin-related receptor kinase B activates the BDNF signaling pathway. Its increased levels can reduce brain issues like depression (Zborowski et al., 2019).

Neurogenesis, angiogenesis, and synaptogenesis drive neuroplasticity. Neurotrophins are key to these processes. They enhance neuronal survival, growth, repair, and differentiation (Huang & Reichardt, 2001). BDNF affects neurons in both the peripheral and central nervous systems. It promotes survival and aids in the formation and maturation of synapses (Song et al., 2020).

BDNF is crucial for learning and memory. It boosts long-term potentiation and neurogenesis in key brain areas (Chen & Russo-Neustadt, 2009; Voss et al., 2011).

BDNF also plays a key role in motor learning (Penna et al., 2021). It activates Akt and CREB in the rat hippocampus. This enhances neuroplasticity during aerobic exercise (Chen & Russo-Neustadt, 2009; Voss et al., 2011).

Exercise benefits the brain by improving blood flow and raising neurotransmitters like serotonin and norepinephrine (Ploughman, 2008). It also boosts neurotrophins, aiding in synaptogenesis and neuronal survival (Schinder & Poo, 2000).

Exercise boosts brain function and increases BDNF levels (Erickson et al., 2009; Griffin et al., 2011). It also improves temporal lobe function. In adult mice, it promotes cell growth and development in the hippocampus, enhancing memory and processing (Liang et al., 2021).

In rats with brain damage, exercise aids neuron recovery and improves movement (Zhang et al., 2013). It also benefits cognition and raises neurofilament levels in older adults (Desai et al., 2022).

Research shows athletes benefit in mood and cognition from various physical activities (Aly et al., 2019; Basso & Suzuki, 2017).

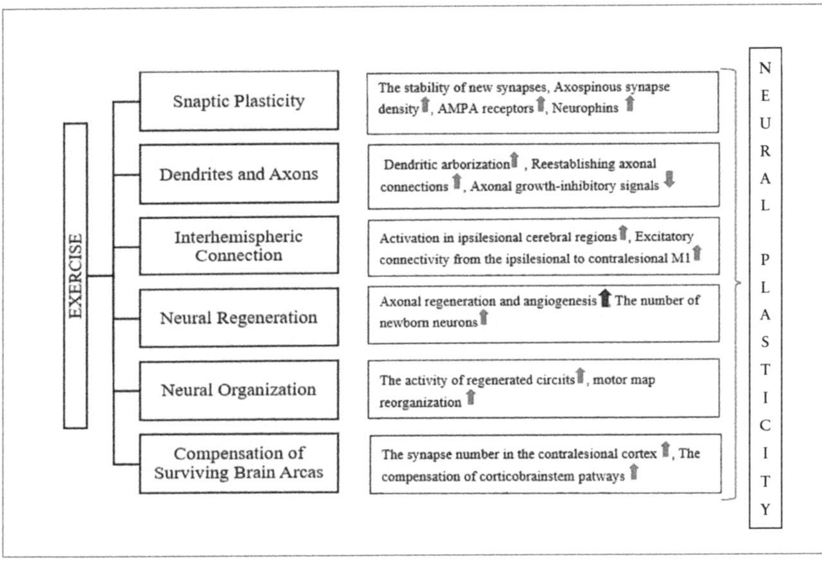

Figure 1: Relationship between exercise and neuroplasticity (Xing & Bai, 2020)

According to Figure 1, exercise boosts brain plasticity by enhancing neurotrophins and synaptic activity. It also aids in neuronal repair and encourages healthy brain areas to adapt. Research shows potential in exercise rehab for stroke recovery, yet more studies are needed. Robot-assisted thera py proves beneficial for stroke patients, improving arm movement, learning, and memory, likely by promoting neuroplasticity. Such therapies should be widely adopted in clinics (Xing & Bai, 2020).

Short-term exercise boosts learning, memory, brain plasticity, and new neuron growth. These benefits are linked to increased BDNF in the hippocampus (Vaynman et al., 2004; Vaynman & Gomez-Pinilla, 2005).

3. Exercise and Neuroplasticity

3.1. Aerobic Exercise and Neuroplasticity

Aerobic exercise is a key promoter of neuroplasticity. It benefits nearly all organ systems and improves cardiovascular, neurological, motor, and cognitive health (Gallardo-Gómez et al., 2022). It likely boosts brain function by enhancing the brain's ability to form and reorganize connections (T.-W. Lin et al., 2018). However, the effect of exercise intensity on neuroplasticity is still unclear (Garber et al., 2011).

Studies show aerobic exercise increases hippocampus volume in young and older adults (Thomas et al., 2016). However, there is insufficient evidence to confirm a direct link between aerobic exercise and structural changes in the adult hippocampus. Moreover, these findings have not always been consistent (Firth et al., 2018; Jonasson et al., 2017).

Aerobic exercise boosts gray and white matter in the prefrontal and temporal brain regions (Colcombe et al., 2006) and improves white matter integrity in these areas (Voss et al., 2013).

Exercise can enhance cognitive performance in young, healthy individuals. High-impact running boosts vocabulary, while cycling improves performance in the Stroop task and map recognition (Griffin et al., 2011).

Moderate aerobic exercise boosts brain responses for motor recovery via plasticity. This likely increases brain-derived neurotrophic factor (BDNF), aiding neuron growth and protection (Penna et al., 2021). Combining aerobic exercise with cognitive training enhances walking speed, fitness, and cognitive scores. However, aerobic exercise alone shows different results. Thus, combining methods boosts neuroplasticity (Triviño-Paredes et al., 2016).

In chronic stroke patients, aerobic exercise significantly improved motor learning. Specifically, three sessions a week for eight weeks of moderate-intensity exercise enhanced cognitive functions, especially motor learning (Quaney et al., 2009). In healthy individuals, aerobic exercise raises serum levels of neurotrophic factors, crucial for neuron and glial cell survival and plasticity (Saunders et al., 2014). Additionally, it enhances executive function and BDNF levels (El-Tamawy et al., 2014).

In animal studies, high-intensity aerobic exercise was found to harm neuroplasticity and impair memory in mice by disrupting key brain pathways (Aguiar et al., 2010).

For humans, aerobic exercise affects gray matter as we age, especially in cognitive areas. Its benefits are closely linked to hippocampal size (Erickson et al., 2014). Furthermore, running activates the dorsolateral prefrontal cortex, linked to improved executive function (Damrongthai et al., 2021).

Aerobic exercise, especially forced ones, aids stroke recovery (Linder et al., 2019). Doing it three times a week for a year prevents hippocampus aging (Erickson et al., 2009). It also increases hippocampus and entorhinal sizes (Duzel et al., 2016; Whiteman et al., 2016).

Many studies highlight the long-term benefits of physical activity on brain function, especially in the prefrontal areas, which affect decision-making (Basso & Suzuki, 2017). It also improves mood, working memory, attention, cognitive flexibility, and control. These benefits last up to two hours after exercise (Basso & Suzuki, 2017; Chiu et al., 2017).

3.1.1. Running

Running significantly boosts neurogenesis (Ra et al., 2002). Endurance runners, compared to non-runners, show better brain connectivity and structure. Specifically, they have more extensive gray matter and increased thickness in the left precentral gyrus. Moreover, their hippocampal volume is greater, and connectivity with other brain regions improves (Cao et al., 2021). Regular running enhances brain structure and function (Serra et al., 2022).

Additionally, running promotes growth in the dentate gyrus of the hippocampus, which continues into adulthood (Spalding et al., 2013).It also increases neurotrophic factors and synaptic modulators, aiding learning and attention early in stroke rehab (Ploughman et al., 2007).According to Yong et al., rats on a treadmill or trained to reach may have higher amounts of neurotrophic factors. The amounts of NGF and BDNF in the animals given the two types of exercise do not appear to differ much (Yong et al., 2017).

After cerebral ischemia, skilled reaching training is more effective than forced running wheel exercise at promoting motor recovery and axonal plasticity in the corticofugal tract (Zhang et al., 2018). Running and reach training together can significantly increase a rat's competent reaching ability, although postural support and gait do not improve following an ischemic stroke (Yong et al., 2017). According to Van Praag et al., voluntary wheel running increases cell proliferation and survival in mice (Van Praag et al., 1999).

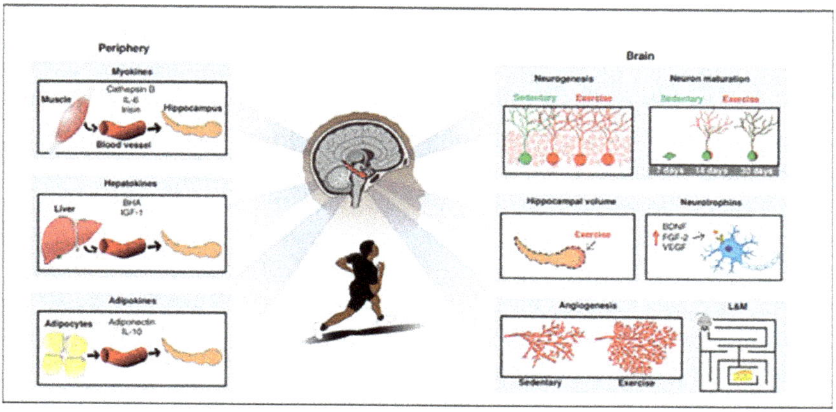

Figure 2: Exercise and brain (Fischer et al., 2014; Steib et al., 2014)

Running alters the hippocampus (Figure 2). It boosts hippocampal volume, speeds up new neuron development, and promotes blood vessel growth. Factors from the brain and body might enhance brain adaptability and memory. Hepatokines, myokines, and adipokines could increase growth factors in the brain. L&M, learning and memory; VEGF, vascular endothelial growth factor; BHA, b-hydroxybutyrate; BDNF, brain-derived neurotrophic factor; FGF-2, fibroblast growth factor 2; IGF-1, insulin-like growth factor 1; IL-6, interleukin 6; IL-10, interleukin 10 (Cooper et al., 2018).

Running boosts the creation of new neurons in the hippocampal dentate gyrus (DG) (Patten et al., 2016). Age, genetics, and running conditions can increase this increase 2–3 times (Cooper et al., 2018). This effect is limited to the dorsal DG, not the ventral DG (Vivar et al., 2016). Additionally, running is linked to better spatial memory (Patten et al., 2016).

The rise in new neurons from running might explain some cognitive benefits. However, neuron development, shape, and connection improvements are also key. Running could speed up neuron maturation and reduce the growth cycle of progenitor cells (Fischer et al., 2014; Steib et al., 2014).

Running activates key brain areas that provide inputs for spatial and temporal processing (Vivar et al., 2016). These areas also help with memory and initiate hippocampal rhythms (Dillingham et al., 2015). Furthermore, running reduces certain types of neuron connections in the hippocampus (Vivar et al., 2016).

3.1.2. Swimming

Master swimmers age better than non-athletes, showing improvements in various functions. This study confirmed aerobic exercise builds reserves (Nørregaard et al., 2014). Young, competitive swimmers showed better motor skills and brain changes than novices (Sato et al., 2020). Water immersion boosts blood flow and reduces muscle activity (Sato et al., 2013). It also enhances venous return. Sensorimotor learning involves many brain areas. Elite swimmers outperform non-experts in sensorimotor skills. However, non-experts have a more excitable primary motor cortex (Serra et al., 2022).

For stroke survivors, water training aids in daily tasks and strength (Mehrholz et al., 2011). Additionally, it boosts memory, attention, executive function, and fitness in older adults (Fedor et al., 2015).

A study showed that swimming improves short-term memory in old rats (Cechella et al., 2014) and boosts new brain cell growth (Ra et al., 2002).

Swimming affects brain functions. One session can reduce stress and improve thinking. Long-term swimming prevents age-related changes in specific brain areas. However, it doesn't increase cell growth in the mouse brain (Ra et al., 2002).

Most research on exercise for stroke recovery focuses on duration and intensity. Currently, swimming sessions are categorized by duration: short (5 min), moderate (10 min), and long (20 min). Each duration's impact on behavior and brain growth was assessed. Findings suggest exercise intensity boosts brain adaptability. In rats after a stroke, regular swimming improved behavior and increased brain growth factors like VEGF and BDNF (Song et al., 2019).

For seniors, water exercise enhances heart health, memory, attention, and decision-making (Fedor et al., 2015). In stroke-affected rats, swimming boosted brain cell growth in the hippocampus (Chae et al., 2014).

This study found that a week of water aerobics significantly improved decision-making, attention, and memory. Land exercises also benefit seniors, enhancing decision-making, attention, and memory (Angevaren et al., 2008; Kara et al., 2005; Langlois et al., 2013).

3.1.3. Treadmill Exercise

Treadmill exercise can boost recovery from stroke-related brain damage in rats. It likely improves dendritic and synaptic plasticity by activating the caveolin-1/VEGF pathway (Xie et al., 2019). Additionally, Trejo et al. showed that such exercise raises cell numbers in the dentate gyrus of rats (Trejo et al., 2001).

3.2. Skilled Reaching Training

Running exercise may not have the same impact on improving motor function as skilled reaching training. Exercise, whether forced or voluntary, may not substantially impact neuroplasticity. Numerous forms of exercise therapy, including video games, biofeedback balance training, and device-assisted mirror symmetrical bimanual movements, have been effective for post-stroke patients in clinical settings (Xing & Bai, 2020).

3.3. High-, Moderate-, and Low-Intensity Exercise and Neuroplasticity

Exercise intensity affects brain recovery after a stroke. There are three levels: high, moderate, and low (Xing & Bai, 2020). High-intensity training (HIT) is best for restoring brain function post-stroke, surpassing moderate and low exercises (Luo et al., 2019).

Moderate-intensity training (MIT) can also help. It improves mobility and exercise ability by managing neural plasticity and inflammation (Oliveira et al., 2019). After a stroke, moderate aerobic exercise is advised. It boosts walking ability, peak oxygen uptake (VO2peak), and maximum running speed (Smax), all key to a good quality of life (Marsden et al., 2013).

High-intensity exercise might enhance brain recovery more due to its greater impact on neuroplasticity. Meanwhile, low-intensity exercise could improve spatial memory (Xing & Bai, 2020).

Low-intensity treadmill exercise is better than high-intensity walking. Moderate treadmill exercise boosts NGF levels (Chae et al., 2014; Ploughman et al., 2007). Moreover, low to moderate exercise might encourage brain cell growth. In contrast, high-intensity exercise could harm it (Aguiar et al., 2010; Gegentonglaga et al., 2013).

In a study with rats, high-intensity interval training proved beneficial. It increased vascular endothelial growth factor and improved stroke outcomes (Rezaei et al., 2018).

The latency of the motor evoked potential increased with exercise intensity. This means that moderate-intensity training reduced it more than high-intensity interval training (HIIT) (Abraha et al., 2018). After training, high-intensity activities raised BDNF levels. This was likely due to the sample's age and neurological issues, which lessened intervention effects (Charalambous et al., 2018). Conversely, high-intensity interval training boosted VEGF1 and IgF1 levels significantly. However, moderate exercise did not (Boyne et al., 2020).

High-intensity interval training significantly boosted BDNF more than moderate training. It also linked reduced intracortical inhibition to increased BDNF. This suggests BDNF levels might indicate neuroplasticity, as reduced inhibition is tied to motor relearning (Penna et al., 2021).

The impact of exercise intensity on neuroplasticity and its link to motor and cognitive performance remains unclear in healthy people and those with multiple sclerosis, Parkinson's disease, or after a stroke (Hortobágyi et al., 2022). Meanwhile, low-intensity activities like yoga and tai chi have been shown to improve motor and cognitive function at any age (Hoy et al., 2021; Ren et al., 2021).

Walking at low intensity improved aerobic fitness. It increased gray matter in the brain and white matter integrity. It also reduced white matter lesions (F. T. Chen et al., 2020; Mendez Colmenares et al., 2021). Furthermore, low-intensity exercise and cognitive training boosted NP markers and cognitive performance (Chow et al., 2021).

Exercise intensity affects neuroplasticity markers. Higher intensity boosts cognitive and motor performance more. It increases blood flow, glucose, and oxygen in the brain. It also makes blood vessels stiffer and raises nerve growth factors. These factors include brain-derived neurotrophic factor, insulin-like growth factor, and vascular endothelial growth factor. Furthermore, exercise reduces neuroinflammation markers like C-reactive protein and cytokines. It also enhances gait, balance, attention, executive function, processing speed, and memory. This improvement happens after both aerobic and resistance training (Hortobágyi et al., 2022).

3.4. Motor Coordination Training and Balance Training

Like aerobic exercise, balance and coordination training can change the brain's structure. For instance, such training has been linked to a larger hippocampus (Niemann et al., 2014). Moreover, short-term balance training increases gray matter in several brain areas (Taubert et al., 2010). Notably, better balance is associated with increased frontal brain volume, suggesting learning-induced changes. Interestingly, the putamen's volume decreased during six weeks of training (Taubert et al., 2010). Additionally, those who improved in balance training showed less connectivity between certain brain areas, unlike the inactive group (Magon et al., 2016).

A study on balance showed that experts in ballet or slacklining outperformed control participants. The controls had no balance training. This was in a task reliant on the hippocampus for learning. Additionally, the experts had larger volumes in the posterior hippocampus and putamen. (Hüfner et al., 2011), but smaller anterior hippocampal and putamen volumes (Hänggi et al., 2010). Moreover, the balance specialists' occipital gray matter volume was larger and their insular cortex had less gray matter (Hüfner et al., 2011).

Exercise engages the whole body. When balancing skills are tested without high metabolic demands, the brain shows unique changes. These changes are both structural and functional. During any physical activity, vestibular input is key. It aids in neuroplasticity, as shown by Smith in 2017. Balance control requires combining proprioceptive, visual, and motor inputs. This process starts in the brain stem's vestibular nuclei, according to Cullen in 2012. Moreover, these nuclei connect to the hippocampus and other brain areas. This link suggests the vestibular system is vital for higher cognitive functions, as noted by El-Sayes and colleagues in 2019.

Thus, whole-body exercise relying on the vestibular system seems to adapt this system structurally. In contrast to a group that only relaxed, our recent study showed something remarkable. After 12 weeks of balance training, healthy individuals improved. Their balance, associative memory, and spatial cognition all got better (Rogge et al., 2017).

El-Sayes' study involved 37 healthy adults aged 19 to 65. They were divided into two groups: one for balance training and the other for relaxation. Both groups trained twice a week for 12 weeks. Balance tests and brain scans were done before and after training. Only the balance group showed significant improvement. Their brains showed more cortical thickness in areas linked to balance and less volume in the putamen. Better balance was linked to thicker cortex and smaller putamen. This suggests balance training boosts brain areas related to movement perception. It may also enhance the cognitive benefits of exercise by activating visual-vestibular pathways (El-Sayes et al., 2019).

3.5. Exergame

Exercise positively affects the brain's structure and function. It boosts neurotrophic factors like BDNF. This leads to better synapses, new neurons in the hippocampus, and more blood vessels in the brain (Geerlings et al., 2015; Lange-Asschenfeldt & Kojda, 2008). Studies show that combining exercise with cognitive tasks improves older adults' cognitive abilities more than exercise alone (Rathore & Lom, 2017; P. J. Smith et al., 2010).

Recent technological advancements have introduced "exergames" to support combined training. These games are available on platforms like "Nintendo Wii," "Xbox," and "Playstation" (Konstantinidis et al., 2016; Manera et al., 2017). Exergames enable exercises that engage both physical and cognitive skills, as players interact with a virtual world (Pichierri et al., 2012). Research has widely supported exergames. They are seen as enjoyable therapy for the elderly, boosting motivation and improving cognitive and sensorimotor skills (Anderson-Hanley et al., 2018; Green & Seitz, 2015).

4. Conclusion

Exercise is the key to unlocking a healthy brain. By improving blood flow, exercise delivers oxygen and essential nutrients to the brain, fueling its proper function. This increased blood flow also plays a critical role in removing waste products that can harm neurons, the building blocks of the brain. Furthermore, exercise has been shown to boost the growth of new neurons in the hippocampus, a crucial region responsible for forming and storing memories, as well as facilitating learning.

Regular aerobic exercise is particularly beneficial, as it increases the size of the hippocampus, leading to enhanced cognitive function and a greater ability to process and retain information. Exercise also improves synaptic plasticity, the brain's ability to adapt and change in response to new experiences, which is essential for learning and personal growth.

The benefits of exercise on brain health are largely attributed to its ability to raise levels of brain-derived neurotrophic factor (BDNF), a protein that promotes the growth and survival of neurons. Higher BDNF levels have been linked to improved learning and memory, as well as a reduced risk of neurodegenerative diseases such as Alzheimer's and Parkinson's. Moreover, different types of exercise have been found to activate various regions of the brain, creating a more interconnected and resilient brain network. For example, exercises that require coordination, such as dancing or playing a musical instrument, can improve the motor and sensory areas of the brain, further highlighting the importance of regular physical activity for overall brain health.

5. Takeaways

1. Exercise stimulates brain activity by increasing blood flow and the release of neurotrophic factors, which support neuronal health and function.
2. Physical activity boosts the activity of neurotransmitters and growth factors that enhance synaptic transmission and plasticity, improving the efficiency of communication between neurons.
3. Exercise promotes the growth of dendritic spines, which are essential for synaptic connections and plasticity, thereby enhancing cognitive and motor functions.
4. Neuroplastic changes collectively contribute to improved cognitive and motor functional recovery. Exercise helps the brain and body adapt to injuries or neurological conditions, facilitating rehabilitation and enhancing overall functional outcomes.
5. Regular physical activity boosts neuronal firing rates and enhances brain-wide connectivity. This increased activity helps stimulate the release of neurotrophic factors, such as BDNF, which are vital for neuronal survival and growth.
6. Regular physical activity supports the formation of myelin, the insulating layer around axons that speeds up electrical impulses and improves neural communication.
7. Aerobic exercise, such as running, swimming, or cycling, has been shown to positively impact various aspects of cognitive function.
8. Aerobic exercise promotes neurogenesis in the hippocampus, a brain region crucial for learning and memory.
9. Regular aerobic exercise can enhance verbal memory, which involves the ability to recall and use language-related information.

10. Aerobic activities also improve spatial memory, which is essential for navigation and understanding spatial relationships, and facilitate overall learning processes.

Abbreviations

BDNF: Brain-Derived Neurotrophic Factors
BHA: b-Hydroxybutyrate
CREB: cAMP-Response Element Binding Protein
DG: Dentate Gyrus
FGF-2: Fibroblast Growth Factor 2
GDNF: Glial Cell Line
HIIT: High-Intensity Interval Training
HIT: High-Intensity Training
IGF-1: Insulin-Like Growth Factor
IL-10: Interleukin
IL-6: Interleukin 6
L&M: Learning and Memory
MIT: Moderate Interval Training
NGF: Nerve Growth Factor
NP: Neuroplasticity
NT3: Neurotrophin 3
NT4: Neurotrophin 4
VEGF: Vascular Endothelial Growth Factor
VEGT: Vascular Endothelial Growth Factor

Bibliography

1. Abraha, B., Chaves, A. R., Kelly, L. P., Wallack, E. M., Wadden, K. P., McCarthy, J., & Ploughman, M. (2018). A bout of high intensity interval training lengthened nerve conduction latency to the non-exercised affected limb in chronic stroke. *Frontiers in Physiology*, 9(JUL), 1–15. https://doi.org/10.3389/fphys.2018.00827
2. Aguiar, A. S., Boemer, G., Rial, D., Cordova, F. M., Mancini, G., Walz, R., de Bem, A. F., Latini, A., Leal, R. B., Pinho, R. A., & Prediger, R. D. S. (2010). High-intensity physical exercise disrupts

implicit memory in mice: Involvement of the striatal glutathione antioxidant system and intracellular signaling. *Neuroscience*, *171*(4), 1216–1227. https://doi.org/10.1016/j.neuroscience.2010.09.053

3. Aly, M., Ahmed, M. A., Hasan, A., Kojima, H., & Abdelhakem, A. R. (2019). Sport experience and physical activity: Event-related brain potential and task performance indices of attention in young adults. *Journal of Functional Morphology and Kinesiology*, *4*(2). https://doi.org/10.3390/jfmk4020033
4. Anderson-Hanley, C., Barcelos, N. M., Zimmerman, E. A., Gillen, R. W., Dunnam, M., Cohen, B. D., Yerokhin, V., Miller, K. E., Hayes, D. J., Arciero, P. J., Maloney, M., & Kramer, A. F. (2018). The Aerobic and Cognitive Exercise Study (ACES) for community-dwelling older adults with or at-risk for mild cognitive impairment (MCI): Neuropsychological, neurobiological and neuroimaging outcomes of a randomized clinical trial. *Frontiers in Aging Neuroscience*, *10*(MAY). https://doi.org/10.3389/fnagi.2018.00076
5. Angevaren, M., Aufdemkampe, G., Verhaar, H. J. J., Aleman, A., & Vanhees, L. (2008). Physical activity and enhanced fitness to improve cognitive function in older people without known cognitive impairment. *Cochrane Database of Systematic Reviews*, *2*. https://doi.org/10.1002/14651858.CD005381.pub2
6. Basso, J. C., & Suzuki, W. A. (2017). The Effects of Acute Exercise on Mood, Cognition, Neurophysiology, and Neurochemical Pathways: A Review. *Brain Plasticity*, *2*(2), 127–152. https://doi.org/10.3233/bpl-160040
7. Boyne, P., Meyrose, C., Westover, J., Whitesel, D., Hatter, K., Reisman, D. S., Carl, D., Khoury, J. C., Gerson, M., Kissela, B., & Dunning, K. (2020). Effects of Exercise Intensity on Acute Circulating Molecular Responses Poststroke. *Neurorehabilitation and Neural Repair*, *34*(3), 222–234. https://doi.org/10.1177/1545968319899915
8. Bozelos, P., Stefanou, S. S., Bouloukakis, G., Melachrinos, C., & Poirazi, P. (2016). REMOD: A tool for analyzing and remodeling the dendritic architecture of neural cells. *Frontiers in Neuroanatomy*, 9(JAN2016), 1–11. https://doi.org/10.3389/fnana.2015.00156
9. Cao, L., Zhang, Y., Huang, R., Li, L., Xia, F., Zou, L., Yu, Q., Lin, J., Herold, F., Perrey, S., Mueller, P., Dordevic, M., Loprinzi, P. D., Wang, Y., Ma, Y., Zeng, H., Qu, S., Wu, J., & Ren, Z. (2021). Structural and functional brain signatures of endurance runners. *Brain Structure and Function*, *226*(1), 93–103. https://doi.org/10.1007/s00429-020-02170-y

10. Cechella, J. L., Leite, M. R., Rosario, A. R., Sampaio, T. B., & Zeni, G. (2014). Diphenyl diselenide-supplemented diet and swimming exercise enhance novel object recognition memory in old rats. *Age*, *36*(4). https://doi.org/10.1007/s11357-014-9666-8
11. Chae, C. H., Jung, S. L., An, S. H., Park, B. Y., Kim, T. W., Wang, S. W., Kim, J. H., Lee, H. C., & Kim, H. T. (2014). Swimming exercise stimulates neuro-genesis in the subventricular zone via increase in synapsin i and nerve growth factor levels. *Biology of Sport*, *31*(4), 309–314. https://doi.org/10.5604/20831862.1132130
12. Charalambous, C. C., Helm, E. E., Lau, K. A., Morton, S. M., & Reisman, D. S. (2018). The feasibility of an acute high-intensity exercise bout to promote locomotor learning after stroke. *Topics in Stroke Rehabilitation*, *25*(2), 83–89. https://doi.org/10.1080/10749357.2017.1399527
13. Chen, F. T., Hopman, R. J., Huang, C. J., Chu, C. H., Hillman, C. H., Hung, T. M., & Chang, Y. K. (2020). The effect of exercise training on brain structure and function in older adults: A systematic review based on evidence from randomized control trials. *Journal of Clinical Medicine*, *9*(4). https://doi.org/10.3390/jcm9040914
14. Chen, K., Zheng, Y., Wei, J. An, Ouyang, H., Huang, X., Zhang, F., Wan Lai, C. S., Ren, C., So, K. F., & Zhang, L. (2019). Exercise training improves motor skill learning via selective activation of mTOR. *Science Advances*, *5*(7), 1–12. https://doi.org/10.1126/sciadv.aaw1888
15. Chen, M. J., & Russo-Neustadt, A. A. (2009). Running exercise-induced up-regulation of hippocampal brain-derived neurotrophic factor is CREB-dependent. *Hippocampus*, *19*(10), 962–972. https://doi.org/10.1002/hipo.20579
16. Chiu, C. N., Chen, C. Y., & Muggleton, N. G. (2017). Sport, time pressure, and cognitive performance. In *Progress in Brain Research* (1st ed., Vol. 234). Elsevier B.V. https://doi.org/10.1016/bs.pbr.2017.06.007
17. Chow, Z. S., Moreland, A. T., Macpherson, H., & Teo, W. P. (2021). The Central Mechanisms of Resistance Training and Its Effects on Cognitive Function. *Sports Medicine*, *51*(12), 2483–2506. https://doi.org/10.1007/s40279-021-01535-5
18. Citri, A., & Malenka, R. C. (2008). Synaptic plasticity: Multiple forms, functions, and mechanisms. *Neuropsychopharmacology*, *33*(1), 18–41. https://doi.org/10.1038/sj.npp.1301559
19. Clark, T. A., Sullender, C., Jacob, D., Zuo, Y., Dunn, A. K., & Jones, T. A. (2019). Rehabilitative training interacts with ischemia-instigated

spine dynamics to promote a lasting population of new synapses in peri-infarct motor cortex. *Journal of Neuroscience*, *39*(43), 8471–8483. https://doi.org/10.1523/JNEUROSCI.1141-19.2019

20. Colcombe, S. J., Erickson, K. I., Scalf, P. E., Kim, J. S., Prakash, R., McAuley, E., Elavsky, S., Marquez, D. X., Hu, L., & Kramer, A. F. (2006). Aerobic exercise training increases brain volume in aging humans. *Journals of Gerontology - Series A Biological Sciences and Medical Sciences*, *61*(11), 1166–1170. https://doi.org/10.1093/gerona/61.11.1166
21. Cooper, C., Moon, H. Y., & Van Praag, H. (2018). On the run for hippocampal plasticity. *Cold Spring Harbor Perspectives in Medicine*, *8*(4), 1–30. https://doi.org/10.1101/cshperspect.a029736
22. Cotman, C. W., Berchtold, N. C., & Christie, L. A. (2007). Exercise builds brain health: key roles of growth factor cascades and inflammation. *Trends in Neurosciences*, *30*(9), 464–472. https://doi.org/10.1016/j.tins.2007.06.011
23. Cullen, K. E. (2012). The vestibular system: Multimodal integration and encoding of self-motion for motor control. *Trends in Neurosciences*, *35*(3), 185–196. https://doi.org/10.1016/j.tins.2011.12.001
24. Damrongthai, C., Kuwamizu, R., Suwabe, K., Ochi, G., Yamazaki, Y., Fukuie, T., Adachi, K., Yassa, M. A., Churdchomjan, W., & Soya, H. (2021). Benefit of human moderate running boosting mood and executive function coinciding with bilateral prefrontal activation. *Scientific Reports*, *11*(1), 1–12. https://doi.org/10.1038/s41598-021-01654-z
25. De Sousa Fernandes, M. S., Ordônio, T. F., Santos, G. C. J., Santos, L. E. R., Calazans, C. T., Gomes, D. A., & Santos, T. M. (2020). Effects of Physical Exercise on Neuroplasticity and Brain Function: A Systematic Review in Human and Animal Studies. *Neural Plasticity*, *2020*. https://doi.org/10.1155/2020/8856621
26. Desai, P., Dhana, K., Decarli, C., Wilson, R. S., McAninch, E. A., Evans, D. A., & Rajan, K. B. (2022). Examination of Neurofilament Light Chain Serum Concentrations, Physical Activity, and Cognitive Decline in Older Adults. *JAMA Network Open*, *5*(3), 1–11. https://doi.org/10.1001/jamanetworkopen.2022.3596
27. Dillingham, C. M., Frizzati, A., Nelson, A. J. D., & Vann, S. D. (2015). How do mammillary body inputs contribute to anterior thalamic function? *Neuroscience and Biobehavioral Reviews*, *54*, 108–119. https://doi.org/10.1016/j.neubiorev.2014.07.025

28. Duzel, E., Van Praag, H., & Sendtner, M. (2016). Can physical exercise in old age improve memory and hippocampal function? *Brain*, *139*(3), 662–673. https://doi.org/10.1093/brain/awv407
29. El-Sayes, J., Harasym, D., Turco, C. V., Locke, M. B., & Nelson, A. J. (2019). Exercise-Induced Neuroplasticity: A Mechanistic Model and Prospects for Promoting Plasticity. *Neuroscientist*, *25*(1), 65–85. https://doi.org/10.1177/1073858418771538
30. El-Tamawy, M. S., Abd-Allah, F., Ahmed, S. M., Darwish, M. H., & Khalifa, H. A. (2014). Aerobic exercises enhance cognitive functions and brain derived neurotrophic factor in ischemic stroke patients. *NeuroRehabilitation*, *34*(1), 209–213. https://doi.org/10.3233/NRE-131020
31. Erickson, K. I., Leckie, R. L., & Weinstein, A. M. (2014). Physical activity, fitness, and gray matter volume. *Neurobiology of Aging*, *35*(SUPPL.2), S20–S28. https://doi.org/10.1016/j.neurobiolaging.2014.03.034
32. Erickson, K. I., Prakash, R. S., Voss, M. W., Chaddock, L., Hu, L., Morris, K. S., White, S. M., Wójcicki, T. R., McAuley, E., & Kramer, A. F. (2009). Aerobic fitness is associated with hippocampal volume in elderly humans. *Hippocampus*, *19*(10), 1030–1039. https://doi.org/10.1002/hipo.20547
33. Fedor, A., Garcia, S., & Gunstad, J. (2015). The effects of a brief, water-based exercise intervention on cognitive function in older adults. *Archives of Clinical Neuropsychology*, *30*(2), 139–147. https://doi.org/10.1093/arclin/acv001
34. Ferris, L. T., Williams, J. S., & Shen, C. L. (2007). The effect of acute exercise on serum brain-derived neurotrophic factor levels and cognitive function. *Medicine and Science in Sports and Exercise*, *39*(4), 728–734. https://doi.org/10.1249/mss.0b013e31802f04c7
35. Firth, J., Stubbs, B., Vancampfort, D., Schuch, F., Lagopoulos, J., Rosenbaum, S., & Ward, P. B. (2018). Effect of aerobic exercise on hippocampal volume in humans: A systematic review and meta-analysis. *NeuroImage*, *166*, 230–238. https://doi.org/10.1016/j.neuroimage.2017.11.007
36. Fischer, T. J., Walker, T. L., Overall, R. W., Brandt, M. D., & Kempermann, G. (2014). Acute effects of wheel running on adult hippocampal precursor cells in mice are not caused by changes in cell cycle length or S phase length. *Frontiers in Neuroscience*, *8*(SEP), 1–6. https://doi.org/10.3389/fnins.2014.00314

37. Gallardo-Gómez, D., del Pozo-Cruz, J., Noetel, M., Álvarez-Barbosa, F., Alfonso-Rosa, R. M., & del Pozo Cruz, B. (2022). Optimal dose and type of exercise to improve cognitive function in older adults: A systematic review and Bayesian model-based network meta-analysis of RCTs. *Ageing Research Reviews*, *76*. https://doi.org/10.1016/j.arr.2022.101591
38. Garber, C. E., Blissmer, B., Deschenes, M. R., Franklin, B. A., Lamonte, M. J., Lee, I. M., Nieman, D. C., & Swain, D. P. (2011). Quantity and quality of exercise for developing and maintaining cardiorespiratory, musculoskeletal, and neuromotor fitness in apparently healthy adults: Guidance for prescribing exercise. *Medicine and Science in Sports and Exercise*, *43*(7), 1334–1359. https://doi.org/10.1249/MSS.0b013e318213fefb
39. Geerlings, M. I., Sigurdsson, S., Eiriksdottir, G., Garcia, M. E., Harris, T. B., Gudnason, V., & Launer, L. J. (2015). Salivary cortisol, brain volumes, and cognition in community-dwellinfile:///C:/Users/Asus/Desktop/Alzheimer's disease, cerebrovascular dysfunction and the benefits of exercise from vessels to neurons.pdf elderly without dementia. *Neurology*, *85*(11), 976–983. https://doi.org/10.1212/WNL.0000000000001931
40. Gegentonglaga, Yoshizato, H., Higuchi, Y., Toyota, Y., Hanai, Y., Ando, Y., & Yoshimura, A. (2013). Variable alteration of regional tissue oxygen pressure in rat hippocampus by acute swimming exercise. *Life Sciences*, *93*(21), 773–777. https://doi.org/10.1016/j.lfs.2013.09.022
41. Ghalandari-Shamami, M., Nourizade, S., Yousefi, B., Vafaei, A. A., Pakdel, R., & Rashidy-Pour, A. (2019). Beneficial Effects of Physical Activity and Crocin Against Adolescent Stress-Induced Anxiety or Depressive-Like Symptoms and Dendritic Morphology Remodeling in Prefrontal Cortex in Adult Male Rats. *Neurochemical Research*, *44*(4), 917–929. https://doi.org/10.1007/s11064-019-02727-2
42. Green, C. S., & Seitz, A. R. (2015). The Impacts of Video Games on Cognition (and How the Government Can Guide the Industry). *Policy Insights from the Behavioral and Brain Sciences*, *2*(1), 101–110. https://doi.org/10.1177/2372732215601121
43. Griffin, É. W., Mullally, S., Foley, C., Warmington, S. A., O'Mara, S. M., & Kelly, Á. M. (2011). Aerobic exercise improves hippocampal function and increases BDNF in the serum of young adult males. *Physiology and Behavior*, *104*(5), 934–941. https://doi.org/10.1016/j.physbeh.2011.06.005

44. Hänggi, J., Koeneke, S., Bezzola, L., & Jäncke, L. (2010). Structural neuroplasticity in the sensorimotor network of professional female ballet dancers. *Human Brain Mapping*, *31*(8), 1196–1206. https://doi.org/10.1002/hbm.20928
45. Hortobágyi, T., Vetrovsky, T., Balbim, G. M., Sorte Silva, N. C. B., Manca, A., Deriu, F., Kolmos, M., Kruuse, C., Liu-Ambrose, T., Radák, Z., Váczi, M., Johansson, H., dos Santos, P. C. R., Franzén, E., & Granacher, U. (2022). The impact of aerobic and resistance training intensity on markers of neuroplasticity in health and disease. *Aging Research Reviews*, *80*(April 2022). https://doi.org/10.1016/j.arr.2022.101698
46. Hoy, S., Östh, J., Pascoe, M., Kandola, A., & Hallgren, M. (2021). Effects of yoga-based interventions on cognitive function in healthy older adults: A systematic review of randomized controlled trials. *Complementary Therapies in Medicine*, *58*(February). https://doi.org/10.1016/j.ctim.2021.102690
47. Huang, E. J., & Reichardt, L. F. (2001). Neurotrophins: Roles in neuronal development and function. *Annual Review of Neuroscience*, *24*, 677–736. https://doi.org/10.1146/annurev.neuro.24.1.677
48. Hüfner, K., Binetti, C., Hamilton, D. A., Stephan, T., Flanagin, V. L., Linn, J., Labudda, K., Markowitsch, H., Glasauer, S., Jahn, K., Strupp, M., & Brandt, T. (2011). Structural and functional plasticity of the hippocampal formation in professional dancers and slackliners. *Hippocampus*, *21*(8), 855–865. https://doi.org/10.1002/hipo.20801
49. Jonasson, L. S., Nyberg, L., Kramer, A. F., Lundquist, A., Riklund, K., & Boraxbekk, C. J. (2017). Aerobic exercise intervention, cognitive performance, and brain structure: Results from the Physical Influences on Brain in Aging (PHIBRA) Study. *Frontiers in Aging Neuroscience*, *8*(JAN), 1–15. https://doi.org/10.3389/fnagi.2016.00336
50. Kara, B., Pinar, L., Uğur, F., & Oğuz, M. (2005). Correlations between aerobic capacity, pulmonary and cognitive functioning in the older women. *International Journal of Sports Medicine*, *26*(3), 220–224. https://doi.org/10.1055/s-2004-820955
51. Kempermann, G., Gage, F. H., Aigner, L., Song, H., Curtis, M. A., Thuret, S., Kuhn, H. G., Jessberger, S., Frankland, P. W., Cameron, H. A., Gould, E., Hen, R., Abrous, D. N., Toni, N., Schinder, A. F., Zhao, X., Lucassen, P. J., & Frisén, J. (2018). Human Adult Neurogenesis: Evidence and Remaining Questions. *Cell Stem Cell*, *23*(1), 25–30. https://doi.org/10.1016/j.stem.2018.04.004

52. Kleemeyer, M. M., Kühn, S., Prindle, J., Bodammer, N. C., Brechtel, L., Garthe, A., Kempermann, G., Schaefer, S., & Lindenberger, U. (2016). Changes in fitness are associated with changes in hippocampal microstructure and hippocampal volume among older adults. *NeuroImage, 131*, 155–161. https://doi.org/10.1016/j.neuroimage.2015.11.026
53. Konstantinidis, E. I., Billis, A. S., Mouzakidis, C. A., Zilidou, V. I., Antoniou, P. E., & Bamidis, P. D. (2016). Design, implementation, and wide pilot deployment of FitForAll: An Easy to use exergaming platform improving physical fitness and life quality of senior citizens. *IEEE Journal of Biomedical and Health Informatics, 20*(1), 189–200. https://doi.org/10.1109/JBHI.2014.2378814
54. Kozorovitskiy, Y., & Gould, E. (2010). Journal of Clinical and Experimental Neuropsychology Adult Neurogenesis : A Mechanism for Brain Repair ? Adult Neurogenesis : A Mechanism for Brain Repair ? *Journal of Clinical and Experimental Neuropsychology, February 2012*, 37–41.
55. Lange-Asschenfeldt, C., & Kojda, G. (2008). Alzheimer's disease, cerebrovascular dysfunction and the benefits of exercise: From vessels to neurons. *Experimental Gerontology, 43*(6), 499–504. https://doi.org/10.1016/j.exger.2008.04.002
56. Langlois, F., Vu, T. T. M., Chassé, K., Dupuis, G., Kergoat, M. J., & Bherer, L. (2013). Benefits of physical exercise training on cognition and quality of life in frail older adults. *Journals of Gerontology - Series B Psychological Sciences and Social Sciences, 68*(3), 400–404. https://doi.org/10.1093/geronb/gbs069
57. Liang, J., Wang, H., Zeng, Y., Qu, Y., Liu, Q., Zhao, F., Duan, J., Jiang, Y., Li, S., Ying, J., Li, J., & Mu, D. (2021). Physical exercise promotes brain remodeling by regulating epigenetics, neuroplasticity and neurotrophins. *Reviews in the Neurosciences, 32*(6), 615–629. https://doi.org/10.1515/revneuro-2020-0099
58. Lin, T.-W., Tsai, S.-F., & Kuo, Y.-M. (2018). Physical Exercise Enhances Neuroplasticity and Delays Alzheimer's Disease. *Brain Plasticity, 4*(1), 95–110. https://doi.org/10.3233/bpl-180073
59. Lin, Y., Dong, J., Yan, T., He, X., Zheng, X., Liang, H., & Sui, M. (2015). Involuntary, forced and voluntary exercises are equally capable of inducing hippocampal plasticity and the recovery of cognitive function after stroke. *Neurological Research, 37*(10), 893–901. https://doi.org/10.1179/1743132815Y.0000000074

60. Linder, S. M., Rosenfeldt, A. B., Davidson, S., Zimmerman, N., Penko, A., Lee, J., Clark, C., & Alberts, J. L. (2019). Forced, Not Voluntary, Aerobic Exercise Enhances Motor Recovery in Persons With Chronic Stroke. *Neurorehabilitation and Neural Repair*, *33*(8), 681–690. https://doi.org/10.1177/1545968319862557
61. Liu-Ambrose, T., Nagamatsu, L. S., Voss, M. W., Khan, K. M., & Handy, T. C. (2012). Resistance training and functional plasticity of the aging brain: A 12-month randomized controlled trial. *Neurobiology of Aging*, *33*(8), 1690–1698. https://doi.org/10.1016/j.neurobiolaging.2011.05.010
62. Luo, L., Zhu, S., Shi, L., Wang, P., Li, M., & Yuan, S. (2019). High Intensity Exercise for Walking Competency in Individuals with Stroke: A Systematic Review and Meta-Analysis. *Journal of Stroke and Cerebrovascular Diseases*, *28*(12). https://doi.org/10.1016/j.jstrokecerebrovasdis.2019.104414
63. Maass, A., Düzel, S., Goerke, M., Becke, A., Sobieray, U., Neumann, K., Lövden, M., Lindenberger, U., Bäckman, L., Braun-Dullaeus, R., Ahrens, D., Heinze, H. J., Müller, N. G., & Düzel, E. (2015). Vascular hippocampal plasticity after aerobic exercise in older adults. *Molecular Psychiatry*, *20*(5), 585–593. https://doi.org/10.1038/mp.2014.114
64. Magon, S., Donath, L., Gaetano, L., Thoeni, A., Radue, E. W., Faude, O., & Sprenger, T. (2016). Striatal functional connectivity changes following specific balance training in elderly people: MRI results of a randomized controlled pilot study. *Gait and Posture*, *49*, 334–339. https://doi.org/10.1016/j.gaitpost.2016.07.016
65. Manera, V., Ben-Sadoun, G., Aalbers, T., Agopyan, H., Askenazy, F., Benoit, M., Bensamoun, D., Bourgeois, J., Bredin, J., Bremond, F., Crispim-Junior, C., David, R., Schutter, B. De, Ettore, E., Fairchild, J., Foulon, P., Gazzaley, A., Gros, A., Hun, S., … Robert, P. (2017). Recommendations for the use of serious games in neurodegenerative disorders: 2016 Delphi Panel. *Frontiers in Physiology*, *8*(JUL), 1–10. https://doi.org/10.3389/fpsyg.2017.01243
66. Marsden, D. L., Dunn, A., Callister, R., Levi, C. R., & Spratt, N. J. (2013). Characteristics of exercise training interventions to improve cardiorespiratory fitness after stroke: A systematic review with meta-analysis. *Neurorehabilitation and Neural Repair*, *27*(9), 775–788. https://doi.org/10.1177/1545968313496329

67. Masley, S., Roetzheim, R., & Gualtieri, T. (2009). Aerobic exercise enhances cognitive flexibility. *Journal of Clinical Psychology in Medical Settings*, *16*(2), 186–193. https://doi.org/10.1007/s10880-009-9159-6
68. Mehrholz, J., Kugler, J., & Pohl, M. (2011). Water-based exercises for improving activities of daily living after stroke. *Cochrane Database of Systematic Reviews*, *1*. https://doi.org/10.1002/14651858.cd008186.pub2
69. Mendez Colmenares, A., Voss, M. W., Fanning, J., Salerno, E. A., Gothe, N. P., Thomas, M. L., McAuley, E., Kramer, A. F., & Burzynska, A. Z. (2021). White matter plasticity in healthy older adults: The effects of aerobic exercise. *NeuroImage*, *239*(June), 118305. https://doi.org/10.1016/j.neuroimage.2021.118305
70. Nesin, S. M., Sabitha, K. R., Gupta, A., & Laxmi, T. R. (2019). Constraint Induced Movement Therapy as a Rehabilitative Strategy for Ischemic Stroke—Linking Neural Plasticity with Restoration of Skilled Movements. *Journal of Stroke and Cerebrovascular Diseases*, *28*(6), 1640–1653. https://doi.org/10.1016/j.jstrokecerebrovasdis.2019.02.028
71. Niemann, C., Godde, B., & Voelcker-Rehage, C. (2014). Not only cardiovascular, but also coordinative exercise increases hippocampal volume in older adults. *Frontiers in Aging Neuroscience*, *6*(JUN), 1–24. https://doi.org/10.3389/fnagi.2014.00170
72. Nørregaard, A. J., Gram, M., Vigelsoe, A., Wiuff, C., & Birk, A. (2014). *Masters Athletes: Exemplars of Successful Aging?*
73. Oliveira, D. M. G., Aguiar, L. T., de Oliveira Limones, M. V., Gomes, A. G., da Silva, L. C., de Morais Faria, C. D. C., & Scalzo, P. L. (2019). Aerobic Training Efficacy in Inflammation, Neurotrophins, and Function in Chronic Stroke Persons: A Randomized Controlled Trial Protocol. *Journal of Stroke and Cerebrovascular Diseases*, *28*(2), 418–424. https://doi.org/10.1016/j.jstrokecerebrovasdis.2018.10.016
74. Patten, A. R., Yau, S. Y., Fontaine, C. J., Meconi, A., Wortman, R. C., & Christie, B. R. (2016). The Benefits of Exercise on Structural and Functional Plasticity in the Rodent Hippocampus of Different Disease Models. *Brain Plasticity*, *1*(1), 97–127. https://doi.org/10.3233/bpl-150016
75. Penna, L. G., Pinheiro, J. P., Ramalho, S. H. R., & Ribeiro, C. F. (2021). Effects of aerobic physical exercise on neuroplasticity after stroke: Systematic review. *Arquivos de Neuro-Psiquiatria*, *79*(9), 832–843. https://doi.org/10.1590/0004-282X-ANP-2020-0551

76. Pichierri, G., Murer, K., & De Bruin, E. D. (2012). A cognitive-motor intervention using a dance video game to enhance foot placement accuracy and gait under dual task conditions in older adults: A randomized controlled trial. *BMC Geriatrics*, *12*(1), 1. https://doi.org/10.1186/1471-2318-12-74
77. Ploughman, M. (2008). Exercise is brain food: The effects of physical activity on cognitive function. *Developmental Neurorehabilitation*, *11*(3), 236–240. https://doi.org/10.1080/17518420801997007
78. Ploughman, M., Attwood, Z., White, N., Doré, J. J. E., & Corbett, D. (2007). Endurance exercise facilitates relearning of forelimb motor skill after focal ischemia. *European Journal of Neuroscience*, *25*(11), 3453–3460. https://doi.org/10.1111/j.1460-9568.2007.05591.x
79. Pontifex, M. B., Hillman, C. H., Fernhall, B., Thompson, K. M., & Valentini, T. A. (2009). The effect of acute aerobic and resistance exercise on working memory. *Medicine and Science in Sports and Exercise*, *41*(4), 927–934. https://doi.org/10.1249/MSS.0b013e3181907d69
80. Quaney, B. M., Boyd, L. A., McDowd, J. M., Zahner, L. H., Jianghua He, Mayo, M. S., & MacKo, R. F. (2009). Aerobic exercise improves cognition and motor function poststroke. *Neurorehabilitation and Neural Repair*, *23*(9), 879–885. https://doi.org/10.1177/1545968309338193
81. Ra, S. M., Kim, H., Jang, M. H., Shin, M. C., Lee, T. H., Lim, B. V., Kim, C. J., Kim, E. H., Kim, K. M., & Kim, S. S. (2002). Treadmill running and swimming increase cell proliferation in the hippocampal dentate gyrus of rats. *Neuroscience Letters*, *333*(2), 123–126. https://doi.org/10.1016/S0304-3940(02)01031-5
82. Rathore, A., & Lom, B. (2017). The effects of chronic and acute physical activity on working memory performance in healthy participants: A systematic review with meta-analysis of randomized controlled trials. *Systematic Reviews*, *6*(1), 1–16. https://doi.org/10.1186/s13643-017-0514-7
83. Ren, F. F., Chen, F. T., Zhou, W. S., Cho, Y. M., Ho, T. J., Hung, T. M., & Chang, Y. K. (2021). Effects of Chinese Mind-Body Exercises on Executive Function in Middle-Aged and Older Adults: A Systematic Review and Meta-Analysis. *Frontiers in Psychology*, *12*(May). https://doi.org/10.3389/fpsyg.2021.656141
84. Rezaei, R., Nasoohi, S., Haghparast, A., Khodagholi, F., Bigdeli, M. R., & Nourshahi, M. (2018). High intensity exercise preconditioning provides differential protection against brain injury following

experimental stroke. *Life Sciences*, *207*(March), 30–35. https://doi.org/10.1016/j.lfs.2018.03.007
85. Rogge, A. K., Röder, B., Zech, A., Nagel, V., Hollander, K., Braumann, K. M., & Hötting, K. (2017). Balance training improves memory and spatial cognition in healthy adults. *Scientific Reports*, 7(1), 1–10. https://doi.org/10.1038/s41598-017-06071-9
86. Sato, D., Yamashiro, K., Yoshida, T., Onishi, H., Shimoyama, Y., & Maruyama, A. (2013). Effects of water immersion on short- and long-latency afferent inhibition, short-interval intracortical inhibition, and intracortical facilitation. *Clinical Neurophysiology*, *124*(9), 1846–1852. https://doi.org/10.1016/j.clinph.2013.04.008
87. Sato, D., Yamazaki, Y., Yamashiro, K., Onishi, H., Baba, Y., Ikarashi, K., & Maruyama, A. (2020). Elite competitive swimmers exhibit higher motor cortical inhibition and superior sensorimotor skills in a water environment. *Behavioural Brain Research*, *395*(July), 112835. https://doi.org/10.1016/j.bbr.2020.112835
88. Saunders, D. H., Greig, C. A., & Mead, G. E. (2014). Physical activity and exercise after stroke: Review of multiple meaningful benefits. *Stroke*, *45*(12), 3742–3747. https://doi.org/10.1161/STROKEAHA.114.004311
89. Schinder, A. F., & Poo, M. ming. (2000). The neurotrophin hypothesis for synaptic plasticity. *Trends in Neurosciences*, *23*(12), 639–645. https://doi.org/10.1016/S0166-2236(00)01672-6
90. Serra, L., Petrosini, L., Mandolesi, L., Bonarota, S., Balsamo, F., Bozzali, M., Caltagirone, C., & Gelfo, F. (2022). Walking, Running, Swimming: An Analysis of the Effects of Land and Water Aerobic Exercises on Cognitive Functions and Neural Substrates. *International Journal of Environmental Research and Public Health*, *19*(23). https://doi.org/10.3390/ijerph192316310
91. Silva-Batista, C., de Lima-Pardini, A. C., Nucci, M. P., Coelho, D. B., Batista, A., Piemonte, M. E. P., Barbosa, E. R., Teixeira, L. A., Corcos, D. M., Amaro, E., Horak, F. B., & Ugrinowitsch, C. (2020). A Randomized, Controlled Trial of Exercise for Parkinsonian Individuals With Freezing of Gait. *Movement Disorders*, *35*(9), 1607–1617. https://doi.org/10.1002/mds.28128
92. Smith, P. F. (2017). Is hippocampal neurogenesis modulated by the sensation of self-motion encoded by the vestibular system? *Neuroscience and Biobehavioral Reviews*, *83*, 489–495. https://doi.org/10.1016/j.neubiorev.2017.09.013

93. Smith, P. J., Blumenthal, J. A., Hoffman, B. M., Cooper, H., Strauman, T. A., Welsh-Bohmer, K., Browndyke, J. N., & Sherwood, A. (2010). Aerobic exercise and neurocognitive performance: A meta-analytic review of randomized controlled trials. *Psychosomatic Medicine*, 72(3), 239–252. https://doi.org/10.1097/PSY.0b013e3181d14633
94. Solstad, T., Moser, E. I., & Einevoll, G. T. (2006). Running induces widespread structural alterations in the hippocampus and entorhinal cortex. *Hippocampus*, *1031*, 1026–1031. https://doi.org/10.1002/hipo
95. Song, M. K., Kim, E. J., Kim, J. K., & Lee, S. G. (2020). Effects of exercise timing and intensity on neuroplasticity in a rat model of cerebral infarction. *Brain Research Bulletin*, *160*(April), 50–55. https://doi.org/10.1016/j.brainresbull.2020.04.002
96. Song, M. K., Kim, E. J., Kim, J. K., Park, H. K., & Lee, S. G. (2019). Effect of regular swimming exercise to duration-intensity on neurocognitive function in cerebral infarction rat model. *Neurological Research*, *41*(1), 37–44. https://doi.org/10.1080/01616412.2018.1524087
97. Spalding, K. L., Bergmann, O., Alkass, K., Bernard, S., Salehpour, M., Huttner, H. B., Boström, E., Westerlund, I., Vial, C., Buchholz, B. A., Possnert, G., Mash, D. C., Druid, H., & Frisén, J. (2013). XDynamics of hippocampal neurogenesis in adult humans. *Cell*, *153*(6), 1219. https://doi.org/10.1016/j.cell.2013.05.002
98. Steib, K., Schäffner, I., Jagasia, R., Ebert, B., & Chichung Lie, D. (2014). Mitochondria modify exercise-induced development of stem cell-derived neurons in the adult brain. *Journal of Neuroscience*, *34*(19), 6624–6633. https://doi.org/10.1523/JNEUROSCI.4972-13.2014
99. Taubert, M., Draganski, B., Anwander, A., Müller, K., Horstmann, A., Villringer, A., & Ragert, P. (2010). Dynamic properties of human brain structure: Learning-related changes in cortical areas and associated fiber connections. *Journal of Neuroscience*, *30*(35), 11670–11677. https://doi.org/10.1523/JNEUROSCI.2567-10.2010
100. Ten Brinke, L. F., Bolandzadeh, N., Nagamatsu, L. S., Hsu, C. L., Davis, J. C., Miran-Khan, K., & Liu-Ambrose, T. (2015). Aerobic exercise increases hippocampal volume in older women with probable mild cognitive impairment: A 6-month randomised controlled trial. *British Journal of Sports Medicine*, *49*(4), 248–254. https://doi.org/10.1136/bjsports-2013-093184
101. Thomas, A. G., Dennis, A., Rawlings, N. B., Stagg, C. J., Matthews, L., Morris, M., Kolind, S. H., Foxley, S., Jenkinson, M., Nichols, T.

E., Dawes, H., Bandettini, P. A., & Johansen-Berg, H. (2016). Multi-modal characterization of rapid anterior hippocampal volume increase associated with aerobic exercise. *NeuroImage*, *131*, 162–170. https://doi.org/10.1016/j.neuroimage.2015.10.090

102. Trejo, J. L., Carro, E., & Torres-Alemán, I. (2001). Circulating insulin-like growth factor I mediates exercise-induced increases in the number of new neurons in the adult hippocampus. *Journal of Neuroscience*, *21*(5), 1628–1634. https://doi.org/10.1523/jneurosci.21-05-01628.2001
103. Triviño-Paredes, J., Patten, A. R., Gil-Mohapel, J., & Christie, B. R. (2016). The effects of hormones and physical exercise on hippocampal structural plasticity. *Frontiers in Neuroendocrinology*, *41*, 23–43. https://doi.org/10.1016/j.yfrne.2016.03.001
104. Tymofiyeva, O., & Gaschler, R. (2021). Training-Induced Neural Plasticity in Youth: A Systematic Review of Structural and Functional MRI Studies. *Frontiers in Human Neuroscience*, *14*(January), 1–24. https://doi.org/10.3389/fnhum.2020.497245
105. Van Praag, H., Christie, B. R., Sejnowski, T. J., & Gage, F. H. (1999). Running enhances neurogenesis, learning, and long-term potentiation in mice. *Proceedings of the National Academy of Sciences of the United States of America*, *96*(23), 13427–13431. https://doi.org/10.1073/pnas.96.23.13427
106. Van Praag, H., Shubert, T., Zhao, C., & Gage, F. H. (2005). Exercise enhances learning and hippocampal neurogenesis in aged mice. *Journal of Neuroscience*, *25*(38), 8680–8685. https://doi.org/10.1523/JNEUROSCI.1731-05.2005
107. Vaynman, S., & Gomez-Pinilla, F. (2005). License to run: Exercise impacts functional plasticity in the intact and injured central nervous system by using neurotrophins. *Neurorehabilitation and Neural Repair*, *19*(4), 283–295. https://doi.org/10.1177/1545968305280753
108. Vaynman, S., Ying, Z., & Gomez-Pinilla, F. (2004). Hippocampal BDNF mediates the efficacy of exercise on synaptic plasticity and cognition. *European Journal of Neuroscience*, *20*(10), 2580–2590. https://doi.org/10.1111/j.1460-9568.2004.03720.x
109. Vilela, T. C., Muller, A. P., Damiani, A. P., Macan, T. P., da Silva, S., Canteiro, P. B., de Sena Casagrande, A., Pedroso, G. dos S., Nesi, R. T., de Andrade, V. M., & de Pinho, R. A. (2017). Strength and Aerobic Exercises Improve Spatial Memory in Aging Rats Through Stimulating Distinct Neuroplasticity Mechanisms.

Molecular Neurobiology, *54*(10), 7928–7937. https://doi.org/10.1007/s12035-016-0272-x

110. Vivar, C., Peterson, B. D., & van Praag, H. (2016). Running rewires the neuronal network of adult-born dentate granule cells. *NeuroImage*, *131*, 29–41. https://doi.org/10.1016/j.neuroimage.2015.11.031
111. Voss, M. W., Heo, S., Prakash, R. S., Erickson, K. I., Alves, H., Chaddock, L., Szabo, A. N., Mailey, E. L., Wójcicki, T. R., White, S. M., Gothe, N., Mcauley, E., Sutton, B. P., & Kramer, A. F. (2013). The influence of aerobic fitness on cerebral white matter integrity and cognitive function in older adults: Results of a one-year exercise intervention. *Human Brain Mapping*, *34*(11), 2972–2985. https://doi.org/10.1002/hbm.22119
112. Voss, M. W., Nagamatsu, L. S., Liu-Ambrose, T., & Kramer, A. F. (2011). Exercise, brain, and cognition across the life span. *Journal of Applied Physiology*, *111*(5), 1505–1513. https://doi.org/10.1152/japplphysiol.00210.2011
113. Westneat, D. F., Potts, L. J., Sasser, K. L., & Shaffer, J. D. (2019). Causes and Consequences of Phenotypic Plasticity in Complex Environments. *Trends in Ecology and Evolution*, *34*(6), 555–568. https://doi.org/10.1016/j.tree.2019.02.010
114. Whiteman, A. S., Young, D. E., Budson, A. E., Stern, C. E., & Schon, K. (2016). Entorhinal volume, aerobic fitness, and recognition memory in healthy young adults: A voxel-based morphometry study. *NeuroImage*, *126*, 229–238. https://doi.org/10.1016/j.neuroimage.2015.11.049
115. Winter, B., Breitenstein, C., Mooren, F. C., Voelker, K., Fobker, M., Lechtermann, A., Krueger, K., Fromme, A., Korsukewitz, C., Floel, A., & Knecht, S. (2007). High impact running improves learning. *Neurobiology of Learning and Memory*, *87*(4), 597–609. https://doi.org/10.1016/j.nlm.2006.11.003
116. Xie, Q., Cheng, J., Pan, G., Wu, S., Hu, Q., Jiang, H., Wang, Y., Xiong, J., Pang, Q., & Chen, X. (2019). Treadmill exercise ameliorates focal cerebral ischemia/reperfusion-induced neurological deficit by promoting dendritic modification and synaptic plasticity via upregulating caveolin-1/VEGF signaling pathways. *Experimental Neurology*, *313*(September 2018), 60–78. https://doi.org/10.1016/j.expneurol.2018.12.005
117. Xing, Y., & Bai, Y. (2020). A Review of Exercise-Induced Neuroplasticity in Ischemic Stroke: Pathology and Mechanisms.

Molecular Neurobiology, *57*(10), 4218–4231. https://doi.org/10.1007/s12035-020-02021-1

118. Yamaguchi, M., Seki, T., Imayoshi, I., Tamamaki, N., Hayashi, Y., Tatebayashi, Y., & Hitoshi, S. (2016). Neural stem cells and neuro/gliogenesis in the central nervous system: understanding the structural and functional plasticity of the developing, mature, and diseased brain. *Journal of Physiological Sciences*, *66*(3), 197–206. https://doi.org/10.1007/s12576-015-0421-4
119. Yau, S. Y., Lau, B. W. M., Tong, J. Bin, Wong, R., Ching, Y. P., Qiu, G., Tang, S. W., Lee, T. M. C., & So, K. F. (2011). Hippocampal neurogenesis and dendritic plasticity support running-improved spatial learning and depression-like behaviour in stressed rats. *PLoS ONE*, *6*(9). https://doi.org/10.1371/journal.pone.0024263
120. Yepes, M. (2020). Urokinase-type plasminogen activator is a modulator of synaptic plasticity in the central nervous system: Implications for neurorepair in the ischemic brain. *Neural Regeneration Research*, *15*(4), 620–624. https://doi.org/10.4103/1673-5374.266904
121. Yong, M. S., Kim, S. G., & Cheon, S. H. (2017). Effects of skilled reach training with affected forelimb and treadmill exercise on the expression of neurotrophic factor following ischemia-induced brain injury in rats. *Journal of Physical Therapy Science*, *29*(4), 647–650. https://doi.org/10.1589/jpts.29.647
122. Zborowski, V. A., Heck, S. O., Sari, M. H. M., Bastos, N. K., Neto, J. S. S., & Nogueira, C. W. (2019). (p-ClPhSe)2 modulates hippocampal BDNF/TrkB signaling and reverses memory impairment induced by diabetes in mice. *Progress in Neuro-Psychopharmacology and Biological Psychiatry*, *94*(May), 109660. https://doi.org/10.1016/j.pnpbp.2019.109660
123. Zhang, C. J., Zou, Y., Li, K., Li, C., Jiang, Y. P., Sun, J., Sun, R., & Wen, H. M. (2018). Different effects of running wheel exercise and skilled reaching training on corticofugal tract plasticity in hypertensive rats with cortical infarctions. *Behavioural Brain Research*, *336*(September 2017), 166–172. https://doi.org/10.1016/j.bbr.2017.09.002
124. Zhang, Q. W., Deng, X. X., Sun, X., Xu, J. X., & Sun, F. Y. (2013). Exercise promotes axon regeneration of newborn striatonigral and corticonigral projection neurons in rats after ischemic stroke. *PLoS ONE*, *8*(11). https://doi.org/10.1371/journal.pone.0080139

Motor Learning and Neuroplasticity: From Theory to Practice

Aygül Köseoğlu Kurt and Mehmet Kaan Altunok

1. Introduction

Learning relies on the brain's ability to reshape itself from our experiences. The learning process must be able to record, store, and retrieve information. Memory helps us recall cherished moments and learn skills like biking.

Functional neuroimaging studies are key. They identify the brain networks in various motor learning practices (Mancini et al., 2022). However, the literature has no consensus. It is unclear which brain regions activate during motor skill acquisition. The studies used different methods (Hardwick et al., 2013). Neuroimaging studies have investigated many sensorimotor tasks. They include complex skills, like playing a musical instrument (Buccino et al., 2004). They also include simple tasks, like knot-tying (Tracy et al., 2003). They also include adapting movements to changes in visuomotor tasks (Inoue et al., 2000; Nezafat et al., 2001).

Neuroplasticity is the ability of neurons and their synapses in the CNS. They can change in structure and function due to environmental stimuli. "Neuroplasticity" is the ability of brain neural pathways to change. This helps an organism adapt to its environment (Johnson & Cohen, 2023). There is a strong link between motor skills and brain plasticity. This occurs at both the cortical and subcortical levels. This plasticity evolves. It shows changes that help activate interconnected, distant brain regions. Many studies have used various paradigms to learn motor skills. Regardless of the paradigm's form or stage, the learning process changes neural connections in the brain.

Neurological rehabilitation, particularly in physical therapy, involves restoring lost motor skills. This is about neuroplasticity, the nervous system's reorganization in response to internal and external experiences (Cramer et al., 2011).

Understanding the principles of motor learning is crucial for developing rehabilitation protocols. This chapter will discuss motor learning techniques and highlight insights for practical use.

2. Motor Learning and Skill Acquisition

Motor learning is acquiring motor skills. It causes lasting improvements in speed and accuracy from practice or experience. Stable motor skills adapt to diverse stressors, enabling broad application across varied contexts. This flexibility allows consistent performance in new environments and unfamiliar scenarios. As motor skills become automated, cognitive load decreases. This reduction allows for smoother task performance. Schmidt and Lee (2005) note that advanced automatization enables more effortless execution. The brain, less taxed by routine movements, can allocate resources elsewhere.

The acquisition of motor skills and their long-term retention in the brain plays a fundamental role in our ability to carry out daily activities. Functional skills such as writing, riding a bicycle, playing the piano, or playing tennis are all acquired through repeated practice. Learning motor skills is the process of improving movement through practice. It makes movements faster and more accurate (Willingham, 1998). Recent advances in neuroimaging and non-invasive brain stimulation have increased our understanding of the brain areas that control motor skills. Scientists have made dramatic discoveries in this field. Animal studies in vivo and in vitro provide insights into the assignment of specific neuronal circuits during motor skill learning. They cover various stages of this process. They add to the scientific knowledge in the community.

Motor skills are typically acquired gradually over multiple training sessions, eventually reaching a point where performance approaches a near-maximum level. At first, motor skills improve rapidly, with significant gains made in a single training session. As practice continues, progress slows down, with only minor improvements resulting from repeated practice (Doyon & Benali, 2005; Doyon et al., 2003). The learning process for motor skills can vary depending on the skill's difficulty level. Relatively simple skills may reach a near-maximum level of performance in a short time. In contrast, more complex motor skills like playing a musical instrument may take much longer to reach this level.

Motor skills can be learned during a training session, whether online or offline, but they can also continue to improve after the session has ended. Offline, the skill can stabilize in the CNS and further develop. This process is referred to as motor memory consolidation. During consolidation, the motor skill becomes permanent and reinforced. This process goes beyond the initial learning phase of motor skills, making it more robust, efficient, and easier to recall over time. Additionally, this process represents an intermediate stage between fast and slow learning (Dayan & Cohen, 2011).

2.1. Fast Motor Skill Learning

Brain imaging techniques, such as Positron Emission Tomography (PET) and functional magnetic resonance imaging (fMRI), have provided insight into how humans acquire new skills. These tools reveal our learning processes' neural foundations, offering glimpses into the brain's internal workings as we absorb and master new information during rapid mastery of sequential motor skills, brain regions like the dorsolateral prefrontal cortex (DLPFC), primary motor cortex (M1), and pre-supplementary motor area (preSMA) display altered activity (Figure 1). As individuals swiftly acquire complex movement patterns, these areas modulate, highlighting the neural plasticity involved in motor learning. The brain's adaptability enables rapid skill development, reshaping neural circuits to support new motor sequences. As learning progresses, activation in these regions decreases, while increased activation is observed in areas such as the premotor cortex, supplementary motor area (SMA), parietal regions, striatum, and brainstem. Thus, learning is associated with different regional activity modulations, including blood-oxygen-level-dependent (BOLD) signals and regional cerebral blood flow (rCBF). When learning spatial coordinates in motor skills, the frontal-parietal-associative striatum-cerebellum circuit supports the process, whereas learning motor coordinates activates the M1-sensorimotor striatum-cerebellum circuit. The SMA, pre-SMA, and premotor cortices are involved in the alternating activation of spatial and motor coordinates. Spatial coordinates are learned more quickly, and when attention and executive functions are engaged in the learning process, prefrontal cortical regions become activated.

Rapid learning hinges on M1, a key brain region. This area swiftly processes and integrates new information, enabling quick skill acquisition and adaptation. Additionally, studies using transcranial magnetic stimulation (TMS) in humans have demonstrated the initiation of neuroplastic activity resembling Long-Term Potentiation (LTP) during the learning of a motor task (Rosenkranz et al., 2007). Moreover, BOLD activity in M1 has been reported to gradually decrease as motor skill learning progresses within a single training session (Karni et al., 1998). The plasticity related to learning in M1 seems to rely on protein synthesis in this area and is thought to be primarily connected to brain-derived neurotrophic factors (Kleim et al., 2003).

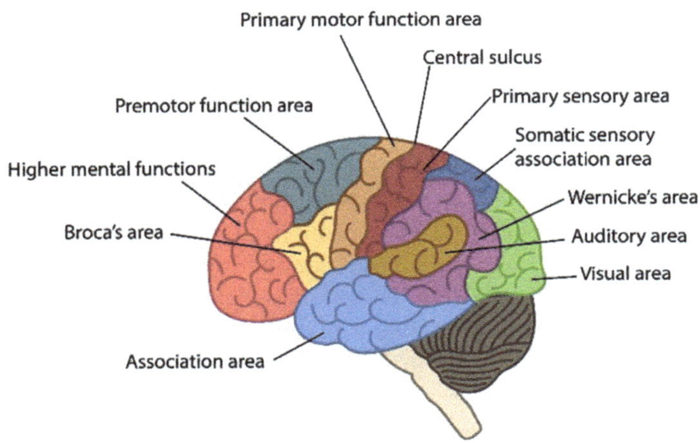

Figure 1: Functional Areas of the Cerebral Cortex

2.2. Slow Motor Skill Learning

In the later stages of motor skill learning, improvements in behavior are generally more modest, and progress is slower than seen during the initial rapid learning phase. Research shows slow learning links to heightened activation in the primary somatosensory cortex, supplementary motor area (SMA), and putamen in the M1. It also relates to reduced activation in cerebellar lobule VI (Dayan & Cohen, 2011). Animal model studies have demonstrated the involvement of neurons in the sensorimotor striatum during later stages of learning. This suggests that slow learning underpins the automation and habituation of motor skills. Ex vivo recordings from the sensorimotor striatum after motor training have revealed persistent changes in glutamatergic neurotransmission (Yin et al., 2009).

Reis et al. (2009) found that administering anodal transcranial direct current stimulation (tDCS) to M1 over five consecutive days notably improved the learning of a sequential visuomotor task. Additionally, the performance benefits seen in the stimulated group compared to the sham control group continued for 90 days following the intervention (Reis et al., 2009).

2.3. Structural Plasticity Associated with Slow Learning

Research has shown that training focused on slow learning increases gray matter in the MT/V5 region, which is involved in motion perception. It also increases gray matter in the left posterior intraparietal sulcus, which is linked to visuomotor processing (May & Gaser, 2006). However, following a decline in skill levels post-training, gray matter volume decreased to near baseline levels over three months. Similarly, gray matter expansion in the MT/V5 area has been reported in older volunteers learning the same task. This suggests that gray matter reorganization can also occur in the aging human brain (Boyke et al., 2008).

Magnetic resonance imaging (MRI) techniques, especially voxel-based morphometry (VBM) (Ashburner & Friston, 2000), assess gray matter changes linked to experience and learning. Studies suggest gray matter volumes differ with motor skill levels. Musicians have more gray matter in the auditory, sensorimotor, and premotor cortices, and cerebellum than non-musicians. Similarly, experts in tasks like typing, basketball, or golf show gray matter differences in specific brain areas compared to novices (Dayan & Cohen, 2011).

A study by Draganski et al. (2004) showed how learning affects brain structure. Participants trained for three months to juggle three balls. Researchers scanned their brains before training, after, and three months later. They found that juggling increased gray matter in specific brain areas. However, this increase disappeared without practice. Such studies are key in linking skill learning to brain changes. Slow learning also affects white matter structure (Doyon & Benali, 2005; Fields, 2008).

2.4. Offline Motor Skill Learning

The shift from quick to slow motor skill learning is called consolidation. It's been noted that sleep boosts skills significantly (Diekelmann & Born, 2010; Fischer et al., 2002). This is especially true for non-rapid eye movement (NREM) sleep, particularly stage II (M. P. Walker et al., 2002). Moreover, learning during sleep, mainly implicit sequential learning, happens without awareness (Robertson et al., 2004; Song et al., 2007).

As fMRI studies show, learning sequential finger movements reduces M1 BOLD activity after sleep. This change highlights sleep's role in acquiring motor skills. Moreover, motor memory consolidation during sleep is linked to time spent in NREM-stage II (Matthew P Walker et al., 2002). Sleep aids explicit motor sequence learning but not implicit learning (Song et al., 2007).

The striatum is key in consolidating these skills. Post-sleep tests show increased striatal activity in humans (Debas et al., 2010). Early improvements in oculomotor learning also show increased BOLD activity in the ventral striatum and hippocampus after sleep (Albouy et al., 2008).

A study found increased activity in the prefrontal, superior, and inferior parietal cortices and the Crus II region of the cerebellum after rest in a visuomotor tracking task (Albert et al., 2009). This rest period likely boosts learning. It may activate key brain regions, refine connections, and provide timing support. Such downtime could be vital for enhancing our ability to learn new physical skills.

2.5. Long-Term Retention of Motor Skills

Motor skills, once learned, stay in the nervous system. They can last or fade. Studies show people retain these skills for up to 12 months. This proves the brain's ability to keep learned tasks (Romano et al., 2010). In real life, skills last even longer. A small quantity of practice over days helps keep explicit skills (Savion-Lemieux & Penhune, 2005).

Rewards during practice boost long-term skill retention. fMRI scans show increased brain activity in key areas with rewards (Abe et al., 2011; Adcock et al., 2006; Wittmann et al., 2005). This is linked to dopamine in the brain (Calabresi et al., 2007). The predictability and size of rewards affect memory (Wittmann et al., 2011) (Figure 2).

It has been shown that training sessions conducted in mixed or random order rather than in blocks can enhance motor learning skills (Shea & Morgan, 1979). Randomly ordered practice is more stable in memory and is recorded more quickly. Some theories suggest that motor learning practices conducted in random order are coded with faster shifts from the SMA to other brain regions, such as the striatum or parietal cortex

(Tanaka et al., 2010). Based on this information, incorporating random practice into a training session may be more beneficial for motor learning. Sleep following motor learning (Stickgold et al., 2000) or even short naps (Nishida & Walker, 2007) can help facilitate the long-term retention of motor skills.

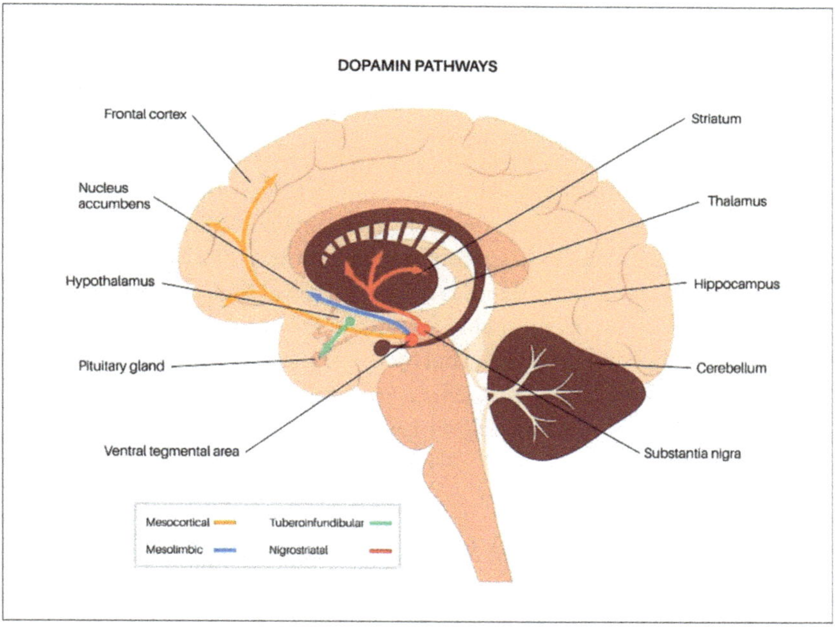

Figure 2: Dopamine Pathways

Adopting theoretical knowledge regarding the consolidation and retention of motor skills can potentially impact clinical outcomes. These principles can be applied in neurorehabilitation interventions focused on motor skill acquisition following brain lesions.

3. Learning Concepts

Research into human neuromuscular design has sped up, seeking motor learning and performance principles. These depend on the adaptability of muscles and nerves for movement. A broad understanding of how central systems work together is crucial. The central nervous system plays a key role in learning movements and acquiring motor skills. After neurological damage, relearning and reorganizing lost skills depend on the brain's neuroplasticity, vital for rehabilitation. Key concepts in learning are crucial for skill acquisition.

3.1. Attention Focus

In rehabilitation for motor skills, practitioners often provide specific movement instructions. These instructions cover aspects like the sequence, timing, speed, and appearance of movements.

An *internal focus of attention* happens when someone concentrates on their body or limbs. Meanwhile, an *external focus of attention* occurs when the focus is on the movement's impact on the environment. For example, a stroke patient relearning to grasp might focus on their shoulder, elbow, wrist, fingers, and muscle actions. This leads to an internal focus. However, if they are asked to carry a cup by following floor lines, they shift to an external focus.

Research indicates that an external focus aids in more automatic learning. It uses unconscious, quick, and reflexive processes. Conversely, an internal focus involves conscious control, which can prevent automatic movement (Wulf et al., 2001).

3.2. Implicit Learning

Motor learning involves two pathways: explicit and implicit. These pathways work together. Explicit learning requires awareness and strategic thinking, often aided by verbal instructions. It involves cognitive stages and working memory. Implicit learning, however, happens without conscious effort. People learn skills without knowing the rules or strategies. This occurs through trial and error or practice. Implicit learning is ideal when the focus is limited or tasks are complex (Masters, 1992).

Automating motor skills is key in rehabilitation. People face complex tasks daily. Stress or multitasking can reduce conscious control over movement and decisions. In these cases, relying on automatic skills is crucial.

Learning motor skills explicitly can be challenging. People may struggle to keep movements fluid and high-quality while remembering instructions. This cognitive effort can cause fatigue, harming performance further (Masters et al., 2008).

3.3. Differential Learning

Differential learning aims to adapt movement patterns for stability in different conditions. It involves practicing movements under varied scenarios to maintain quality. The focus is on learning the movement, not correcting it, no matter the repetitions.

At this stage, people try different movements in various conditions to build motor skills. They learn to adapt movements to changing conditions and develop new strategies.

Virtual reality aids in differential learning for neurological rehab. It exposes individuals to different conditions, improving their adaptability. This method fosters flexible movement patterns, which is crucial for better performance and recovery (Levin & Demers, 2021).

3.4. Self-Directed Learning

In this learning theory, the individual or patient is allowed to learn independently. In doing so, the person plays a more active role in the learning process, which can reinforce the effectiveness of motor learning by providing vital internal feedback, motivation, and enjoyment (Bjork et al., 2013).

3.5. Motor Learning Process

Procedural learning is the most used form of implicit learning in acquiring motor skills. This learning method enables the acquisition of functional skills through task repetition without using high cognitive abilities. Thus, movements become automatic without the need for conscious awareness. For example, when an individual learns to ride a bicycle, the ability to pedal and maintain balance becomes automatic through repetition without conscious thought. This is an example of procedural learning. This process plays a vital role in the formation of motor memory.

For the task to be performed at its best, the movement must be repeated frequently and stored in procedural memory. Motor skills transferred to procedural memory are very difficult to forget.

The first step in the motor learning process is the acquisition phase, during which practice is carried out to acquire motor skills. The target motor skill is learned through repetition, and skill acquisition is attempted, also called the online process. The consolidation phase, the offline period, follows this. This learning phase occurs between the first training session in which the movement is learned and the subsequent training session. During this period, learning is reinforced without additional practice, increasing motor function skills. Studies have shown that motor performance can spontaneously improve within six hours without further practice (Masaki & Sommer, 2012).

In the acquisition phase, motor skills are initially stored in short-term memory. In contrast, during consolidation, these skills are transferred from short-term to long-term memory to maintain motor skills for years. However, acquiring motor skills is not always achievable through a single practice session. As the difficulty level of the skill increases, this process may extend over a longer period. With each practice session, the consolidated information gradually becomes automatic and can be retained for years. When the acquired motor skill is performed with high coordination, permanent changes in the CNS occur, and the skill becomes firmly established for long-term retention.

4. Definition of the Principles of Neurorehabilitation in Motor Learning

4.1. Intensive and Repetitive Practice

Intensive practice involves little to no rest between work phases. Skills can be repeated continuously or in blocks. In rehabilitation, this means often using a damaged limb. Repetition boosts learning and helps predict movements based on sensory input.

Animal studies show that skill movements change brain structure. However, this does not happen with non-skill movements. Block training is faster for humans, but varied practice is better for long-term learning and transfer.

There is no set rule for repetition in therapies or studies. However, more sessions or longer ones are key. Many studies mix protocols, difficulty, and tasks, making pure repetition rare. This complexity shows the varied effects of repetitive training. Intensive practice is vital, but often combined with other methods. Counting repetitions in sessions and therapies is crucial for comparisons. This helps analyze training intensity and duration. It aids in understanding the impact of repetition on motor learning (Maier et al., 2019).

4.2. Spaced Practice

Spaced practice in motor learning involves repeating tasks with breaks in between. Research shows that these breaks can improve test scores. However, if the breaks are too long, learning might suffer. The first learning session activates a memory trace. Later attempts require less effort, potentially weakening the memory. This is known as "repetition suppression." It suggests we do not fully understand how spaced practice works.

TMS studies show that certain brain areas are crucial for motor memory. These areas benefit from spaced training. Moreover, learning and exercise are linked to the growth of new brain cells. In animals, spaced practice helps memory by supporting cell survival in a key area. It also enhances processes that aid memory.

Few studies have examined spaced practice for stroke recovery. One study found that stretching out Constraint-Induced Therapy (CIT) over more days, with shorter sessions, led to a similar recovery as traditional CIT and improved quality of life (Dettmers et al., 2005).

4.3. Dose and Duration

In therapy, "dose" is less clear than in pharmacology. It often refers to therapy length, session frequency, and training needed to learn. High-intensity training achieves greater doses. Motor learning sessions last 22 to 60 minutes (Veerbeek et al., 2014). As learning improves, session length may reduce. More therapy time typically leads to better outcomes (Lohse et al., 2014; Veerbeek et al., 2014). Neurological rehab needs serious commitment. Research shows that over 6 months, 16+ extra training hours are vital. This means over 71 daily minutes for 3 months. Such effort in the first 6 months is crucial for recovery (Veerbeek et al., 2014). Longer sessions can change brain function. High-dose training boosts brain activity and motor skills. It reorganizes networks, enhancing function. Evidence backs this method for lasting change (Kwakkel et al., 2015; Veerbeek et al., 2014).

4.4. Task-Specific Practice

Task-specific practice helps learn motor skills and adapt them to new conditions. It also makes the skill more automatic and improves sensorimotor abilities (Rizzolatti & Sinigaglia, 2010). Once sensorimotor skills are gained, performance can become highly specialized. Rehabilitation often includes meaningful daily activities to support independent living. Boyd et al. (2010) showed that targeted training boosts motor learning and grip strength. Furthermore, a study by Singer et al. (2013) found that interhemispheric inhibition decreased after targeted practice.

4.5. Goal-Oriented Application

Learning to throw a ball into a basket involves linking movement patterns. In such practice, the aim is not just to focus on muscles or movements. It is about helping patients find the best ways to complete the task (Wu et al., 2000). Setting specific, challenging goals boosts motor learning. These targets lead to better skill development than vague ones (Guadagnoli & Lee, 2004). Research shows different brain patterns for tool use in animals and humans. Movement targets are coded uniquely, showing cognitive differences. In the anterior intraparietal sulcus and M1, targets are linked to specific body parts. Meanwhile, the ventral intraparietal sulcus and premotor cortex are more general (Gallivan & Culham, 2015). Additionally, goal-directed movements ramp up activity in sensorimotor areas of the brain (Nathan et al., 2012).

4.6. Diversifying (Enriching) Practice

Practice can be improved in two ways. First, adding variability, called *practice variability* (Schmidt, 1975). Second, by randomizing training sequences, known as *random practice* or *contextual interference* (Shea & Morgan, 1979). Studies using fMRI and TMS have shown that varied practice boosts brain activity and connectivity. This is especially true in areas linked to motor learning. It also improves recall (Lage et al., 2015). The M1 area of the brain is more active during recall. This suggests that varied practice makes it easier to retrieve motor memories (Lin et al., 2011). Innovation seems to trigger systems that enhance brain adaptability. This could explain why varied practice is so effective in learning.

4.7. Increasing the Difficulty Level

The best learning happens when task difficulty, processing, and performance are balanced. Training tailored to individual abilities outperforms fixed difficulty increases (Wickens et al., 2013). Participants perform better when they set their difficulty (Andrieux et al., 2012). However, too much difficulty can hurt performance. Neuroimaging shows more activity in specific brain areas under increased difficulty (Winstein et al., 1997). This suggests these areas are key in handling task difficulty. Such increases in difficulty are effective in standard care, robot therapy (Winstein et al., 1997), and virtual reality systems (Ballester et al., 2016). All these methods aid motor recovery.

4.8. Multisensory Stimulation

The brain integrates multiple senses and enhances perception through predictions. Key areas for this are the temporal sulcus, intraparietal cortex, and frontal cortex. Recent studies reveal that once thought to handle only one sense, some areas also process multiple senses (Ghazanfar & Schroeder, 2006). For example, touch can affect how we see things. This means multisensory feedback improves our sensory interpretation. Using multiple senses during movements aids motor learning (McGann, 2010). In rehabilitation, combining vision and proprioception eases movement. Both are crucial for motor planning, allowing for multisensory approaches in therapy (Sober & Sabes, 2003). Adding tactile feedback during motor imagery improves brain-computer interface accuracy and strengthens sensorimotor links (Gomez-Rodriguez et al., 2011). Training with multisensory stimulation helps those with sensory difficulties, like vision or hearing impairments, to improve (Làdavas, 2008).

4.9. Rhythmic Cueing

Neuroentrainment links body movements with rhythmic environmental stimuli (Rossignol & Jones, 1976). It applies to auditory, visual, tactile, and vestibular skills. Auditory cues match movements to rhythmic patterns. These cues boost activity in motor networks and the cerebellum. They also enhance the brain's perception of rhythm and tempo. Creating a mental model of the rhythm aids in anticipating movements (Schaefer, 2014).

Research by Grahn (2012) and Nombela et al. (2013) shows interactions between the auditory and motor systems. Motor training with auditory cues can change these connections (Moore et al., 2017).

4.10. Explicit Feedback/Knowledge of Results

Knowledge of results (KR) feedback is verbal and final and adds to goal achievement. Research shows that external feedback fosters simple reactions, especially in animals. However, KR feedback goes beyond rewards. It involves cognitive processing, not just conditioning (Salmoni et al., 1984).

KR feedback can be quantitative or qualitative, covering success or failure. It does not have to be verbal. For instance, a patient missing a goal might get feedback through tone changes or color shifts (Taylor et al., 2014). Kinematic feedback is another type where movement errors are recorded and reviewed (Salmoni et al., 1984). Explicit feedback boosts learning mechanisms and affects implicit ones (Taylor et al., 2014).

Research shows that feedback can impact motor learning differently. Punishment may speed up learning. Meanwhile, rewards help retain long-term knowledge (Abe et al., 2011; Galea et al., 2015). Positive feedback boosts dopamine, preventing information loss (Wickens et al., 2013). Reward-driven neuron activity is linked to neuroplasticity. Dopamine builds up, aiding memory (Abe et al., 2011). In therapies like Constraint-Induced Motor Therapy (CIMT), KR boosts adherence (Taub et al., 1994).

4.11. Implicit Feedback/Knowledge of Performance

Advancements in technology now allow performance feedback to be delivered online, implicitly, and during movement. This feedback can be about internal body processes and movement mechanics, both verbally and nonverbally. EMGs and biofeedback give patients real-time signals about their movements (Huang et al., 2006).

Virtual reality (VR) applications offer similar benefits. For instance, stroke patients who trained for four weeks using a VR system with implicit feedback on arm movements showed notable improvements in motor function. Additionally, there was increased activity in the primary sensorimotor cortex on the same side of the brain (Jang et al., 2005).

4.12. Modulate Effector Selection

After a stroke, patients often avoid using the affected limb. They tend to overuse the unaffected limb instead. This can lead to problems, a phenomenon called *learned non-use*. CIMT is the most researched therapy to prevent this, showing the best results (Kwakkel et al., 2015).

In a study, stroke patients completed a two-week CIMT program at home. The results showed increased brain activity when using the paretic arm (Johansen-Berg et al., 2002).

4.13. Action Observation/Embodied Practice

Action Observation/Embodied Practice has gained attention since mirror neurons were discovered (Rizzolatti & Sinigaglia, 2010). Studies in monkeys showed that specific neurons fired when a monkey acted and watched another act. In human studies, those who watched a task before trying it performed better than those who did not watch or saw a different task (Mattar & Gribble, 2005). Watching a movement activates the visual cortex. Doing it also activates the M1 (Hardwick et al., 2018).

TMS studies found that watching actions boosts the activation of relevant muscles (Fadiga et al., 1995). This implies that action observation therapy could help patients with hemiparesis or paralysis.

Mirror therapy has shown notable benefits in therapy (Ramachandran & Rogers-Ramachandran, 1996). Unlike action observation, it significantly activates the M1 and visual areas related to the reflected movement. Moreover, it boosts connectivity between motor areas in the brain, enhancing links between hemispheres (Arya, 2016; Hamzei et al., 2012).

4.14. Mental Practice/Motor Imagery

Mental practice and motor imagery allow people to simulate movements mentally. As explained by simulation theory (Jeannerod, 2001), this doesn't require physical execution. Motor imagery serves as a rehearsal for future actions and aids in planning. It's vital for learning movements (Di Rienzo et al., 2016). However, physical practice is more effective than mental practice (Hird et al., 1991).

A study showed similar brain activity regarding mental imagery and actual movement. Both activate premotor areas, the somatosensory cortex, and subcortical regions. However, motor imagery focuses more on cognitive control. In contrast, physical movement engages areas related to essential motor functions (Hardwick et al., 2018).

Motor imagery's benefits have been noted in sports and rehabilitation. It is beneficial for those with severe disabilities, like stroke survivors (Mulder, 2007). For early stage recovery patients who are bed-bound, motor imagery can be crucial. It may improve outcomes and aid in learning movements.

4.15. Social Interaction

Many daily activities require social interaction. Lacking these skills can make people depend on others. Comments from others and self-belief affect learning motor skills. An fMRI study showed that brain areas for social skills, attention, and rewards are active during social interactions. This shows that social context is vital in learning and rehabilitation (Redcay et al., 2010).

In studies with animals, those who socialized after a blockage showed better recovery and lower death rates. This was especially true when paired with a healthy partner. These findings call for more research on social interaction in rehab and learning motor skills. There is a lack of trials in this area (Venna et al., 2014).

Social interaction seems key in learning about and treating impairments. Group exercise for similar patients could be helpful. However, more controlled studies are needed to explore this.

4.16. Principles and Outcomes in Motor Learning

- Studies on intensive practice vary in repetition counts, complicating the assessment of its effects on motor learning.
- The reliability of these studies is hindered by inconsistent measures for practice types and session lengths.
- Task-specific practice aims to master a skill, while task-oriented practice involves any suitable movement.
- Variable practice is common, but its effects need more study.
- Strategies like rhythmic stimuli and multisensory integration are under-researched and need better protocols.
- Some principles target specific brain areas, while others engage broader networks.
- The best neuroplastic activity should form the basis of effective rehabilitation protocols.

- Selection of principles should consider the patient's stage and the extent of damage.
- Techniques like action observation, mirror therapy, and motor imagery are ideal for those with severe damage.
- These principles apply beyond stroke patients. They are relevant for Parkinson's, Alzheimer's (for more information on this topic, please refer to *Chapter 13: Neuroplasticity's Impact on Neurodegenerative Disorders*), orthopedic rehabilitation, and sports studies (for more information on this topic, please refer to *Chapter 5: Neuroplasticity in Sports Medicine: Injury Prevention And Performance Enhancement*).

taVNS is a promising rehabilitation method that boosts motor learning. For more details, see *Chapter 10: Neuroplasticity and Neuromodulation*. It explores neuroplasticity and discusses neuromodulation techniques like transcranial stimulation and taVNS. These methods aid brain recovery and improve function in rehabilitation.

5. Vagus Nerve Stimulation and Motor Learning

Various neuromodulation techniques are used to enhance or improve motor learning. Transcutaneous auricular vagus nerve stimulation (taVNS) has gained more interest in practice due to its non-invasive nature, cost-effectiveness, and ease of application. Studies in the literature have shown that taVNS modulates motor learning. Information regarding its enhancement of motor learning is related to changes in hormone and neurotransmitter concentrations. Research has shown that taVNS can increase noradrenergic activity by activating the nucleus tractus solitarius and locus coeruleus (Jongkees et al., 2018). A common neurochemical effect of taVNS is increased brain γ-aminobutyric acid (GABA) levels. GABA is associated with behavioral decision-making (Paine et al., 2015) and implicit motor learning (Frangou et al., 2019). A study on participants with upper extremity dysfunction following ischemic stroke showed that taVNS improved rehabilitation scores and positively affected motor recovery (Frangou et al., 2019). In another study on healthy people, the taVNS group outperformed the sham group in learning. They had greater cortical potential amplitude related to motor skills, a reduced alpha-gamma modulation index, and stronger connectivity to the frontal lobe. These results suggest that taVNS can enhance motor learning by facilitating simple movement learning and strengthening cognitive and memory functions (Chen et al., 2024).

taVNS is a promising method that facilitates rehabilitation due to its undeniable positive effects on improving motor learning skills. For more information on this topic, please refer to *Chapter 10: Neuroplasticity and Neuromodulation*. This chapter will likely delve into the mechanisms of neuroplasticity and how neuromodulation techniques, such as transcranial stimulation or taVNS, are used to promote brain recovery and functional improvement in various rehabilitation settings.

6. Music Therapy and Motor Learning

There is evidence that music therapy and rhythmic auditory stimulation significantly improve a range of symptoms in both neurological and non-neurological conditions. Studies have shown that music and rhythm-based motor learning training can improve gait disorders and motor symptoms (particularly freezing symptoms) in patients with Parkinson's disease (Devlin et al., 2019). Music therapy regulates "on" and "off" dopaminergic activities, which are crucial for one of the critical motor skills: walking. Thus, combining music therapy with neurological rehabilitation could be valuable in light of the existing evidence.

Harrison et al. (2019) demonstrated that singing loudly at different tempos could improve outcomes in older adults and Parkinson's patients (Harrison et al., 2019). Similarly, Clabrò et al. (2019) emphasized that rhythmic auditory stimulation training could help compensate for the loss of automatic and rhythmic movements. The study found that rhythmic auditory cues during gait training reshaped sensorimotor rhythms and frontal-temporoparietal/temporal connectivity (Calabrò et al., 2019). For more information on this topic, please refer to *Chapter 13: Neuroplasticity's Impact on Neurodegenerative Disorders.* This chapter will provide a detailed exploration of how neuroplasticity influences neurodegenerative diseases such as Alzheimer's, Parkinson's, and Huntington's diseases. It will examine the potential therapeutic approaches that utilize neuroplastic mechanisms to slow disease progression, enhance cognitive function, and improve overall quality of life for patients with these conditions.

Different types of music have various effects on motor skill acquisition. For example, classical music reduces speed and body inclination, while rock, motivational, and heavy metal songs increase the range of pelvic obliquity movements (Murgia et al., 2018). Moreover, auditory rhythms strengthen movement's temporal and spatial stability (Véron-Delor et al., 2020). In light of this information, music not only aids motor learning by addressing movement disorders but also contributes to psychological, social, and well-being aspects. Thus, music therapy could be used in training aimed at motor learning to assist patients.

7. Virtual Reality (VR) and Motor Learning

For effective outcomes in motor learning, sustainability and repetition are crucial. Continuous repetitions may hinder patient engagement and motivation in long-term motor learning rehabilitation approaches. Also, when working with children in motor learning, adding entertainment and novelty to the therapy is critical. This keeps them engaged and motivated during rehab. In this context, VR applications allow tasks to be completed meaningfully and enjoyably, especially in pediatric populations. VR-based applications allow the integration of various motor experiences and provide an opportunity to practice tasks that would be difficult or impossible to perform in the real world. Tasks that are hard to practice in real life can be safely simulated through VR applications. Moreover, VR applications can be implemented at home, ensuring that the necessary practice for long-term motor learning can continue in the long term (Palaus et al., 2017).

For more information on this topic, please refer to *Chapter 12: Virtual Reality Applications in Neuroplasticity Research and Therapy*. This chapter will explore how VR technologies are utilized in research and therapeutic contexts to stimulate neuroplasticity. It will explore how VR enhances motor learning, helps patients with neurological conditions like stroke or cerebral palsy recover, and triggers the brain's adaptive mechanisms for functional recovery. The chapter will also cover case studies and examples of VR applications in neurorehabilitation, providing a deeper understanding of its impact on neural recovery and motor function.

Current studies (Chen et al., 2022; Zhang et al., 2021) indicate that VR applications in rehabilitation processes targeting motor learning improve motor skills in both upper and lower extremities following a stroke. Similarly, VR-based interventions in children with cerebral palsy have been shown to increase participation rates in therapy and improve motor functions (Tobaiqi et al., 2023). For more information on this topic, please refer to *Chapter 7: Stroke Rehabilitation: Harnessing Neuroplasticity for Functional Recovery.* This chapter will likely provide a deeper exploration of how neuroplasticity can be leveraged in stroke rehabilitation, including the role of modern technologies.

8. Conclusion

Motor learning and neuroplasticity work together to improve brain function. As patients learn new motor skills during rehab, their neural connections strengthen and reorganize. This process leads to lasting changes, improving their abilities and helping them develop new ones. The brain's ability to adapt, combined with focused learning, is key to effective neurological recovery and improving skills.

Neuroplasticity is crucial for recovering from brain disorders like stroke, Parkinson's disease, and cerebral palsy. By applying motor learning principles, rehab outcomes can be significantly improved by promoting brain changes that restore function. Clinicians need to understand these concepts to create effective treatment plans. By using targeted exercises to tap into the brain's adaptability, therapists can improve patient outcomes and offer hope to those facing neurological challenges.

9. Takeaways

1. Reward-based practices during motor skill training improve the long-term retention of motor skills.
2. Random or variable training is better than blocked practice. It enhances motor learning.
3. Sleep or short naps following motor learning significantly aid in the long-term retention of motor skills.
4. Intensive practice is a critical factor in improving motor learning.
5. Task-specific practice plays a crucial role in the acquisition of sensorimotor skills.

6. Goal-oriented exercises during motor learning yield better performance compared to non-goal-oriented activities.
7. The inclusion of tasks with optimal difficulty levels leads to improved motor learning outcomes.
8. Diverse practice variations in training boost neuroplasticity. So, varying practices is important.
9. Training personalized to an individual's abilities yields better results. It is more effective than using fixed difficulty increments.
10. Excessive task difficulty that surpasses an individual's capacity for success can negatively affect performance.
11. Multisensory stimulation during goal-oriented movements facilitates motor skill acquisition.
12. Providing tactile feedback during motor imagery strengthens sensorimotor connectivity.
13. Consistent perception of sounds and rhythms boosts motor networks' and cerebellum activity.
14. Punishment-based feedback can accelerate motor learning, while reward-based feedback supports long-term retention.
15. Neuronal activity is driven by rewards resulting from neuroplastic changes, and the gradual accumulation of dopamine contributes to long-term memory formation.
16. CIMT is among the most effective and widely used techniques with high success rates in motor learning.
17. TMS studies have demonstrated increased muscle activation patterns during action observation, highlighting the potential benefits of action observation therapy in treating individuals with hemiparesis or complete paralysis.
18. Motor imagery, regarded as a mental rehearsal of future movements and motor planning, plays a significant role in motor learning. It has been reported as advantageous in motor learning strategies for individuals with severe physical impairments, particularly in stroke rehabilitation.
19. Social interactions stimulate brain functions and facilitate activation in areas associated with goal-directed attention, visual attention, and reward processing.
20. taVNS enhances GABA activity in the brain, improving behavioral decision-making and implicit motor learning functions.

21. Music- and rhythm-based motor learning programs have been shown to improve gait disturbances and motor symptoms, especially freezing episodes, in patients with Parkinson's disease.
22. VR-based practice is increasingly used in motor learning strategies, particularly for children with stroke or cerebral palsy. VR interventions are vital in developing upper and lower extremity motor skills.

Abbreviations

BOLD: Blood-Oxygen-Level-Dependent
CIMT: Constraint-Induced Motor Therapy
CNS: Central Nervous System
DLPFC: Dorsolateral Prefrontal Cortex
fMRI: Functional Magnetic Resonance Imaging
GABA: γ-aminobütirik asit
KR: Knowledge Of Results
LTP: Long-Term Potentiation
M1: Primary Motor Cortex
MRI: Magnetic Resonance Imaging
NREM: Non-Rapid Eye Movement
PET: Positron Emission Tomography
preSMA: Pre-supplementary Motor Area
rCBF: Cerebral Blood Flow
SMA: Supplementary Motor Area
taVNS: Transcutaneous Auricular Vagus Nerve Stimulation
tDCS: Transcranial Direct Current Stimulation
TMS: Transcranial Magnetic Stimulation
VBM: Voxel-Based Morphometry
VR: Virtual Reality

Bibliography

1. Abe, M., Schambra, H., Wassermann, E. M., Luckenbaugh, D., Schweighofer, N., & Cohen, L. G. (2011). Reward improves long-term retention of a motor memory through induction of offline memory gains. *Current Biology, 21*(7), 557–562. https://doi.org/ 10.1016/j.cub.2011.02.030

2. Adcock, R. A., Thangavel, A., Whitfield-Gabrieli, S., Knutson, B., & Gabrieli, J. D. (2006). Reward-motivated learning: mesolimbic activation precedes memory formation. *Neuron, 50*(3), 507–517. https://doi.org/ 10.1016/j.neuron.2006.03.036
3. Albert, N. B., Robertson, E. M., & Miall, R. C. (2009). The resting human brain and motor learning. *Current Biology, 19*(12), 1023–1027. https://doi.org/10.1016/j.cub.2009.04.028
4. Albouy, G., Sterpenich, V., Balteau, E., Vandewalle, G., Desseilles, M., Dang-Vu, T., … Degueldre, C. (2008). Both the hippocampus and striatum are involved in consolidation of motor sequence memory. *Neuron, 58*(2), 261–272. https://doi.org/10.1016/j.neuron.2008.02.008
5. Andrieux, M., Danna, J., & Thon, B. (2012). Self-control of task difficulty during training enhances motor learning of a complex coincidence-anticipation task. *Research quarterly for exercise and sport, 83*(1), 27–35. https://doi.org/10.1080/02701367.2012.10599822
6. Arya, K. N. (2016). Underlying neural mechanisms of mirror therapy: Implications for motor rehabilitation in stroke. *Neurology India, 64*(1), 38–44. https://doi.org/10.4103/0028-3886.173622
7. Ashburner, J., & Friston, K. J. (2000). Voxel-based morphometry--the methods. *Neuroimage, 11*(6 Pt 1), 805–821. https://doi.org/10.1006/nimg.2000.0582
8. Ballester, B. R., Maier, M., San Segundo Mozo, R. M., Castañeda, V., Duff, A., & MJ Verschure, P. F. (2016). Counteracting learned non-use in chronic stroke patients with reinforcement-induced movement therapy. *Journal of neuroengineering and rehabilitation, 13*, 1–15. https://doi.org/10.1186/s12984-016-0178-x
9. Bjork, R. A., Dunlosky, J., & Kornell, N. (2013). Self-regulated learning: Beliefs, techniques, and illusions. *Annual review of psychology, 64*(1), 417–444. https://doi.org/10.1146/annurev-psych-113011-143823
10. Boyd, L. A., Vidoni, E. D., & Wessel, B. D. (2010). Motor learning after stroke: is skill acquisition a prerequisite for contralesional neuroplastic change? *Neuroscience letters, 482*(1), 21–25. https://doi.org/10.1016/j.neulet.2010.06.082
11. Boyke, J., Driemeyer, J., Gaser, C., Büchel, C., & May, A. (2008). Training-induced brain structure changes in the elderly. *J Neurosci, 28*(28), 7031–7035. https://doi.org/10.1523/jneurosci.0742-08.2008
12. Buccino, G., Vogt, S., Ritzl, A., Fink, G. R., Zilles, K., Freund, H. J., & Rizzolatti, G. (2004). Neural circuits underlying imitation learning

of hand actions: an event-related fMRI study. *Neuron, 42*(2), 323–334. https://doi.org/10.1016/s0896-6273(04)00181-3

13. Calabresi, P., Picconi, B., Tozzi, A., & Di Filippo, M. (2007). Dopamine-mediated regulation of corticostriatal synaptic plasticity. Trends in neurosciences, 30(5), 211–219. https://doi.org/10.1016/j.tins.2007.03.001
14. Calabrò, R. S., Naro, A., Filoni, S., Pullia, M., Billeri, L., Tomasello, P., … Bramanti, P. (2019). Walking to your right music: a randomized controlled trial on the novel use of treadmill plus music in Parkinson's disease. *Journal of neuroengineering and rehabilitation, 16*, 1–14. https://doi.org/10.1186/s12984-019-0533-9
15. Chen, J., Or, C. K., & Chen, T. (2022). Effectiveness of Using Virtual Reality-Supported Exercise Therapy for Upper Extremity Motor Rehabilitation in Patients With Stroke: Systematic Review and Meta-analysis of Randomized Controlled Trials. *J Med Internet Res, 24*(6), e24111. https://doi.org/10.2196/24111
16. Chen, L., Tang, C., Wang, Z., Zhang, L., Gu, B., Liu, X., & Ming, D. (2024). Enhancing Motor Sequence Learning via Transcutaneous Auricular Vagus Nerve Stimulation (taVNS): An EEG Study. *IEEE J Biomed Health Inform, 28*(3), 1285–1296. https://doi.org/10.1109/jbhi.2023.3344176
17. Cramer, S. C., Sur, M., Dobkin, B. H., O'Brien, C., Sanger, T. D., Trojanowski, J. Q., … Chen, D. (2011). Harnessing neuroplasticity for clinical applications. *Brain, 134*(6), 1591–1609. https://doi.org/10.1093/brain/awr039
18. Dayan, E., & Cohen, L. G. (2011). Neuroplasticity subserving motor skill learning. *Neuron, 72*(3), 443–454. https://doi.org/10.1016/j.neuron.2011.10.008
19. Debas, K., Carrier, J., Orban, P., Barakat, M., Lungu, O., Vandewalle, G., … Ungerleider, L. G. (2010). Brain plasticity is related to the consolidation of motor sequence learning and motor adaptation. *Proceedings of the National Academy of Sciences, 107*(41), 17839–17844. https://doi.org/10.1073/pnas.1013176107
20. Dettmers, C., Teske, U., Hamzei, F., Uswatte, G., Taub, E., & Weiller, C. (2005). Distributed form of constraint-induced movement therapy improves functional outcome and quality of life after stroke. *Archives of Physical Medicine and Rehabilitation, 86*(2), 204–209. https://doi.org/10.1016/j.apmr.2004.05.007

21. Devlin, K., Alshaikh, J. T., & Pantelyat, A. (2019). Music therapy and music-based interventions for movement disorders. *Current Neurology and Neuroscience Reports, 19*, 1–13. https://doi.org/10.1007/s11910-019-1005-0
22. Di Rienzo, F., Debarnot, U., Daligault, S., Saruco, E., Delpuech, C., Doyon, J., … Guillot, A. (2016). Online and offline performance gains following motor imagery practice: a comprehensive review of behavioral and neuroimaging studies. *Frontiers in human neuroscience, 10*, 315. https://doi.org/10.3389/fnhum.2016.00315
23. Diekelmann, S., & Born, J. (2010). The memory function of sleep. *Nat Rev Neurosci, 11*(2), 114–126. https://doi.org/10.1038/nrn2762
24. Doyon, J., & Benali, H. (2005). Reorganization and plasticity in the adult brain during learning of motor skills. *Curr Opin Neurobiol, 15*(2), 161–167. https://doi.org/10.1016/j.conb.2005.03.004
25. Doyon, J., Penhune, V., & Ungerleider, L. G. (2003). Distinct contribution of the cortico-striatal and cortico-cerebellar systems to motor skill learning. *Neuropsychologia, 41*(3), 252–262. https://doi.org/10.1016/s0028-3932(02)00158-6
26. Draganski, B., Gaser, C., Busch, V., Schuierer, G., Bogdahn, U., & May, A. (2004). Neuroplasticity: changes in grey matter induced by training. *Nature, 427*(6972), 311–312. https://doi.org/10.1038/427311a
27. Fadiga, L., Fogassi, L., Pavesi, G., & Rizzolatti, G. (1995). Motor facilitation during action observation: a magnetic stimulation study. *Journal of Neurophysiology, 73*(6), 2608–2611. https://doi.org/10.1152/jn.1995.73.6.2608
28. Fields, R. D. (2008). White matter in learning, cognition, and psychiatric disorders. *Trends Neurosci, 31*(7), 361–370. https://doi.org/10.1016/j.tins.2008.04.001
29. Fischer, S., Hallschmid, M., Elsner, A. L., & Born, J. (2002). Sleep forms memory for finger skills. *Proc Natl Acad Sci U S A, 99*(18), 11987–11991. https://doi.org/10.1073/pnas.182178199
30. Frangou, P., Emir, U. E., Karlaftis, V. M., Nettekoven, C., Hinson, E. L., Larcombe, S., … Kourtzi, Z. (2019). Learning to optimize perceptual decisions through suppressive interactions in the human brain. *Nature communications, 10*(1), 474. https://doi.org/10.1038/s41467-019-08313-y
31. Galea, J. M., Mallia, E., Rothwell, J., & Diedrichsen, J. (2015). The dissociable effects of punishment and reward on motor learning. *Nature Neuroscience, 18*(4), 597–602. https://doi.org/10.1038/nn.3956

32. Gallivan, J. P., & Culham, J. C. (2015). Neural coding within human brain areas involved in actions. *Current opinion in neurobiology, 33*, 141–149. https://doi.org/10.1016/j.conb.2015.03.012
33. Ghazanfar, A. A., & Schroeder, C. E. (2006). Is neocortex essentially multisensory? *Trends in cognitive sciences, 10*(6), 278–285. https://doi.org/10.1016/j.tics.2006.04.008
34. Gomez-Rodriguez, M., Peters, J., Hill, J., Schölkopf, B., Gharabaghi, A., & Grosse-Wentrup, M. (2011). Closing the sensorimotor loop: Haptic feedback facilitates the decoding of motor imagery. *Journal of neural engineering, 8*(3), 036005.
35. Grahn, J. A. (2012). Neural mechanisms of rhythm perception: current findings and future perspectives. *Topics in cognitive science, 4*(4), 585–606. https://doi.org/10.1111/j.1756-8765.2012.01213.x
36. Guadagnoli, M. A., & Lee, T. D. (2004). Challenge point: a framework for conceptualizing the effects of various practice conditions in motor learning. *Journal of motor behavior, 36*(2), 212–224. https://doi.org/10.3200/JMBR.36.2.212-224
37. Hamzei, F., Läppchen, C. H., Glauche, V., Mader, I., Rijntjes, M., & Weiller, C. (2012). Functional plasticity induced by mirror training: the mirror as the element connecting both hands to one hemisphere. *Neurorehabilitation and neural repair, 26*(5), 484–496. https://doi.org/10.1177/1545968311427917
38. Hardwick, R. M., Caspers, S., Eickhoff, S. B., & Swinnen, S. P. (2018). Neural correlates of action: Comparing meta-analyses of imagery, observation, and execution. *Neuroscience & Biobehavioral Reviews, 94*, 31–44. https://doi.org/10.1016/j.neubiorev.2018.08.003
39. Hardwick, R. M., Rottschy, C., Miall, R. C., & Eickhoff, S. B. (2013). A quantitative meta-analysis and review of motor learning in the human brain. *NeuroImage, 67*, 283–297. https://doi.org/10.1016/j.neuroimage.2012.11.020
40. Harrison, E. C., Horin, A. P., & Earhart, G. M. (2019). Mental Singing Reduces Gait Variability More Than Music Listening for Healthy Older Adults and People With Parkinson's Disease. *J Neurol Phys Ther, 43*(4), 204–211. https://doi.org/10.1097/npt.0000000000000288
41. Hird, J. S., Landers, D. M., Thomas, J. R., & Horan, J. J. (1991). Physical practice is superior to mental practice in enhancing cognitive and motor task performance. *Journal of Sport and Exercise Psychology, 13*(3), 281–293. https://doi.org/10.1123/jsep.13.3.281

42. Huang, H., Wolf, S. L., & He, J. (2006). Recent developments in biofeedback for neuromotor rehabilitation. *Journal of neuroengineering and rehabilitation, 3*, 1–12. https://doi.org/10.1186/1743-0003-3-11
43. Inoue, K., Kawashima, R., Satoh, K., Kinomura, S., Sugiura, M., Goto, R., … Fukuda, H. (2000). A PET Study of Visuomotor Learning under Optical Rotation. *NeuroImage, 11*(5), 505–516. https://doi.org/10.1006/nimg.2000.0554
44. Jang, S. H., You, S. H., Hallett, M., Cho, Y. W., Park, C.-M., Cho, S.-H., … Kim, T.-H. (2005). Cortical reorganization and associated functional motor recovery after virtual reality in patients with chronic stroke: an experimenter-blind preliminary study. *Archives of Physical Medicine and Rehabilitation, 86*(11), 2218–2223. https://doi.org/10.1016/j.apmr.2005.04.015
45. Jeannerod, M. (2001). Neural simulation of action: a unifying mechanism for motor cognition. *NeuroImage, 14*(1), S103-S109. https://doi.org/10.1006/nimg.2001.0832
46. Johansen-Berg, H., Dawes, H., Guy, C., Smith, S. M., Wade, D. T., & Matthews, P. M. (2002). Correlation between motor improvements and altered fMRI activity after rehabilitative therapy. *Brain, 125*(12), 2731–2742. https://doi.org/10.1093/brain/awf282
47. Johnson, B. P., & Cohen, L. G. (2023). Applied strategies of neuroplasticity. *Handb Clin Neurol, 196*, 599–609. https://doi.org/10.1016/b978-0-323-98817-9.00011-9
48. Jongkees, B. J., Immink, M. A., Finisguerra, A., & Colzato, L. S. (2018). Transcutaneous vagus nerve stimulation (tVNS) enhances response selection during sequential action. *Frontiers in Psychology, 9*, 1159. https://doi.org/10.3389/fpsyg.2018.01159
49. Karni, A., Meyer, G., Rey-Hipolito, C., Jezzard, P., Adams, M. M., Turner, R., & Ungerleider, L. G. (1998). The acquisition of skilled motor performance: fast and slow experience-driven changes in primary motor cortex. *Proc Natl Acad Sci U S A, 95*(3), 861–868. https://doi.org/10.1073/pnas.95.3.861
50. Kleim, J. A., Bruneau, R., Calder, K., Pocock, D., VandenBerg, P. M., MacDonald, E., … Nader, K. (2003). Functional organization of adult motor cortex is dependent upon continued protein synthesis. *Neuron, 40*(1), 167–176. https://doi.org/10.1016/s0896-6273(03)00592-0
51. Kwakkel, G., Veerbeek, J. M., van Wegen, E. E., & Wolf, S. L. (2015). Constraint-induced movement therapy after stroke.

The Lancet Neurology, 14(2), 224–234. https://doi.org 10.1016/S1474-4422(14)70160-7
52. Làdavas, E. (2008). Multisensory-based Approach to the recovery of unisensory deficit. *Annals of the New York Academy of Sciences, 1124*(1), 98–110. https://doi.org/10.1196/annals.1440.008
53. Lage, G. M., Ugrinowitsch, H., Apolinário-Souza, T., Vieira, M. M., Albuquerque, M. R., & Benda, R. N. (2015). Repetition and variation in motor practice: A review of neural correlates. *Neuroscience & Biobehavioral Reviews, 57*, 132–141. https://doi.org/10.1016/j.neubiorev.2015.08.012
54. Levin, M. F., & Demers, M. (2021). Motor learning in neurological rehabilitation. Disability and Rehabilitation, 43(24), 3445–3453. https://doi.org/10.1080/09638288.2020.175231
55. Lin, C.-H. J., Knowlton, B. J., Chiang, M.-C., Iacoboni, M., Udompholkul, P., & Wu, A. D. (2011). Brain–behavior correlates of optimizing learning through interleaved practice. *NeuroImage, 56*(3), 1758–1772. https://doi.org/10.1016/j.neuroimage.2011.02.066
56. Lohse, K. R., Lang, C. E., & Boyd, L. A. (2014). Is more better? Using metadata to explore dose–response relationships in stroke rehabilitation. *Stroke, 45*(7), 2053–2058. https://doi.org/10.1161/STROKEAHA.114.004695
57. Maier, M., Ballester, B. R., & Verschure, P. F. (2019). Principles of neurorehabilitation after stroke based on motor learning and brain plasticity mechanisms. *Frontiers in systems neuroscience, 13*, 74. https://doi.org/10.3389/fnsys.2019.00074
58. Mancini, A., de Iure, A., & Picconi, B. (2022). Chapter 2 - Basic mechanisms of plasticity and learning. In A. Quartarone, M. F. Ghilardi, & F. Boller (Eds.), *Handbook of Clinical Neurology* (Vol. 184, pp. 21–34): Elsevier. https://doi.org/10.1016/B978-0-12-819410-2.00002-3
59. Masaki, H., & Sommer, W. (2012). Cognitive neuroscience of motor learning and motor control. *The Journal of Physical Fitness and Sports Medicine, 1*(3), 369–380. https://doi.org/10.7600/jpfsm.1.369
60. Masters, R., Poolton, J. M., Maxwell, J. P., & Raab, M. (2008). Implicit motor learning and complex decision making in time-constrained environments. *Journal of Motor Behavior, 40*(1), 71–79. https://doi.org/10.3200/JMBR.40.1.71-80
61. Masters, R. S. (1992). Knowledge, knerves and know-how: The role of explicit versus implicit knowledge in the breakdown of a complex

motor skill under pressure. *British journal of psychology, 83*(3), 343–358. https://doi.org/10.1111/j.2044-8295.1992.tb02446.x

62. Mattar, A. A., & Gribble, P. L. (2005). Motor learning by observing. *Neuron, 46*(1), 153-160. https://doi.org/10.1016/j.neuron.2005.02.009
63. May, A., & Gaser, C. (2006). Magnetic resonance-based morphometry: a window into structural plasticity of the brain. *Curr Opin Neurol, 19*(4), 407–411. https://doi.org/10.1097/01.wco.0000236622.91495.21
64. McGann, M. (2010). Perceptual modalities: modes of presentation or modes of interaction? *Journal of Consciousness Studies, 17*(1–2), 72–94.
65. Moore, E., Schaefer, R. S., Bastin, M. E., Roberts, N., & Overy, K. (2017). Diffusion tensor MRI tractography reveals increased fractional anisotropy (FA) in arcuate fasciculus following music-cued motor training. *Brain and Cognition, 116*, 40–46. https://doi.org/10.1016/j.bandc.2017.05.001
66. Mulder, T. (2007). Motor imagery and action observation: cognitive tools for rehabilitation. *Journal of neural transmission, 114*, 1265–1278. https://doi.org/10.1007/s00702-007-0763-z
67. Murgia, M., Pili, R., Corona, F., Sors, F., Agostini, T. A., Bernardis, P., … Pau, M. (2018). The use of footstep sounds as rhythmic auditory stimulation for gait rehabilitation in Parkinson's disease: a randomized controlled trial. *Frontiers in neurology, 9*, 348. https://doi.org/10.3389/fneur.2018.00348
68. Nathan, D. E., Prost, R. W., Guastello, S. J., Jeutter, D. C., & Reynolds, N. C. (2012). Investigating the neural correlates of goal-oriented upper extremity movements. *NeuroRehabilitation, 31*(4), 421–428. https://doi.org/10.3233/NRE-2012-00812
69. Nezafat, R., Shadmehr, R., & Holcomb, H. H. (2001). Long-term adaptation to dynamics of reaching movements: a PET study. *Experimental Brain Research, 140*(1), 66–76. https://doi.org/10.1007/s002210100787
70. Nishida, M., & Walker, M. P. (2007). Daytime naps, motor memory consolidation and regionally specific sleep spindles. *PloS one, 2*(4), e341. https://doi.org/10.1371/journal.pone.0000341
71. Nombela, C., Hughes, L. E., Owen, A. M., & Grahn, J. A. (2013). Into the groove: can rhythm influence Parkinson's disease? *Neuroscience & Biobehavioral Reviews, 37*(10), 2564–2570. https://doi.org/10.1016/j.neubiorev.2013.08.003

72. Paine, T. A., O'hara, A., Plaut, B., & Lowes, D. C. (2015). Effects of disrupting medial prefrontal cortex GABA transmission on decision-making in a rodent gambling task. *Psychopharmacology, 232*, 1755–1765. https://doi.org/10.1007/s00213-014-3816-7
73. Palaus, M., Marron, E. M., Viejo-Sobera, R., & Redolar-Ripoll, D. (2017). Neural Basis of Video Gaming: A Systematic Review. *Front Hum Neurosci, 11*, 248. https://doi.org/10.3389/fnhum.2017.00248
74. Ramachandran, V. S., & Rogers-Ramachandran, D. (1996). Synaesthesia in phantom limbs induced with mirrors. *Proceedings of the Royal Society of London. Series B: Biological Sciences, 263*(1369), 377–386. https://doi.org/10.1098/rspb.1996.0058
75. Redcay, E., Dodell-Feder, D., Pearrow, M. J., Mavros, P. L., Kleiner, M., Gabrieli, J. D., & Saxe, R. (2010). Live face-to-face interaction during fMRI: a new tool for social cognitive neuroscience. *NeuroImage, 50*(4), 1639–1647. https://doi.org/10.1016/j.neuroimage.2010.01.052
76. Reis, J., Schambra, H. M., Cohen, L. G., Buch, E. R., Fritsch, B., Zarahn, E., … Krakauer, J. W. (2009). Noninvasive cortical stimulation enhances motor skill acquisition over multiple days through an effect on consolidation. *Proceedings of the National Academy of Sciences, 106*(5), 1590–1595. https://doi.org/10.1073/pnas.0805413106
77. Rizzolatti, G., & Sinigaglia, C. (2010). The functional role of the parieto-frontal mirror circuit: interpretations and misinterpretations. *Nature Reviews Neuroscience, 11*(4), 264–274. https://doi.org/10.1038/nrn2805
78. Robertson, E. M., Pascual-Leone, A., & Press, D. Z. (2004). Awareness modifies the skill-learning benefits of sleep. *Curr Biol, 14*(3), 208–212. https://doi.org/10.1016/j.cub.2004.01.027
79. Romano, J. C., Howard Jr., J. H., & Howard, D. V. (2010). One-year retention of general and sequence-specific skills in a probabilistic, serial reaction time task. *Memory, 18*(4), 427–441. https://doi.org/10.1080/09658211003742680
80. Rosenkranz, K., Kacar, A., & Rothwell, J. C. (2007). Differential modulation of motor cortical plasticity and excitability in early and late phases of human motor learning. *J Neurosci, 27*(44), 12058–12066. https://doi.org/10.1523/jneurosci.2663-07.2007
81. Rossignol, S., & Jones, G. M. (1976). Audio-spinal influence in man studied by the H-reflex and its possible role on rhythmic

movements synchronized to sound. *Electroencephalography and clinical neurophysiology, 41*(1), 83–92. https://doi.org/10.1016/0013-4694(76)90217-0

82. Salmoni, A. W., Schmidt, R. A., & Walter, C. B. (1984). Knowledge of results and motor learning: a review and critical reappraisal. *Psychological bulletin, 95*(3), 355. https://doi.org/10.1037/0033-2909.95.3.355
83. Savion-Lemieux, T., & Penhune, V. B. (2005). The effects of practice and delay on motor skill learning and retention. *Experimental Brain Research, 161*, 423–431. https://doi.org/10.1007/s00221-004-2085-9
84. Schaefer, R. S. (2014). Auditory rhythmic cueing in movement rehabilitation: findings and possible mechanisms. *Philosophical Transactions of the Royal Society B: Biological Sciences, 369*(1658), 20130402. https://doi.org/10.1098/rstb.2013.0402
85. Schmidt, R., & Lee, T. (2005). Motor Learning and Performance: A behavioral emphasis. *Champaign, IL: Human Kinetics*, 130–146.
86. Schmidt, R. A. (1975). A schema theory of discrete motor skill learning. *Psychological review, 82*(4), 225. https://doi.org/10.1037/h0076770
87. Shea, J. B., & Morgan, R. L. (1979). Contextual interference effects on the acquisition, retention, and transfer of a motor skill. *Journal of Experimental psychology: Human Learning and memory, 5*(2), 179. https://doi.org/10.1037/0278-7393.5.2.179
88. Singer, B. J., Vallence, A.-M., Cleary, S., Cooper, I., & Loftus, A. M. (2013). The effect of EMG triggered electrical stimulation plus task practice on arm function in chronic stroke patients with moderate-severe arm deficits. *Restorative neurology and neuroscience, 31*(6), 681–691. https://doi.org/10.3233/RNN-130319
89. Sober, S. J., & Sabes, P. N. (2003). Multisensory integration during motor planning. *Journal of Neuroscience, 23*(18), 6982–6992. https://doi.org/10.1523/JNEUROSCI.23-18-06982.2003
90. Song, S., Howard, J. H., & Howard, D. V. (2007). Sleep does not benefit probabilistic motor sequence learning. *Journal of Neuroscience, 27*(46), 12475–12483. https://doi.org/10.1523/JNEUROSCI.2062-07.2007
91. Stickgold, R., Whidbee, D., Schirmer, B., Patel, V., & Hobson, J. A. (2000). Visual discrimination task improvement: A multi-step process occurring during sleep. *Journal of cognitive neuroscience, 12*(2), 246–254. https://doi.org/10.1162/089892900562075

92. Tanaka, S., Honda, M., Hanakawa, T., & Cohen, L. G. (2010). Differential contribution of the supplementary motor area to stabilization of a procedural motor skill acquired through different practice schedules. *Cerebral Cortex, 20*(9), 2114–2121.
93. Taub, E., Crago, J. E., Burgio, L. D., Groomes, T. E., Cook III, E. W., DeLuca, S. C., & Miller, N. E. (1994). An operant approach to rehabilitation medicine: overcoming learned nonuse by shaping. *Journal of the experimental analysis of behavior, 61*(2), 281–293. https://doi.org/10.1901/jeab.1994.61-281
94. Taylor, J. A., Krakauer, J. W., & Ivry, R. B. (2014). Explicit and implicit contributions to learning in a sensorimotor adaptation task. *Journal of Neuroscience, 34*(8), 3023–3032. https://doi.org/10.1523/JNEUROSCI.3619-13.2014
95. Tobaiqi, M. A., Albadawi, E. A., Fadlalmola, H. A., & Albadrani, M. S. (2023). Application of Virtual Reality-Assisted Exergaming on the Rehabilitation of Children with Cerebral Palsy: A Systematic Review and Meta-Analysis. *Journal of Clinical Medicine, 12*(22), 7091. https://doi.org/10.3390/jcm12227091
96. Tracy, J., Flanders, A., Madi, S., Laskas, J., Stoddard, E., Pyrros, A., … DelVecchio, N. (2003). Regional Brain Activation Associated with Different Performance Patterns during Learning of a Complex Motor Skill. *Cerebral Cortex, 13*(9), 904–910. https://doi.org/10.1093/cercor/13.9.904
97. Veerbeek, J. M., van Wegen, E., van Peppen, R., Van der Wees, P. J., Hendriks, E., Rietberg, M., & Kwakkel, G. (2014). What is the evidence for physical therapy poststroke? A systematic review and meta-analysis. *PloS one, 9*(2), e87987. https://doi.org/10.1371/journal.pone.0087987
98. Venna, V., Xu, Y., Doran, S., Patrizz, A., & McCullough, L. (2014). Social interaction plays a critical role in neurogenesis and recovery after stroke. *Translational psychiatry, 4*(1), e351-e351. https://doi.org/10.1038/tp.2013.128
99. Véron-Delor, L., Pinto, S., Eusebio, A., Azulay, J. P., Witjas, T., Velay, J. L., & Danna, J. (2020). Musical sonification improves motor control in Parkinson's disease: a proof of concept with handwriting. *Annals of the New York Academy of Sciences, 1465*(1), 132–145. https://doi.org/10.1111/nyas.14252
100. Walker, M. P., Brakefield, T., Morgan, A., Hobson, J. A., & Stickgold, R. (2002). Practice with sleep makes perfect: sleep-dependent motor

skill learning. *Neuron, 35*(1), 205–211. 10.1016/S0896-6273(02)00746-8 External Link

101. Wickens, C. D., Hutchins, S., Carolan, T., & Cumming, J. (2013). Effectiveness of part-task training and increasing-difficulty training strategies: a meta-analysis approach. *Human Factors, 55*(2), 461–470. https://doi.org/10.1177/0018720812451994
102. Willingham, D. B. (1998). A neuropsychological theory of motor skill learning. *Psychological review, 105*(3), 558.
103. Winstein, C. J., Grafton, S. T., & Pohl, P. S. (1997). Motor task difficulty and brain activity: investigation of goal-directed reciprocal aiming using positron emission tomography. *Journal of Neurophysiology,* 77(3), 1581–1594. https://doi.org/10.1152/jn.1997.77.3.1581
104. Wittmann, B. C., Dolan, R. J., & Düzel, E. (2011). Behavioral specifications of reward-associated long-term memory enhancement in humans. *Learning & Memory, 18*(5), 296–300.
105. Wittmann, B. C., Schott, B. H., Guderian, S., Frey, J. U., Heinze, H.-J., & Düzel, E. (2005). Reward-related FMRI activation of dopaminergic midbrain is associated with enhanced hippocampus-dependent long-term memory formation. *Neuron, 45*(3), 459–467. https://doi.org/10.1016/j.neuron.2005.01.010
106. Wu, C.-y., Trombly, C. A., Lin, K.-c., & Tickle-Degnen, L. (2000). A kinematic study of contextual effects on reaching performance in persons with and without stroke: influences of object availability. *Archives of Physical Medicine and Rehabilitation, 81*(1), 95–101. https://doi.org/10.1016/S0003-9993(00)90228-4
107. Wulf, G., Shea, C., & Park, J.-H. (2001). Attention and motor performance: Preferences for and advantages of an external focus. *Research quarterly for exercise and sport,* 72(4), 335–344. https://doi.org/10.1080/02701367.2001.10608970
108. Yin, H. H., Mulcare, S. P., Hilário, M. R., Clouse, E., Holloway, T., Davis, M. I., … Costa, R. M. (2009). Dynamic reorganization of striatal circuits during the acquisition and consolidation of a skill. *Nat Neurosci, 12*(3), 333–341. https://doi.org/10.1038/nn.2261
109. Zhang, B., Li, D., Liu, Y., Wang, J., & Xiao, Q. (2021). Virtual reality for limb motor function, balance, gait, cognition and daily function of stroke patients: A systematic review and meta-analysis. *J Adv Nurs,* 77(8), 3255–3273. https://doi.org/10.1111/jan.14800

Neuroplasticity in Sports Medicine: Injury Prevention and Performance Enhancement

Pelin Pişirici and Büşra Tamgüç

1. Introduction

Sports activities are vital for a healthy lifestyle and well-being. They are even suggested as alternatives to medication (Latino & Tafuri, 2024). However, as more people join sports, injuries also rise. This trend could harm participation and health (Swanik et al., 2007). In the United States, about 4.3 million people seek emergency help for sports injuries each year (Stephenson et al., 2021). Knee and ankle injuries comprise over 60% of these (Emery & Tyreman, 2009). Noncontact anterior cruciate ligament (ACL) injuries are the most common (about 70%) among knee injuries (Mancino et al., 2024).

The risk of injury increases with professionalism. For professional athletes, ACL injuries occur between 0.21% and 3.67%. In the general population, it's only 0.03% (Bram et al., 2021; Mancino et al., 2024). Ankle injuries are the second most common, following knee injuries (Stephenson et al., 2021). They account for about 15% of sports injuries (Achenbach, 2018). Athletes with previous ankle injuries are three times more likely to reinjure themselves (Al Attar WSA, 2019).

Injuries vary by sport (Gottschalk, 2011). Overhead athletes often get shoulder injuries. The shoulder, prone to injury, accounts for 17% of all injuries (Cools, 2020). Overhead athletes see shoulder injury rates between 18% and 61% (Cools 2020, Asker Aslan 2018, Wright, 2021).

Post-injury, professional athletes face sensorimotor pain, reduced strength, and muscle imbalances. They also struggle with balance, proprioception, and a higher osteoarthritis risk. These issues lead to performance decreases. The recovery process is crucial. It includes time away from the sport, not meeting previous participation levels, and training difficulties (Ardern et al., 2014; Fältström et al., 2023; Mancino et al. 2024; Sugimoto et al., 2015). Loss of musculoskeletal abilities makes returning to the top level hard, increasing re-injury risk. For example, with ACL injuries, 80% return to sport post-surgery. However, only 65% reach pre-injury levels, and just 55% return to competition (Della Villa et al., 2021). After anterior cruciate ligament reconstruction (ACLR) surgery, 15% of patients reinjure their ACL. Of these, 7% injure the same knee, while 8% injure the other (Wiggings et al., 2016).

Musculoskeletal injuries in athletes also lead to central nervous system issues (Rush et al., 2021; Swanik et al., 2007; Lepley et al., 2020; Xue et al., 2022; Zhang et al., 2024). Changes occur in the cerebrum (Nigg & Liu, 1999), corticospinal tract (Perez et al., 2006), and spinal excitability (Lin et al., 2024; Cheriyan & Sheets, 2018) after injury. The primary motor cortex is vital for motor skills in both animals and humans (VandenBerg et al., 2002; Li et al., 2015). Research shows brain changes are linked to joint injuries, especially ACL tears. In people with ACL problems, increased activity is found in areas related to visual processing, motor planning, and pain processing (Nigg & Liu, 1999). A study on chronic ankle instability (CAI) looked at the link between fear of pain and brain changes. It found that shrinking in certain brain areas could be related to this fear (Zhang et al., 2024). The corticospinal tract is key for motor actions. It connects regions involved in movement and sensation (Lemon, 2008). This tract manages muscle contractions and adjustments (Needle et al., 2017). It's seen as the main cause of muscle deficits in knee injuries (Needle et al., 2017). This highlights the need to address both knee joint structure and neuroplasticity issues (Yu et al., 2024). Treatments focusing on the corticospinal tract help with knee injuries (Yu et al., 2024). Spinal excitability is the activation level of spinal cord motor neurons. Knee injuries affect muscle function and activation patterns (Tayfur et al., 2021; Kapreli & Athanasopoulos 2006). These changes occur in the quadriceps, hamstrings, and gastrocnemius muscles. They also affect cortical and spinal excitability, both short and long term. Even two years after surgery, differences in spinal excitability were noted (Rush et al., 2021). The injured leg had lower excitability than the healthy one. The healthy leg was more excitable than the injured one. Research on individuals with CAI suggests reduced neural conduction to ankle stabilizers might cause sensorimotor deficiencies. This could lead to disability due to functional limitations (Kim et al., 2019).

Preventing sports injuries is crucial to avoid long-term issues, especially sensorimotor and central nervous system impairments (Ghanati et al., 2022). Athletes' training includes neuromuscular programs with diverse exercises tailored to their sport.

2. Neuromuscular Training Programs and Motor Learning

2.1. Neuromuscular Training Programs

Over 25 years, neuromuscular training programs have evolved into effective injury prevention tools. For instance, they reduce the injury risk in young female athletes significantly. From 1 in 54, the risk drops to 1 in 111 (Petushek et al., 2019). Moreover, these programs reduce noncontact injuries in female athletes by 67% (Webster & Hewett, 2018).

FIFA 11+, a popular program for football players, showcases impressive results. It reduces hamstring injuries by 60%, hip/groin injuries by 41%, knee injuries by 48%, and ankle injuries by 32% (Thorborg et al., 2017).

Effective programs first identify risk factors. Then, they should be sustainable and match real-world training or competition conditions (Wilk et al., 2023). Current rehabilitation programs mix various exercises: strengthening, plyometric, stability, endurance, agility, and those improving neuromuscular function. They also include postural stabilization exercises with feedback on landing techniques to address muscle activation or imbalances (Huang et al., 2020; Boden & Sheehan, 2022). For best results, programs should last 6–8 weeks (Wong et al., 2023). Each session must be over 20 minutes, several times a week. Total training should exceed 30 minutes weekly. Programs should start before the season and continue throughout. High adherence is crucial. These guidelines are backed by strong evidence (Arundale et al., 2022).

However, these programs can't prevent all injuries. The return-to-sport success rate post-injury is low. After ACLR, 80% of athletes return to some sport. Only 65% get back to their pre-injury level, and 55% return to competition (Della Villa et al., 2021). Yet, 10% of athletes under 25 face new injuries, even after returning to their pre-injury level (Wiggins et al., 2016).

Injuries and re-injuries often happen due to specific reasons. Many programs overlook biomechanical and neuromuscular risk factors (Alentorn-Geli et al., 2014; Ericksen et al., 2013). They also lack high-intensity and sport-specific challenges during rehabilitation. Including single-leg activities and predictable movements in return-to-sport criteria is a mistake. Programs must have objective return-to-sport criteria (Wilk et al., 2023). These programs often fail to prepare athletes for the unpredictable demands of returning to sport (Grindem et al., 2020). So, there's a need for better injury prevention strategies to reduce risks and improve outcomes (Padua et al., 2018; Arundale et al., 2018).

2.2. Motor Learning

Motor learning is the ability to acquire new movement patterns (Richard & Magill, 2001). Engaging in sports and physical activities enhances this ability. It does so by improving movement recall and observation. Sports also aid in better movement control, precision, and coordination. With practice, rough movements become smooth and goal-oriented (Di Palma et al., 2016). This process is crucial for learning and motor skills. It also aids in reorganizing neural networks and enhances brain plasticity (Roberta et al., 2020).

2.2.1. Steps of Motor Learning

Paul Fitts outlines three stages from basic to automatic movement (FITTS, 1964):

1. ***Cognitive Phase:*** In the cognitive phase, basic movement components are learned and coordinated. Beginners aim to understand the action and its execution. In this stage, individuals apply the rules thoughtfully and consciously and use internal verbal expressions to reinforce their actions. Performance is often poor and movements require a lot of energy. An increase in gray matter occurs, but further skill improvement stalls (D'Angelo & Rosa, 2020; Driemeyer et al., 2008).
2. ***Associative Phase:*** Movements become more controlled and purposeful.
3. ***Automation Phase:*** Movements become automatic, even in challenging conditions like competitions. This level is reached through extensive practice. Repetition makes movements automatic, enhancing performance and reducing distraction sensitivity. This phase has two steps: strengthening existing connections and forming new ones (Roberta et al., 2020).

3. Central Nervous System Adaptation in Athletes During Motor Learning

Injury prevention programs for athletes focus on motor learning. They blend awareness, automatic movements, and complex control. These aspects often differ from explicit learning strategies (Beek PJ, 2000; Benjaminse & Otten, 2011). The motor cortex is vital for planning and executing voluntary movements (Svoboda & Li, 2018). Additionally, the cerebellum and striatal nucleus are key in motor learning. The cerebellum is crucial early on and adapts with practice. The striatum then optimizes learning strategies (Rossi et al., 2008). Research links better motor and cognitive performance with increased gray matter (Etgen et al., 2005; Gaser & Schlaug, 2003; Jacini et al., 2009). However, lower gray matter volume sometimes correlates with better performance (Maguire et al., 2000).

Studies on dancers show clear differences between experts and novices. Expert dancers perform faster and more accurately. This improvement links to more synapses in their motor cortex. Training is key to these differences. High-level dancers show increased gray matter in specific areas. Growth can be due to larger cells, new neurons, or denser dendrites. These changes reverse after practice stops, but performance remains high. Experts also display greater "neural efficiency" and different activation patterns than novices. Regular practice is believed to drive these changes (Bengtsson et al., 2005; Cannonieri et al., 2007; Maguire et al., 2000).

Research extends beyond dancers. It looks at motor learning and brain plasticity in other sports too. For instance, Park et al. (2009) studied basketball players and found gray matter changes. Jäncke et al. (2009) confirmed similar findings in golfers. Professional archers showed different brain activation patterns compared to non-archers (Chang et al., 2011).

4. Adding Neurocognitive Tasks to Injury Prevention Programs

Recent studies highlight the need to add neurocognitive tasks to injury prevention programs (Piskin et al., 2021; Wilk et al., 2023). Athletes need to stay aware, monitor their surroundings, filter information, and perform complex movements. These tasks require advanced cognitive processing abilities (Kuo et al., 1983).

Neurocognitive tasks reflect brain function. They are assessed by reaction time, processing speed, and memory (Kennedy et al., 1982). Key traits such as awareness, sensory integration, planning, and coordination are involved (Konishi et al., 2003). Factors such as alertness and attention affect these functions. They also influence the integration of visual, vestibular, and other sensory information needed for control (Butler et al., 1980). This balance is crucial.

After an ACL tear, the central nervous system relies more on visual feedback and spatial awareness. This shift is vital for balance and coordination post-injury. Visual feedback aids interaction with the environment, while spatial awareness helps in body movement control. Such adaptations enhance motor control during recovery and lower re-injury risks (Swanik et al., 2007; Sharma et al., 2013).

Gokeler and colleagues propose viewing ACL injuries as involving biomechanics, neurophysiology, and cognition. They suggest a gradual, constraints-led rehabilitation model for athletes. This model helps athletes meet the desired movement, performance, and psychological readiness goals safely. It allows controlled changes in speed, timing, and complexity. This approach offers insights into handling physical and cognitive stress in training and competition.

Traditional rehabilitation often uses simple protocols, neglecting real-world challenges. This limits athletes' preparation for the game. The authors advocate for a comprehensive view of ACL injuries, integrating motor tasks, sensory input, and cognitive processes. Based on Baumeister's model, it highlights the connection between motor, sensory, and cognitive systems. Small changes in these areas can significantly impact coordination and performance, reducing re-injury risk (Gokeler et al., 2023).

It is suggested that returning to sport should involve multiple tasks. These include neurocognitive challenges, like attention distraction and chaotic movements. Also, it should enhance motor control through intense neuromuscular training (Wilk et al., 2023; Piskin et al., 2021; Grooms et al., 2022a; Grooms et al., 2022b). Adding neurocognitive challenges and unexpected stimuli to hop tests proves more effective than hop tests alone (Simon, 2020). Moreover, the reactive agility test outperforms the pre-planned agility test (Serpell et al., 2010).

5. Integration of Motor Learning Principles into Injury Prevention Programs

5.1. Multiple Tasking During a Movement

The dual-task condition requires performing two tasks at once (Bayot et al., 2018). This is known as "dual-task intervention" (Bayot et al., 2018; Nishimoto et al., 2024). Schaefer et al. (2024) state that complex movements need attention, anticipation, memory, and decision-making. Dual-process theories explain thinking and problem-solving. Type 1 processes allow quick, intuitive actions without thought. In contrast, Type 2 processes need effort and memory, often used by learners. Practicing tasks leads to automatic skills, falling under Type 1 processes for skilled athletes. Beginners rely on Type 2 processes, needing attention and memory. Type 1 processes are default. Type 2 processes kick in when needed. Experienced athletes handle challenges better than beginners (Furley & Memmert, 2015; Schaefer et al., 2024).

Athletes usually focus on opponents, balls, and tasks. However, adding a second task increases cognitive load and injury risk (Olsen et al., 2004; Krosshaug et al., 2007; Bayot et al., 2018). Performing a secondary task while decelerating can increase ACL injury risk (Hughes & Dai, 2023). The Capacity Sharing Theory explains this risk. It states that cognitive resources are limited. These resources get divided between tasks, leading to slower processing and reduced performance (Tombu & Jolicoeur, 2003).

The dual-task concept often uses a primary/secondary task method. Here, participants perform a motor task while a secondary task disrupts it (Abernethy, 2001). For injury prevention, dual-task training combines a primary task (like a drop vertical jump) with a secondary task (like heading a ball while jumping). This training is recommended to improve dual-task performance (Akbari et al., 2023; Ford et al., 2005; Almonroeder et al., 2018; Dai et al., 2018; Mok et al., 2017; Hamoongard et al., 2022). Akbari et al. (2023) studied the impact of heading a soccer ball while jumping on knee injury risk. The results showed significant changes in body mechanics. Heading the ball reduced flexion in the knee, hip, and trunk. It also increased knee stiffness and trunk lateral flexion. These changes raise ACL injury risk. Thus, mastering landings while heading a ball can help prevent ACL injuries (Akbari et al., 2023). Almonroeder et al. (2018) looked at how cognitive demands affect lower body mechanics during a drop vertical jump. They tested four conditions after warming up. The results showed that extra cognitive demands harm lower body mechanics, increasing ACL load (Almonroeder et al., 2018).

5.2. Focus of Attention Feedback

Attention-focused instructions, an important concept in motor learning, can have potentially great effects in terms of supporting exercise-related muscle adaptations and improving athlete performance (Schoenfeld et al., 2016). In rehabilitation processes, when motor skills need to be relearned, athletes are given instructions on correct movement techniques. These instructions focus on the coordination of a player's movements, including basic extremity movements, as well as the form and timing of the movement (Gokeler et al., 2019). In Figure 1, there is an example of the comparison of instructions to reduce knee valgus motion during a balance task. An internal focus attention example is shown in Figure 1A: "Keep your balance by stabilizing your body" and an external focus attention is shown in Figure 1B: "Keep the bar horizontal."

Figure 1: Comparison of instructions to reduce knee valgus motion during a balance task A: Internal focus of attention ("Keep your balance by stabilizing your body"). B: External focus attention ("Keep the bar horizontal") (Benjaminse et al., 2015).

5.2.1. External Focus of Attention

How instructions are given in training or competitions greatly influences learning and performance (Gokeler et al., 2019). Instructions that focus on how movements affect the environment, called "external focus of attention," improve movement efficiency and effectiveness (Lohse et al., 2010). This approach taps into automatic processes in movement. This concept is supported by the Constrained Action Hypothesis, which suggests that directing attention toward movement outcomes facilitates automatic motor control by reducing conscious interference. Constrained Action Hypothesis argues that focusing on movement effects reduces mental load and promotes automatic control (Wulf, 2012).

Recent studies have shed light on the brain mechanisms behind attention focus. They found that external focus instructions boost intracortical inhibition in the motor cortex, regulating motor output (Kuhn et al., 2017). This aligns with the Constrained Action Hypothesis, showing how the brain controls movements in real time. The balance of signals in the motor cortex is key for optimal performance, especially in tasks needing precise timing and coordination. Moreover, the collaboration among the brain's motor, sensory, and insular regions may explain performance boosts in hand-eye coordination tasks and synchronized movements in response to external cues (Kuhn et al., 2017; Gokeler et al., 2019).

These neurophysiological findings regarding the effects of external focus on intracortical inhibition and motor control are particularly relevant for ACL injuries. Crucial for ACL injuries. Such injuries often lead to decreased muscle activation, especially in the quadriceps. This is due to changes in brain inhibition and facilitation (Zarzycki et al., 2018). Reduced muscle activation slows recovery and harms function during rehabilitation (Luc-Harkey et al., 2017). External focus training might help by boosting brain inhibition and restoring quadriceps activity.

Injured athletes often focus too much on their injuries. This happens due to pain, fear of re-injury, and other psychological factors (Singh et al., 2021; Bushnell et al.,2013). In later rehabilitation stages, focusing on external factors improved landing techniques and reduced stiffness in female athletes. This method was better than focusing internally. In a jump performance assessment study, those focusing externally had better knee movement than those focusing internally. These improvements lasted a week and helped in sudden lateral movements (van Weert, 2023). Based on this, Gokeler et al. suggested that focusing on movement effects, not just body mechanics, could boost rehabilitation outcomes. This approach taps into automatic control processes and lowers mental effort. It makes movements more efficient and helps restore normal function by affecting brain mechanisms (Gokeler et al., 2019).

5.2.2. Internal Focus of Attention

Instructions that direct patients to their movements foster an internal focus. This method proves successful in various sports (Chua et al., 2021; Herrebrøden, 2023). While external focus benefits sports like running and cycling, internal focus is key in sports requiring precise body control, such as gymnastics, ski jumping, and diving (Chua et al., 2021). It enhances awareness and performance in these sports (Herrebrøden, 2023). Studies show internal focus training outperforms external focus in movement precision and muscle activation (Gottwald et al., 2020). This suggests it improves control and coordination, which is vital for technically demanding sports. Hence, internal focus instructions are crucial for better performance in these areas.

External focus instructions boost force production. However, internal focus prompts higher muscle activity, as seen in electromyography (EMG) studies (Marchant et al., 2009). Coratella et al. (2022) found more muscle activity during squats with internal focus than external focus. Atalay et al. (2024) compared muscle activation under different focus conditions. They found external focus led to higher activation than internal focus or no focus. Despite extensive research, a clear consensus remains elusive.

5.2.3. Video Feedback

Video feedback uses high-definition cameras, smartphones, or tablets (Etnoyer et al., 2013; Onate et al., 2005; Parsons et al., 2012). It's effective and easy to use in clinical settings, both individually and in groups (Benjaminse et al., 2015). Athletes watch videos of experts demonstrating the correct movement. They learn by observing and imitating. For instance, showing a video of an athlete successfully jumping after catching a basketball can inspire the whole team to mimic that performance (Benjaminse et al., 2015).

Observational learning is key in acquiring motor skills through video feedback (McCullagh et al., 1989). The mirror neuron system is activated, prompting athletes to imitate observed movements (Dallinga et al., 2017). Mirror neurons activate during both action and observation (Rizzolatti & Craighero, 2004). They link visual input to motor output. Imitation occurs when observing similar actions by others (Grooms et al., 2015). This process improves athletic performance (Benjaminse et al., 2015) (Figure 2).

Moreover, video feedback is not limited to expert videos. Recent studies have added sensor-based feedback. Sensors on athletes provide real-time visual feedback during movement, adding tactile and auditory elements (Benjaminse et al., 2017). This feedback significantly boosts performance and motor memory (Myer et al., 2006). It helps athletes improve in future trials. Past research showed immediate benefits in landing techniques using video feedback. Athletes showed better knee and hip angles, reduced ground forces, and quicker reactions (Parsons et al., 2012; Munro & Herrington, 2014; Dallinga et al., 2017). Nyman & Armstrong (2015) tested sensor-based feedback in a 4-week program for female athletes. Those using this feedback saw significant improvements in knee movement during drills (Nyman & Armstrong, 2015). Van Hooren et al. (2024) studied real-time feedback on runners. They found it reduced injuries and improved performance (Van Hooren et al., 2024).

Figure 2: Real-time learning in video feedback (Benjaminse et al., 2015)

5.3. Differential Learning (DL)

Differential learning (DL) is rooted in Dynamic Systems Theory (Wulf & Lewthwaite, 2016; Gokeler et al., 2019). It encourages athletes to explore different movement patterns (Scholhorn et al., 2009). During DL practice, movements are varied intentionally. Athletes seek new ways to execute movements instead of just the "correct" form (Gokeler et al., 2023). The idea is to perform movements in diverse ways. There's no perfect repetition, and feedback is not corrective (Tassignon et al., 2021). DL lacks standard programs with fixed repetitions or sets (Gokeler et al., 2023). No movement is repeated exactly more than once. Training should mix different movement combinations (Gokeler et al., 2023). Since athletes don't follow set patterns in competitions, DL prepares them with coping strategies by introducing movement variations in rehabilitation. Research shows DL outperforms other methods in motor skill learning (Tassignon et al., 2021).

Most studies on motor learning focus on event-related potentials (ERPs). However, they overlook how it affects electroencephalogram (EEG) brain waves (Henz & Schöllhorn, 2016). Research shows learning leaves marks on the brain after tasks (Tanaka et al., 2011; Buschkuehl et al., 2012; Crupi et al., 2013). These marks are specific to tasks and areas in the brain. For example, after a sequence task, increases in theta and alpha waves were noted (Moisello et al., 2013). These changes might indicate learning and reflect brain adaptability (Henz & Schöllhorn, 2016).

Studies show better performance in interleaved practice than repetitive practice. Increased brain activity was found in key areas during interleaved practice (Cross et al., 2007; Wymbs & Grafton, 2009). This includes the dorsal premotor and prefrontal areas. Moreover, better connectivity between these regions was noted, enhancing memory and retrieval (Tanaka et al., 2005). However, the reason for better performance in dual learning remains unclear. Early studies showed increased brain activity in certain areas after DL training (Henz et al., 2013, 2015; Henz et al., 2015). Different processes were observed between DL and repetitive training, with more activity in certain brain areas during DL (Henz & Schöllhorn, 2016).

DL training has shown significant clinical benefits. For instance, it reduced risk factors in ACL injury patients (Ghanati et al., 2022). It also improved motor skills in athletes' post-surgery (Gholami et al., 2023). In football, varied training enhances performance and skills (Santos et al., 2018; Oftadeh et al., 2022). In basketball and golf, competitions with DL training boosted physical performance and adaptability (Arede et al., 2022; Valença et al., 2022).

6. Conclusion

Sports are vital for health but pose injury risks, especially to knees and ankles. These injuries can lead to long-term physical and neurological issues. Neuromuscular training cuts injury risk and aids recovery by boosting strength and control. Yet, re-injuries and performance drops are still problems. A better approach includes prevention and rehabilitation that considers physical, control, and cognitive factors. This should involve tasks that improve planning and awareness. Such methods, backed by brain adaptability, enhance recovery and lower re-injury chances.

Incorporating motor learning strategies in prevention programs can boost athletes' physical and cognitive skills. Techniques like dual-task training and feedback improve efficiency and adaptability. This not only cuts injury risk but also speeds up recovery. Evidence shows combining mental and physical training is the best approach. Programs should be tailored to each sport's needs. Cognitive factors must be included in rehabilitation, and nervous system assessments should track changes. Cognitive tests should also be part of performance reviews and return-to-sport decisions.

7. Takeaways

1. Neuromuscular rehabilitation programs are not completely successful in preventing injuries.
2. Musculoskeletal injuries have been shown to damage various parts of the central nervous system.
3. To prevent injuries and increase performance, motor learning principles should be integrated into training programs.
4. Neurocognitive tasks show how the brain works and are measured by reaction time, processing speed, and memory.
5. Various focus of attention and video feedback, multitasking during a movement, and differential learning are motor learning examples to be added to training programs.

Abbreviations

ACL: Anterior Cruciate Ligament
ACLR: Anterior Cruciate Ligament Reconstruction
CAI: Chronic Ankle Sprain
DL: Differential Learning
EMG: Electromyography
EEG: Electroencephalogram

Bibliography

1. Abernethy, B. (2001). Attention. In R. N. Singer, H. A. Hausenblas, & C. M. Janelle (Eds.), *Handbook of sport psychology* (2nd ed., pp. 53–58). Wiley.
2. Achenbach, L. (2018). The young handball player. In *Handball Sports Medicine: Basic Science, Injury Management and Return to Sport* (pp. 571–582).
3. Akbari, H., Kuwano, S., & Shimokochi, Y. (2023). Effect of heading a soccer ball as an external focus during a drop vertical jump task. *Orthopaedic Journal of Sports Medicine, 11*(4), 23259671231164706. https://doi.org/10.1177/23259671231164706
4. Alentorn-Geli, E., Mendiguchía, J., Samuelsson, K., Musahl, V., Karlsson, J., Cugat, R., & Myer, G. D. (2014). Prevention of non-contact anterior cruciate ligament injuries in sports. Part II: Systematic review of the effectiveness of prevention programmes in male athletes. K*nee Surgery, Sports Traumatology, Arthroscopy, 22*(1), 16–25. https://doi.org/10.1007/s00167-013-2739-x
5. Almonroeder, T. G., Kernozek, T., Cobb, S., Slavens, B., Wang, J., & Huddleston, W. (2018). Cognitive demands influence lower extremity mechanics during a drop vertical jump task in female athletes. *Journal of Orthopaedic Sports Physical Therapy, 48*(5), 381–387. https:// doi.org/ 10.2519/jospt.2018.7739
6. Ardern, C. L., Taylor, N. F., Feller, J. A., & Webster, K. E. (2014). Fifty-five percent return to competitive sport following anterior cruciate ligament reconstruction surgery: An updated systematic review and meta-analysis including aspects of physical functioning and contextual factors. *British Journal of Sports Medicine, 48*(21), 1543–1552. https://doi.org/10.1136/bjsports-2013-093398
7. Arede, J., Fernandes, J. F. T., Schöllhorn, W. I., & Leite, N. (2022). Differential repeated sprinting training in youth basketball players: An analysis of effects according to maturity status. *International Journal of Environmental Research and Public Health, 19*, 12265. https:// doi.org/10.3390/ijerph191912265
8. Arundale, A. J. H., Silvers-Granelli, H. J., & Myklebust, G. (2022). ACL injury prevention: Where have we come from and where are we going? *Journal of Orthopaedic Research, 40*(1), 43–54. https://doi.org/10.1002/jor.25058
9. Arundale, A. J. H., Bizzini, M., Giordano, A., Hewett, T. E., Logerstedt, D. S., Mandelbaum, B., Scalzitti, D. A., Silvers-Granelli,

H., Snyder-Mackler, L., Altman, R. D., et al. (2018). Exercise-based knee and anterior cruciate ligament injury prevention: Clinical practice guidelines linked to the international classification of functioning, disability and health from the Academy of Orthopaedic Physical Therapy and the American Academy of Sports Physical Therapy. *Journal of Orthopaedic Sports Physical Therapy, 48*(1), A1–A42. https://doi.org/10.2519/jospt.2018.0303

10. Asker, M., Brooke, H. L., Walden, M., et al. (2018). Risk factors for, and prevention of, shoulder injuries in overhead sports: A systematic review with best-evidence synthesis. *British Journal of Sports Medicine, 52*(20), 1312–1319. https://doi.org/10.1136/bjsports-2017-098254
11. Atalay, E. S., Soylu, Ç., Kılınç, B., Şahan, T. Y., Türker, D., Demir, P., & Yıldırım, N. Ü. (2024). Comparative effects of different focus methods on muscular activation during single-leg-squat exercise. *Nigerian Journal of Clinical Practice, 27*(10), 1146–1153. https://doi.org/10.4103/njcp.njcp_119_24
12. Bayot, M., Dujardin, K., Tard, C., Defebvre, L., Bonnet, C. T., Allart, E., & Delval, A. (2018). The interaction between cognition and motor control: A theoretical framework for dual-task interference effects on posture, gait initiation, gait and turning. *Neurophysiologie Clinique - Clinical Neurophysiology, 48*(6), 361–375. https://doi.org/10.1016/j.neucli.2018.10.003
13. Beek, P. J. (2000). Toward a theory of implicit learning in the perceptual-motor domain. *International Journal of Sport Psychology, 31*(4), 547–554.
14. Bengtsson, S. L., Nagy, Z., Skare, S., Forsman, L., Forssberg, H., & Ullén, F. (2005). Extensive piano practicing has regionally specific effects on white matter development. *Nature Neuroscience, 8*(9), 1148–1150. https://doi.org/10.1038/nn1516
15. Benjaminse, A., & Otten, E. (2011). ACL injury prevention, more effective with a different way of motor learning? *Knee Surgery, Sports Traumatology, Arthroscopy, 19*(4), 622–627. https://doi.org/10.1007/s00167-010-1313-z
16. Benjaminse, A., Gokeler, A., Dowling, A. V., Faigenbaum, A., Ford, K. R., Hewett, T. E., Onate, J. A., Otten, B., & Myer, G. D. (2015). Optimization of the anterior cruciate ligament injury prevention paradigm: Novel feedback techniques to enhance motor learning and reduce injury risk. *The Journal of Orthopaedic and Sports Physical Therapy, 45*(3), 170–182. https://doi.org/10.2519/jospt.2015.4986

17. Benjaminse, A., Postma, W., Janssen, I., & Otten, E. (2017). Video feedback and 2-dimensional landing kinematics in elite female handball players. *Journal of Athletic Training, 52*(11), 993–1001. https://doi.org/10.4085/1062-6050-52.10.11
18. Boden, B. P., & Sheehan, F. T. (2022). Mechanism of non-contact ACL injury: OREF Clinical Research Award 2021. *Journal of Orthopaedic Research, 40*(3), 531–540. https://doi.org/10.1002/jor.25257
19. Bram, J. T., Magee, L. C., Mehta, N. N., Patel, N. M., & Ganley, T. J. (2021). Anterior cruciate ligament injury incidence in adolescent athletes: A systematic review and meta-analysis. *The American Journal of Sports Medicine, 49*(7), 1962–1972. https://doi.org/10.1177/0363546520959619
20. Buschkuehl, M., Jaeggi, S. M., & Jonides, J. (2012). Neuronal effects following working memory training. *Developmental Cognitive Neuroscience, 2*, S167–S179. https://doi.org/10.1016/j.dcn.2011.10.001
21. Bushnell, M. C., Ceko, M., & Low, L. A. (2013). Cognitive and emotional control of pain and its disruption in chronic pain. *Nature Reviews Neuroscience, 14*(7), 502–511. https://doi.org/10.1038/nrn3516
22. Butler, D. L., Noyes, F. R., & Grood, E. S. (1980). Ligamentous restraints to anterior-posterior drawer in the human knee: A biomechanical study. *Journal of Bone and Joint Surgery. American Volume, 62*, 259–270.
23. Cannonieri, G. C., Bonilha, L., Fernandes, P. T., Cendes, F., & Li, L. M. (2007). Practice and perfect: Length of training and structural brain changes in experienced typists. *Neuroreport, 18*(10), 1063–1066. https://doi.org/10.1097/WNR.0b013e3281a030e5
24. Chang, Y., Lee, J. J., Seo, J. H., Song, H. J., Kim, Y. T., Lee, H. J., Kim, H. J., Lee, J., Kim, W., Woo, M., & Kim, J. G. (2011). Neural correlates of motor imagery for elite archers. *NMR in biomedicine, 24*(4), 366–372. https://doi.org/10.1002/nbm.1600
25. Cheriyan, J., & Sheets, P. L. (2018). Altered excitability and local connectivity of mPFC-PAG neurons in a mouse model of neuropathic pain. *Journal of Neuroscience, 38*(20), 4829–4839. https://doi.org/10.1523/JNEUROSCI.2731-17.2018
26. Chua, L.-K., Jimenez-Diaz, J., Lewthwaite, R., Kim, T., & Wulf, G. (2021). Superiority of external attentional focus for motor performance and learning: Systematic reviews and meta-analyses. *Psychological Bulletin, 147*(6), 618–645. https://doi.org/10.1037/bul0000335

27. Cools, A. M., Maenhout, A. G., Vanderstukken, F., Decleve, P., Johansson, F. R., & Borms, D. (2020). The challenge of the sporting shoulder: From injury prevention through sport-specific rehabilitation toward return to play. *Annals of Physical and Rehabilitation Medicine, 101*, 384. https://doi.org/10.1016/j.physre.2020.101384
28. Coratella, G., Tornatore, G., Longo, S., Borrelli, M., Doria, C., Esposito, F., et al. (2022). The effects of verbal instructions on lower limb muscles' excitation in back-squat. *Research Quarterly for Exercise and Sport, 93*, 429–435. https://doi.org/10.1080/02701367.2020.1840496
29. Cross, E. S., Schmitt, P. J., & Grafton, S. T. (2007). Neural substrates of contextual interference during motor learning support a model of active preparation. *Journal of Cognitive Neuroscience, 19*(11), 1854–1871. https://doi.org/10.1162/jocn.2007.19.11.1854
30. Crupi, D., Cruciata, G., Moisello, C., Green, P., Naro, A., Ricciardi, L., et al. (2013). Protracted exercise without overt neuromuscular fatigue influences cortical excitability. *Journal of Motor Behavior, 45*, 127–138. https://doi.org/10.1080/00222895.2012.760514
31. Dai, B., Cook, R. F., Meyer, E. A., Sciascia, Y., Hinshaw, T. J., Wang, C., & Zhu, Q. (2018). The effect of a secondary cognitive task on landing mechanics and jump performance. *Sports Biomechanics, 17*(2), 192–205. https://doi.org/10.1080/14763141.2016.1265579
32. Dallinga, J., Benjaminse, A., Gokeler, A., Cortes, N., Otten, E., & Lemmink, K. (2017). Innovative video feedback on jump landing improves landing technique in males. *International Journal of Sports Medicine, 38*(2), 150–158. https://doi.org/10.1055/s-0042-106298
33. Della Villa, F., Straub, R. K., Mandelbaum, B., & Powers, C. M. (2021). Confidence to return to play after anterior cruciate ligament reconstruction is influenced by quadriceps strength symmetry and injury mechanism. *Sports Health, 13*(3), 304–309. https://doi.org/10.1177/1941738120976377
34. Denti, M., Monteleone, M., Berardi, A., & Panni, A. S. (1994). Anterior cruciate ligament mechanoreceptors: Histologic studies on lesions and reconstruction. *Clinical Orthopaedics and Related Research, 308*, 29–32.
35. Di Palma, D., Raiola, G., & Tafuri, D. (2016). Disability and Sport Management: a systematic review of the literature. *Journal of Physical Education and Sport, 16*(3), 785. https://doi.org/10.7752/jpes.2016.03125

36. Driemeyer, J., Boyke, J., Gaser, C., Büchel, C., & May, A. (2008). Changes in gray matter induced by learning—revisited. *PLOS ONE, 3*(7), e2669. https://doi.org/10.1371/journal.pone.0002669
37. Emery, C. A., & Tyreman, H. (2009). Sport participation, sport injury, risk factors, and sport safety practices in Calgary and area junior high schools. *Paediatrics & Child Health,* 14(7), 439–444. https://doi.org/10.1093/pch/14.7.439
38. Ericksen, H. M., Gribble, P. A., Pfile, K. R., & Pietrosimone, B. G. (2013). Different modes of feedback and peak vertical ground reaction force during jump landing: A systematic review. *Journal of Athletic Training, 48*(5), 685–695. https://doi.org/10.4085/1062-6050-48.3.02
39. Fältström, A., Hägglund, M., Kvist, J., & Mendonça, L. D. (2023). Risk factors for sustaining a second ACL injury after primary ACL reconstruction in female football players: A study investigating the effects of follow-up time and the statistical approach. *Sports Medicine.* https://doi.org/10.1186/s40798-023-00571-x
40. Fitts, P. M. (1964). Perceptual-motor skill learning. In *Categories of Human Learning* (Vol. 49, Issue 638, pp. 11–31). Academic Press. https://doi.org/10.1016/b978-1-4832-3145-7.50016-9
41. Ford, K. R., Myer, G. D., Smith, R. L., Byrnes, R. N., Dopirak, S. E., & Hewett, T. E. (2005). Use of an overhead goal alters vertical jump performance and biomechanics. *Journal of Strength and Conditioning Research,* 19(2), 394–399. https://doi.org/10.1519/15834.1
42. Furley, P., & Memmert, D. (2015). Creativity and working memory capacity in sports: Working memory capacity is not a limiting factor in creative decision making amongst skilled performers. *Frontiers in Psychology, 6*, 115. https://doi.org/10.3389/fpsyg.2015.00115
43. Gaser, C., & Schlaug, G. (2003). Gray matter differences between musicians and nonmusicians. *Annals of the New York Academy of Sciences, 999*(1), 514–517. https://doi.org/10.1196/annals.1284.062
44. Ghanati, H. A., Letafatkar, A., Shojaedin, S., Hadadnezhad, M., & Schöllhorn, W. I. (2022). Comparing the effects of differential learning, self-controlled feedback, and external focus of attention training on biomechanical risk factors of anterior cruciate ligament (ACL) in athletes: A randomized controlled trial. *International Journal of Environmental Research and Public Health, 19*(16), 10052. https://doi.org/10.3390/ijerph191610052

45. Gholami, F., Letafatkar, A., Moghadas Tabrizi, Y., Gokeler, A., Rossettini, G., Ghanati, H. A., & Schöllhorn, W. I. (2023). Comparing the effects of differential and visuo-motor training on functional performance, biomechanical, and psychological factors in athletes after ACL reconstruction: A randomized controlled trial. *Journal of Clinical Medicine, 12*, 2845. https://doi.org/10.3390/jcm12082845
46. Gokeler, A., Neuhaus, D., Benjaminse, A., Grooms, D. R., & Baumeister, J. (2019). Principles of motor learning to support neuroplasticity after ACL injury: Implications for optimizing performance and reducing risk of second ACL injury. *Sports Medicine, 49*(6), 853–865. https://doi.org/10.1007/s40279-019-01058-0
47. Gokeler, A., Nijmeijer, E. M., Heuvelmans, P., Tak, I., Ramponi, C., & Benjaminse, A. (2023). Motor learning principles during rehabilitation after anterior cruciate ligament injury: Time to create an enriched environment to improve clinical outcome. *Arthroskopie, 36*(3), 164-170. https://doi.org/10.1007/s00142-023-00606-1
48. Gottwald, V. M., Owen, R., Lawrence, G. P., & McNevin, N. (2020). An internal focus of attention is optimal when congruent with afferent proprioceptive task information. *Psychology of Sport and Exercise, 47*, 101634. https://doi.org/10.1016/j.psychsport.2019.101634
49. Grindem, H., Engebretsen, L., Axe, M., Snyder-Mackler, L., & Risberg, M. A. (2020). Activity and functional readiness, not age, are the critical factors for second anterior cruciate ligament injury: The Delaware-Oslo ACL cohort study. *British Journal of Sports Medicine, 54*(18), 1099–1102. https://doi.org/10.1136/bjsports-2019-100623
50. Grooms, D. R., Diekfuss, J. A., Slutsky-Ganesh, A. B., DiCesare, C. A., Bonnette, S., Riley, M. A., Kiefer, A. W., Wohl, T. R., Criss, C. R., Lamplot, J., Thomas, S. M., Foss, K. D. B., Faigenbaum, A. D., Wong, P., Simon, J. E., & Myer, G. D. (2022b). Preliminary report on the Train the Brain Project, Part II: Neuroplasticity of augmented neuromuscular training and improved injury-risk biomechanics. *Journal of Athletic Training, 57*(9–10), 911–920. https://doi.org/10.4085/1062-6050-0548.21
51. Grooms, D. R., Diekfuss, J. A., Slutsky-Ganesh, A. B., Ellis, J. D., Criss, C. R., Thomas, S. M., DiCesare, C. A., Wong, P., Anand, M., Lamplot, J., Simon, J. E., & Myer, G. D. (2022a). Preliminary report on the Train the Brain Project, Part I: Sensorimotor neural correlates of anterior cruciate ligament injury risk biomechanics.

Journal of Athletic Training, 57(9–10), 902–910. https://doi.org/10.4085/1062-6050-0547.21

52. Grooms, D., Appelbaum, G., & Onate, J. (2015). Neuroplasticity following anterior cruciate ligament injury: A framework for visual-motor training approaches in rehabilitation. *Journal of Orthopaedic and Sports Physical Therapy, 45*, 381–393. https://doi.org/10.2519/jospt.2015.5549
53. Hamoongard, M., Hadadnezhad, M., & Abbasi, A. (2022). Effect of combining eight weeks of neuromuscular training with dual cognitive tasks on landing mechanics in futsal players with knee ligament dominance defect: A randomized controlled trial. *BMC Sports Science, Medicine and Rehabilitation, 14*(1), 196. https://doi.org/10.1186/s13102-022-00593-0
54. Henz, D., & Schöllhorn, W. I. (2016). Differential training facilitates early consolidation in motor learning. *Frontiers in Behavioral Neuroscience, 10*,199. https://doi.org/10.3389/fnbeh.2016.00199
55. Henz, D., Hofmann, M., & Schöllhorn, W. I. (2013). Increased EEG theta and alpha activity indicate somatosensory working memory processes in differential badminton serve training. In *Conference Proceedings of the 18th Annual Congress of the European College of Sport Science ECSS Barcelona 2013* (pp. 853). N. Balagué, C. Torrents, A. Vilanova, J. Cadefau, R. Tarragó, & E. Tsolakidis (Eds.). Barcelona: Open Print.
56. Henz, D., Kenville, R., Simon, M., Leinberger, O., & Schöllhorn, W. I. (2015). EEG brain activity in differential, contextual interference and classical repetition based badminton serve training. In *Proceedings of the 20th Annual Congress of the European College of Sports Science* (pp. 502). A. Radmann, S. Hedenborg, & E. Tsolakidis (Eds.). Malmö.
57. Henz, D., Leinberger, O., & Schöllhorn, W. I. (2014). EEG brain activation patterns in differential and mental differential soccer shooting training. In *Book of Abstracts of the 19th Annual Congress of the European College of Sport Science* (pp. 609). A. De Haan, C. J. De Ruiter, & E. Tsolakidis (Eds.). Utrecht: Digital Printing Partners.
58. Herrebrøden, H. (2023). Motor performers need task-relevant information: Proposing an alternative mechanism for the attentional focus effect. *Journal of Motor Behavior, 55*(1), 125–134. https://doi.org/10.1080/00222895.2022.2122920

59. Huang, J., Gadotti, V. M., Chen, L., et al. (2019). A neuronal circuit for activating descending modulation of neuropathic pain. *Nature Neuroscience, 22*(10), 1659–1668. https://doi.org/10.1038/s41593-019-0481-5
60. Huang, Y. L., Jung, J., Mulligan, C. M. S., Oh, J., & Norcross, M. F. (2020). A majority of anterior cruciate ligament injuries can be prevented by injury prevention programs: A systematic review of randomized controlled trials and cluster-randomized controlled trials with meta-analysis. *The American Journal of Sports Medicine, 48*(6), 1505–1515. https://doi.org/10.1177/0363546519870175
61. Hughes, G., & Dai, B. (2023). The influence of decision making and divided attention on lower limb biomechanics associated with anterior cruciate ligament injury: A narrative review. *Sports Biomechanics, 22*(1), 30–45. https://doi.org/10.1080/14763141.2021.1898671
62. Jacini, W. F. S., Cannonieri, G. C., Fernandes, P. T., Bonilha, L., Cendes, F., & Li, L. M. (2009). Can exercise shape your brain? Cortical differences associated with judo practice. *Journal of Science and Medicine in Sport, 12*(6), 688–690. https://doi.org/10.1016/j.jsams.2008.11.004
63. Jäncke, L., Koeneke, S., Hoppe, A., Rominger, C., & Hänggi, J. (2009). The architecture of the golfer's brain. *PLOS ONE, 4*(3), e4785. https://doi.org/10.1371/journal.pone.0004785
64. Kapreli, E., & Athanasopoulos, S. (2006). The anterior cruciate ligament deficiency as a model of brain plasticity. *Medical Hypotheses, 67*(3), 645–650. https://doi.org/10.1016/j.mehy.2006.01.063
65. Kennedy, J. C., Alexander, I. J., & Hayes, K. C. (1982). Nerve supply of the human knee and its functional importance. *American Journal of Sports Medicine, 10*, 329–335. https://doi.org/10.1177/036354658201000601
66. Kim, K. M., Kim, J. S., Cruz-Díaz, D., Ryu, S., Kang, M., & Taube, W. (2019). Changes in spinal and corticospinal excitability in patients with chronic ankle instability: A systematic review with meta-analysis. *Journal of Clinical Medicine, 8*(7), 1037. https://doi.org/10.3390/jcm8071037
67. Konishi, Y., Konishi, H., & Fukubayashi, T. (2003). Gamma loop dysfunction in quadriceps on the contralateral side in patients with ruptured ACL. *Medicine and Science in Sports and Exercise, 35*, 897–900.

68. Krosshaug, T., Nakamae, A., Boden, B. P., Engebretsen, L., Smith, G., Slauterbeck, J. R., Hewett, T. E., & Bahr, R. (2007). Mechanisms of anterior cruciate ligament injury in basketball: Video analysis of 39 cases. *The American Journal of Sports Medicine, 35*(3), 359–367. https://doi.org/10.1177/0363546506293899
69. Kuhn, Y. A., Keller, M., Ruffieux, J., & Taube, W. (2017). Adopting an external focus of attention alters intracortical inhibition within the primary motor cortex. *Acta Physiologica, 220*(2), 289–299. https://doi.org/10.1111/apha.12807
70. Kuo, C. Y., Louie, J. K., & Mote, C. D., Jr. (1983). Field measurements in snow skiing injury research. *Journal of Biomechanics, 16*, 609–624. https://doi.org/10.1016/0021-9290(83)90111-2
71. Latino, F., & Tafuri, F. (2024). Physical activity and cognitive functioning. *Medicina (Kaunas, Lithuania), 60*(2), 216. https://doi.org/10.3390/medicina60020216
72. Lemon, R. N. (2008). Descending pathways in motor control. *Annual Review of Neuroscience, 31*, 195–218. https://doi.org/10.1146/annurev.neuro.31.060407.125547
73. Lepley, A. S., Ly, M. T., Grooms, D. R., Kinsella-Shaw, J. M., & Lepley, L. K. (2020). Corticospinal tract structure and excitability in patients with anterior cruciate ligament reconstruction: A DTI and TMS study. *NeuroImage: Clinical, 25*, 102157. https://doi.org/10.1016/j.nicl.2019.102157
74. Li, N., Chen, T. W., Guo, Z. V., Gerfen, C. R., & Svoboda, K. (2015). A motor cortex circuit for motor planning and movement. *Nature, 519*(7541), 51–56. https://doi.org/10.1038/nature14178
75. Lin, C. W., Jankaew, A., & Lin, C. F. (2024). Physical therapy intervention effects on alteration of spinal excitability in patients with chronic ankle instability: A systematic review and meta-analysis. *Sports Health*. https://doi.org/10.1177/19417381241253248
76. Lohse, K. R., Sherwood, D. E., & Healy, A. F. (2010). How changing the focus of attention affects performance, kinematics, and electromyography in dart throwing. *Human Movement Science, 29*(4), 542–555. https://doi.org/10.1016/j.humov.2010.05.001
77. Luc-Harkey, B. A., Harkey, M. S., Pamukoff, D. N., Kim, R. H., Royal, T. K., Blackburn, J. T., et al. (2017). Greater intracortical inhibition associates with lower quadriceps voluntary activation in individuals with ACL reconstruction. *Experimental Brain Research, 235*(4), 1129–1137. https://doi.org/10.1007/s00221-017-4877-8

78. Maguire, E. A., Gadian, D. G., Johnsrude, I. S., Good, C. D., Ashburner, J., Frackowiak, R. S., & Frith, C. D. (2000). Navigation-related structural change in the hippocampi of taxi drivers. *Proceedings of the National Academy of Sciences of the United States of America, 97*(8), 4398–4403. https://doi.org/10.1073/pnas.070039597
79. Mancino, F., Kayani, B., Gabr, A., Fontalis, A., Plastow, R., & Haddad, F. S. (2024). Anterior cruciate ligament injuries in female athletes: risk factors and strategies for prevention. Bone & joint open, 5(2), 94–100. https://doi.org/10.1302/2633-1462.52.BJO-2023-0166
80. Marchant, D. C., Greig, M., & Scott, C. (2009). Attentional focusing instructions influence force production and muscular activity during isokinetic elbow flexions. *Journal of Strength and Conditioning Research, 23*(8), 2358–2366. https://doi.org/10.1519/JSC.0b013e3181b8d1e5
81. Moisello, C., Meziane, H. B., Kelly, S., Perfetti, B., Kvint, S., Voutsinas, N., et al. (2013). Neural activations during visual sequence learning leave a trace in post-training spontaneous EEG. *PLOS ONE, 8*, e65882. https://doi.org/10.1371/journal.pone.0065882
82. Mok, K. M., Bahr, R., & Krosshaug, T. (2017). The effect of overhead target on the lower limb biomechanics during a vertical drop jump test in elite female athletes. *Scandinavian Journal of Medicine & Science in Sports, 27*(2), 161–166. https://doi.org/10.1111/sms.12640
83. Munro, A., & Herrington, L. (2014). The effect of videotape augmented feedback on drop jump landing strategy: Implications for anterior cruciate ligament and patellofemoral joint injury prevention. *Knee, 21*, 891–895. https://doi.org/10.1016/j.knee.2014.05.011
84. Myer, G. D., Ford, K. R., Brent, J. L., & Hewett, T. E. (2006). The effects of plyometric vs. dynamic stabilization and balance training on power, balance, and landing force in female athletes. *Journal of Strength and Conditioning Research, 20*, 345–353.
85. Needle, A. R., Lepley, A. S., & Grooms, D. R. (2017). Central nervous system adaptation after ligamentous injury: A summary of theories, evidence, and clinical interpretation. *Sports Medicine, 47*(7), 1271–1288. https://doi.org/10.1007/s40279-016-0666-y
86. Nigg, B. M., & Liu, W. (1999). The effect of muscle stiffness and damping on simulated impact force peaks during running. *Journal of Biomechanics, 32*, 849–856. https://doi.org/10.1016/S0021-9290(99)00048-2
87. Nyman, E., Jr., & Armstrong, C. W. (2015). Real-time feedback during drop landing training improves subsequent frontal and

sagittal plane knee kinematics. *Clinical Biomechanics (Bristol, Avon), 30*(9), 988–994. https://doi.org/10.1016/j.clinbiomech.2015.06.018
88. Oftadeh, S., Bahram, A., Yaali, R., Ghadiri, F., & Schöllhorn, W. I. (2022). External focus or differential learning: Is there an additive effect on learning a futsal goal kick? *International Journal of Environmental Research and Public Health, 19*, 317. https://doi.org/10.3390/ijerph19010317
89. Olsen, O. E., Myklebust, G., Engebretsen, L., & Bahr, R. (2004). Injury mechanisms for anterior cruciate ligament injuries in team handball: A systematic video analysis. *The American Journal of Sports Medicine, 32*(4), 1002–1012. https://doi.org/10.1177/0363546503261724
90. Padua, D. A., DiStefano, L. J., Hewett, T. E., Garrett, W. E., Marshall, S. W., Golden, G. M., Shultz, S. J., & Sigward, S. M. (2018). National Athletic Trainers' Association Position Statement: Prevention of anterior cruciate ligament injury. *Journal of Athletic Training, 53*(1), 5–19. https://doi.org/10.4085/1062-6050-99-16
91. Park, I. S., Lee, K. J., Han, J. W., Lee, N. J., Lee, W. T., Park, K. A., & Rhyu, I. J. (2009). Experience-dependent plasticity of cerebellar vermis in basketball players. *Cerebellum (London, England)*, 8(3), 334–339. https://doi.org/10.1007/s12311-009-0100-1
92. Parsons, J. L., & Alexander, M. J. (2012). Modifying spike jump landing biomechanics in female adolescent volleyball athletes using video and verbal feedback. *Journal of Strength and Conditioning Research, 26*, 1076–1084. https://doi.org 10.1519/JSC.0b013e31822e5876
93. Perez, M. A., Lundbye-Jensen, J., & Nielsen, J. B. (2006). Changes in corticospinal drive to spinal motoneurons following visuo-motor skill learning in humans. *The Journal of Physiology, 573*(3), 843–855. https://doi.org/10.1113/jphysiol.2006.105361
94. Petushek, E. J., Sugimoto, D., Stoolmiller, M., Smith, G., & Myer, G. D. (2019). Evidence-based best-practice guidelines for preventing anterior cruciate ligament injuries in young female athletes: A systematic review and meta-analysis. *The American Journal of Sports Medicine, 47*(7), 1744–1753. https://doi.org/10.1177/0363546518782460
95. Piskin, D., Benjaminse, A., Dimitrakis, P., et al. (2021). Neurocognitive and neurophysiological functions related to ACL injury: A framework for neurocognitive approaches in rehabilitation and return-to-sports tests. *Sports Health*. https://doi.org/10.1177/19417381211029265

96. Rizzolatti, G., & Craighero, L. (2004). The mirror-neuron system. *Annual Review of Neuroscience, 27*, 169–192. https://doi.org/10.1146/annurev.neuro.27.070203.144230
97. Roberta, M., Belfiore, P., & Liparoti, M. (2020). Neuroplasticity and motor learning in sport activity. *Journal of Physical Education and Sport, 20*, 2354–2359. https://doi.org/10.7752/jpes.2020.s4318
98. Rossi, S., Mataluni, G., De Bartolo, P., Prosperetti, C., Foti, F., De Chiara, V., Musella, A., Mandolesi, L., Bernardi, G., & Centonze, D. (2008). Cerebellar control of cortico-striatal LTD. *Restorative Neurology and Neuroscience, 26*(6), 475–480. https://doi.org/10.3233/RNN-2008-00433
99. Rush, J. L., Glaviano, N. R., & Norte, G. E. (2021). Assessment of quadriceps corticomotor and spinal-reflexive excitability in individuals with a history of anterior cruciate ligament reconstruction: A systematic review and meta-analysis. *Sports Medicine, 51*, 961–990. https://doi.org/10.1007/s40279-020-01403-8
100. Santos, S., Coutinho, D., Gonçalves, B., Schöllhorn, W., Sampaio, J., & Leite, N. (2018). Differential learning as a key training approach to improve creative and tactical behavior in soccer. *Research Quarterly for Exercise and Sport, 89*(1), 11–24. https://doi.org/10.1080/02701367.2017.1412063
101. Schaefer Nishimoto, R., Inokuchi, H., Fujiwara, S., & Ogata, T. (2024). Implicit learning provides advantage over explicit learning for gait-cognitive dual-task interference. *Scientific Reports, 14*(1), 18336. https://doi.org/10.1038/s41598-024-68284-z
102. Serpell, B. G., Ford, M., & Young, W. B. (2010). The development of a new test of agility for rugby league. *Journal of Strength and Conditioning Research, 24*(12), 3270–3277. https://doi.org/10.1519/JSC.0b013e3181b60430
103. Sharma, N., Classen, J., & Cohen, L. G. (2013). Neural plasticity and its contribution to functional recovery. In *Handbook of Clinical Neurology* (Vol. 110, pp. 3–12). https://doi.org/10.1016/B978-0-444-52901-5.00001-0
104. Simon, J. E., Millikan, N., Yom, J., & Grooms, D. R. (2020). Neurocognitive challenged hops reduced functional performance relative to traditional hop testing. *Physical Therapy in Sport*, 41, 97–102. https://doi.org/10.1016/j.ptsp.2019.12.002
105. Singh, H., Gokeler, A., & Benjaminse, A. (2021). Effective attentional focus strategies after anterior cruciate ligament reconstruction:

A commentary. *International Journal of Sports Physical Therapy, 16*(6), 1575–1585. https://doi.org/10.26603/001c.29848

106. Stephenson, S. D., Kocan, J. W., Vinod, A. V., Kluczynski, M. A., & Bisson, L. J. (2021). A comprehensive summary of systematic reviews on sports injury prevention strategies. *Orthopaedic Journal of Sports Medicine, 9*(10) https://doi.org/10.1177/23259671211035776
107. Sugimoto, D., Myer, G. D., Foss, K. D., & Hewett, T. E. (2015). Specific exercise effects of preventive neuromuscular training intervention on anterior cruciate ligament injury risk reduction in young females: Meta-analysis and subgroup analysis. *British Journal of Sports Medicine, 49*(5), 282–289. https://doi.org/10.1136/bjsports-2014-093461
108. Svoboda, K., & Li, N. (2018). Neural mechanisms of movement planning: Motor cortex and beyond. *Current Opinion in Neurobiology, 49*, 33–41. https://doi.org/10.1016/j.conb.2017.10.023
109. Swanik, C. B., Covassin, T., Stearne, D. J., & Schatz, P. (2007). The relationship between neurocognitive function and noncontact anterior cruciate ligament injuries. *American Journal of Sports Medicine*, 35, 943–948. https://doi.org/10.1177/0363546507299532
110. Tanaka, S., Sandrini, M., & Cohen, L. G. (2011). Modulation of motor learning and memory formation by non-invasive cortical stimulation of the primary motor cortex. *Neuropsychological Rehabilitation, 21*, 650–675. https://doi.org/10.1080/09602011.2011.605589
111. Tanaka, T., Suzuki, K., & Imanaka, K. (2005). An EEG study on contextual interference. *Journal of Sport & Exercise Psychology, 27S*, S151–S152.
112. Tassignon, B., Verschueren, J., Baeyens, J. P., Benjaminse, A., Gokeler, A., Serrien, B., & Clijsen, R. (2021). An exploratory meta-analytic review on the empirical evidence of differential learning as an enhanced motor learning method. *Frontiers in Psychology, 12*, 533033. https://doi.org/10.3389/fpsyg.2021.533033
113. Tayfur, B., Charuphongsa, C., Morrissey, D., & Miller, S. C. (2021). Neuromuscular function of the knee joint following knee injuries: Does it ever get back to normal? A systematic review with meta-analyses. *Sports Medicine (Auckland, N.Z.), 51*(2), 321–338. https://doi.org/10.1007/s40279-020-01386-6
114. Thorborg, K., Krommes, K. K., Esteve, E., Clausen, M. B., Bartels, E. M., & Rathleff, M. S. (2017). Effect of specific exercise-based football injury prevention programmes on the overall injury rate in football: A systematic review and meta-analysis of the FIFA 11 and 11+

programmes. *British Journal of Sports Medicine, 51*(7), 562–571. https://doi.org/10.1136/bjsports-2016-097066
115. Tombu, M., & Jolicoeur, P. (2003). A central capacity sharing model of dual-task performance. *Journal of Experimental Psychology: Human Perception and Performance, 29*(1), 3–18. https://doi.org/10.1037//0096-1523.29.1.3
116. Valença, M., Coutinho, D., Schöllhorn, W., Ribeiro, N., & Santos, S. (2022). Investigating the effects of differential learning on golfers' pitching performance as a function of handicap. *International Journal of Environmental Research and Public Health, 19*, 12550. https://doi.org/10.3390/ijerph191912550
117. Van Hooren, B., Plasqui, G., & Meijer, K. (2024). The effect of wearable-based real-time feedback on running injuries and running performance: A randomized controlled trial. *The American Journal of Sports Medicine, 52*(3), 750–765. https://doi.org/10.1177/03635465231222464
118. van Weert, M. B., Rathleff, M. S., Eppinga, P., Mølgaard, C. M., & Welling, W. (2023). Using a target as external focus of attention results in a better jump-landing technique in patients after anterior cruciate ligament reconstruction: A cross-over study. *The Knee, 42*, 390–399. https://doi.org/10.1016/j.knee.2023.04.016
119. VandenBerg, P. M., Hogg, T. M., Kleim, J. A., & Whishaw, I. Q. (2002). Long-Evans rats have a larger cortical topographic representation of movement than Fischer-344 rats: A microstimulation study of motor cortex in naive and skilled reaching-trained rats. *Brain Research Bulletin, 59*, 197–203. https://doi.org/10.1016/S0361-9230(02)00865-1
120. Webster, K. E., & Hewett, T. E. (2018). Meta-analysis of meta-analyses of anterior cruciate ligament injury reduction training programs. *Journal of Orthopaedic Research, 36*, 2696–2708. https://doi.org/10.1002/jor.24043
121. Wiggins, A. J., Grandhi, R. K., Schneider, D. K., Stanfeld, D., Webster, K. E., & Myer, G. D. (2016). Risk of secondary injury in younger athletes after anterior cruciate ligament reconstruction: A systematic review and meta-analysis. *American Journal of Sports Medicine*, 44, 1861–1876. https://doi.org/10.1177/0363546515621554
122. Wilk, K., Thomas, Z. M., Arrigo, C. A., & Davies, G. J. (2023). The need to change return to play testing in athletes following ACL injury: A theoretical model. *International Journal of Sports Physical Therapy*, 18(1), 272–281. https://doi.org/10.26603/001c.67988

123. Wong, C. Y., Mok, K. M., & Yung, S. H. (2023). Secondary anterior cruciate ligament injury prevention training in athletes: What is the missing link? *International Journal of Environmental Research and Public Health*, 20(6), 4821. https://doi.org/10.3390/ijerph20064821
124. Wright, A. A., Ness, B. M., Donaldson, M., Hegedus, E. J., Salamh, P., & Cleland, J. A. (2021). Effectiveness of shoulder injury prevention programs in an overhead athletic population: A systematic review. *Physical Therapy in Sport*, 52, 189–193. https://doi.org/10.1016/j.ptsp.2021.09.004
125. Wulf, G. (2012). Attentional focus and motor learning: A review of 15 years. *International Review of Sport and Exercise Psychology*, 6(1), 77–104. https://doi.org/10.1080/1750984X.2012.723728
126. Wulf, G., & Lewthwaite, R. (2016). Optimizing performance through intrinsic motivation and attention for learning: The OPTIMAL theory of motor learning. *Psychonomic Bulletin & Review*, 23*(5), 1382–1414. https://doi.org/10.3758/s13423-015-0999-9
127. Wymbs, N. F., & Grafton, S. T. (2009). Neural substrates of practice structure that support future offline learning. *Journal of Neurophysiology*, 102, 2462–2476. https://doi.org/10.1152/jn.00315.2009
128. Xue, X., Li, Q., Wang, Y., et al. (2022). Impaired corticospinal tract in chronic ankle instability: A diffusion tensor imaging (DTI) and neurite orientation dispersion and density imaging (NODDI) study at 7.0 Tesla. *Journal of Science and Medicine in Sport*, 25(8), 649–654. https://doi.org/10.1016/j.jsams.2022.05.003
129. Yu, L., Tao, W., Jin, Z., Li, Y., Xue, X. A., Wang, R., & Hua, Y. (2024). Causal relationships of ligamentous injuries in the knee on corticospinal tract structure: A Mendelian randomization analysis. *Sports Health*. https://doi.org/10.1177/19417381241255342
130. Zarzycki, R., Morton, S. M., Charalambous, C. C., Marmon, A., & Snyder-Mackler, L. (2018). Corticospinal and intracortical excitability differ between athletes early after ACLR and matched controls. *Journal of Orthopaedic Research*. https://doi.org/10.1002/jor.24062
131. Zhang, Y., Xue, X., Guo, G., Cao, R., Yu, L., Tao, W., Pan, S., Hua, Y., & Wang, H. (2024). Association between neural plasticity and pain-related fear in chronic ankle instability: A structural neuroimaging study. *Journal of Athletic Training*. https://doi.org/10.4085/1062-6050-0214.24.

Neuroplasticity in Traumatic Brain Injury: Recovery and Rehabilitation

Öznur Şimşek and Selen Gür Özmen

1. Introduction

Traumatic brain injury (TBI), damage to the brain by an external force, can be one of the most serious and complex healthcare challenges (Rubiano et al., 2015). This damage can be mild, moderate, or severe. Even though many of the TBI cases are considered mild, they can still cause damage to the brain tissue. In moderate and severe cases, damage is more extensive. The brain exhibits a remarkable capacity for recovery, a phenomenon known as neuroplasticity—the brain's ability to change according to intrinsic or extrinsic stimuli (Puderbaugh & Emmady, 2023). This change could take hours to years, depending on the damage level. Sometimes, the damage cannot be repaired thoroughly, and compensatory strategies should be developed to manage everyday life. This chapter will examine neuroplasticity after a traumatic brain injury. In the first section, we will start by investigating TBI. There will be an overview of TBI, including mechanical and chemical changes in the brain and interventions for TBI. Then, neuroplasticity and the factors that influence neuroplasticity will be examined after traumatic brain injuries (TBIs). Following this, therapeutic factors that aim to increase neuroplasticity will be investigated—lastly, the differences between neuroplasticity after TBI in adults and children will be investigated. Please see the relevant chapter for a more comprehensive understanding of neuroplasticity and its mechanisms.

2. Traumatic Brain Injury

Traumatic brain injury refers to pathologies or alterations in brain functions due to external reasons. The most common sources of TBIs are traffic accidents, workplace accidents, and sports injuries (Khellaf et al., 2019).

There are two classifications of injury depending on the course. It can be penetrating or closed. If the trauma is caused by an external force strong enough to breach the skull, it is called a penetrating injury. If there is no breach in the skull, it is called a closed or blunt injury. Different forces can cause damage, including acceleration, deceleration, acceleration-deceleration, rotation, or compression. Such traumas can affect a patient's cognitive, emotional, and physical well-being (Mckee & Daneshvar, 2015; Maas et al., 2022).

According to pathological features, TBIs can also be focal or multifocal. Focal injuries can affect the whole brain by causing compressions in the surrounding regions, leading to edema, elevating intracranial pressure (ICP), and herniation. In focal injuries, symptoms relate to the function of the damaged area. In diffuse or multifocal injuries, multiple parts of the brain are damaged (Conde & Siebner, 2020 Mckee & Daneshvar, 2015).

The severity of the injury is determined using the Glasgow Coma Scale (GCS) with scores between 1 and 15. The smaller number on the scale means more severe injury. This scale measures eye, motor, and verbal responses (Teasdale & Jennett, 1974; Pavlovic et al., 2019).

TBIs should be considered as progressive illnesses rather than an isolated event. TBIs can be analyzed in two stages: primary and secondary. The primary stage occurs during the impact, resulting in focal or diffuse injury. Concussions, contusions, and intracranial hematomas are results of primary-stage TBIs. Continuation of the damage as the result of the primary injury causes secondary brain injuries. Ischemia, hypotension, impaired autoregulation and ICP are some examples of secondary brain injuries (Su et al., 2015). TBIs are associated with dementia and other neurodegenerative diseases even years after the accident (Al-Dahhak et al., 2018; Brett et al., 2022).

2.1. Mechanical and Chemical Changes of TBI

The brain undergoes many chemical and mechanical changes in the acute phase of TBI.

2.1.1. Chemical Changes

Trauma disrupts neurotransmission, causing too much glutamate to be released. Glutamate is vital for synaptic transmission and plasticity. However, excess glutamate activates specific receptors in neurons. This leads to too much calcium entering the cells. Such overload damages cells and can cause cell death (Yi & Hazell, 2006).

Moreover, the surplus of glutamate and calcium strains mitochondria. This reduces ATP (adenosine triphosphate) production. Consequently, it affects processes like neurotransmitter synthesis (Sowers et al., 2021).

High glutamate levels also create reactive oxygen species (ROS), including superoxide anions, hydrogen peroxide, and hydroxyl radicals. ROS levels exceeding the brain's neutralizing capacity harm cellular components and hamper the brain's repair ability (Fesharaki-Zadeh, 2022).

Trauma can breach the blood–brain barrier, letting harmful agents into the brain. This boosts inflammation and triggers the immune system, causing nerve damage (Cash & Theus, 2020).

Traumatic brain injury (TBI) makes glial cells overly active. These cells then release substances that damage nerves and cause chronic inflammation. Additionally, they can form scars (Li et al., 2022; Yi & Hazell, 2006; Mira et al., 2021).

2.1.2. Mechanical Changes

2.1.2.1. Primary Injury Mechanisms

Primary injuries stem from initial impact or force. Key types include coup, contrecoup injuries, diffuse axonal injury (DAI), and penetration.

Coup injuries occur when the brain is hit. Contrecoup injuries happen on the opposite side, as the brain rebounds and hits the skull. Such forces can lead to bruising and bleeding in the brain (Finan et al., 2024; Agrawal, 2021).

Rapid shifts in speed induce rotational forces in the brain. These forces stretch and tear axons, leading to cell death (Meythaler et al., 2001).

A sharp object penetrating the brain causes tissue damage, disrupts circuits, and leads to contamination (Hawryluk et al., 2022).

2.1.2.2. Secondary Injury Mechanisms

Primary impact leads to biochemical and mechanical changes in the brain. These changes cause more prolonged effects called secondary injuries.

After the initial trauma, the brain can accumulate fluid, and edema can develop. Edema leads to ICP, which further suppresses cerebral blood flow. ICP can compress brain structures and lead to herniation (Jha et al., 2019).

The mechanical impact may damage the blood vessels and affect the blood flow, leading to ischemia and hypoxia. Lack of oxygen contributes to cell death in the affected area (Cash & Theus, 2020).

2.3. Interventions to Traumatic Brain Injury (TBI)

TBIs can be examined in both the acute and chronic stages. Early and correct interventions can change the outcome of the injury (Königs et al., 2018). Interventions at different stages of TBI have different goals. While in the acute stage, the aim is to minimize primary and secondary damage, in the chronic stage, the objective is to enhance neuroplasticity and maximize recovery.

Generally, in mild TBI, no medical interventions are necessary. The first aim of moderate to severe traumas is to stabilize the patient's vitals and prevent a secondary injury. ICP and arterial hypotension should be monitored, and appropriate interventions should be made when necessary (Stocchetti et al., 2017). Surgical and/or pharmacological interventions are performed depending on the patient's condition. The patient should be evaluated for seizure risk (Pease et al., 2024)

3. Neuroplasticity After Traumatic Brain Injury

TBI causes cell deaths and a decrease in cortical inhibitory networks. Immediately after a TBI, the brain starts to rehabilitate itself. Hours to days after the injury are considered the immediate stage, and weeks to months after are considered the delayed stage (Zotey et al., 2023). Neurotransmitter release and synapse strength changes set off a recovery. The brain unmasks and recruits new and secondary pathways to keep functioning efficiently. Inhibitory networks start to shift to excitatory. Neuronal proliferation and synaptogenesis begin. Damaged cells start to be replaced; gliotic scar tissue begins to alleviate, and revascularization starts. After a while, remodeling and cortical changes happen through the upregulation of synaptic markers and axonal sprouting (Puderbaugh & Emmady, 2023).

Neurogenesis in humans has been proven by Altman and Das (1965). In nonhuman primates, researchers demonstrated that an injured site can relocate to an adjacent site (Taupin & Gage, 2002).

Neuroplasticity largely depends on post-injury behaviors. Neuroplasticity affects the patient's behavior and extends beyond the injured area (Fraser et al., 2002; Schlaug et al., 2009).

3.1. Factors Influencing Neuroplasticity

3.1.1. Trauma Severity

In cases of mild TBIs, the brain's ability to heal itself works fast and efficiently due to the trauma's less severe nature (Huang et al., 2024). Mild TBI patients seem to recover fully in a year or less. In cases of more severe TBI, recovery is slower and can be incomplete (McCrea et al., 2021). Moderate TBI shows the most extensive recovery in a year (Scholten et al., 2015). In the case of severe TBIs, recovery takes longer time, and the patient's recovery may not be complete.

3.1.2. Location of Injury

The location of the injury can disrupt different brain functions according to the affected area. Different regions have distinct roles in the brain, and the disruption to any of these areas will have a distinct impact. Cortexes have higher cortical plasticity than connective tracts, except for unimodal areas in the cortex (Herbet et al., 2016). While injury to the Broca area results in problems in language, injury to the frontal lobe will result in executive function problems.

3.1.3. Pre-Injury Cognitive Reserve

Higher cognitive reserve and younger age were associated with more significant cognitive recovery after TBI (Fraser et al., 2019). This protective effect of the cognitive reserve diminishes over time after the injury in the older population. Moreover, the time after the accident and the person's age seem to modify the effect of the cognitive reserve (Venkatesan et al., 2022).

3.1.4. Age

Research highlights the dual benefits and drawbacks of age in the context of traumatic brain injuries (TBI). Studies have yielded conflicting results, making it crucial to delve deeper into the relationship between age and TBI outcomes.

On one hand, younger age is linked to faster recovery rates, which can be attributed to the youth's robust adaptive capacity. The Kennard principle, as stated by Elliott (2020), suggests that younger individuals possess a greater ability to recover from brain injuries. This is supported by Kolb's (2014) findings, which indicate that the brain's ongoing development can protect against injuries and lasting damage. For instance, in the case of a 10-year-old who suffers from a TBI, their developing brain is more likely to reorganize and compensate for the damaged areas, leading to a faster recovery.

On the other hand, a TBI sustained during critical periods of growth can have a profoundly damaging impact on normal development, leading to lasting consequences. Levin et al. (2001) emphasize that injuries during these critical periods can disrupt the normal maturation process, resulting in long-term effects on cognitive and motor skills. Furthermore, Schuett and Zihl's (2013) research highlights that age may not be a significant factor in all cases, and other variables may play a more crucial role in determining TBI outcomes.

Interestingly, Levin (2003) notes that age can be a protective factor in focal vascular lesions, where the injury is localized to a specific area of the brain. However, in diffusional axonal injuries, which involve damage to multiple areas of the brain, age does not appear to have a beneficial effect. This suggests that the relationship between age and TBI outcomes is complex and influenced by various factors, including the type and severity of the injury.

3.1.5. Timing of Intervention

Timing of the intervention after TBI can massively affect neuroplasticity and the outcome of the injury. Early onset neurorehabilitation programs show beneficial effects on the recovery rate and/or speed (Königs et al., 2018). The chemical cascade after TBI promotes neuroplasticity and trauma management in the early stage is an important modifier to recovery regardless of neurorehabilitation (Kleim & Jones, 2008). Minimizing secondary damages with early intervention is key to faster and more complete recovery (Su et al., 2015).

3.1.6. Individual Variability

Genetic makeup (Cortes & Pera, 2021), pre-existing brain health, personal resilience, lifestyle, and social support impact the brain's ability to reorganize and adapt (Voss et al., 2017). Individuality in neuroplasticity mechanisms requires individual rehabilitation and treatments for TBI patients. Social support, an active lifestyle, and mental and physical health before the injury can speed up recovery.

3.1.7. Rehabilitation and Environmental Enrichment

Several rehabilitation techniques include cognitive rehabilitation, virtual reality, gamification, and sensorimotor integration training (Zotey et al., 2023). These techniques are important for the brain to reorganize itself. Furthermore, physical activity and social interactions create a positive environment for it (Merzenich et al., 2014; Appelbaum et al., 2023).

4. Therapeutic Approaches Enhancing Neuroplasticity

4.1. Non-Invasive Brain Stimulation

Non-invasive brain stimulation techniques, such as Vagus nerve stimulation (VNS), transcranial magnetic stimulation (TMS), and transcranial direct current stimulation (tDCS), show promise in neurological and psychological areas (Xiong et al., 2022; Hsu et al., 2024).

tDCS aids cognition recovery in traumatic brain injury (TBI) patients (Schwertfeger et al., 2023). It also enhances memory and learning and reduces impulsivity in TBI-affected rodents (Surendrakumar et al., 2023).

Hara and colleagues (2021) reviewed TMS and tDCS effects on executive functioning, attention, and memory. Stimulation of the dorsolateral prefrontal cortex showed mixed results.

Despite their promise, these techniques are new. More research is needed to understand their mechanisms and ensure safety (Ziesel et al., 2023).

4.2. Traumatic Brain Injury (TBI) Rehabilitation

The management of TBI depends on several factors, including the nature of the incident, the patient's age, and pre-injury health. The patient's age, the injury's severity, whether the injury is focal or multifocal, and whether prior or polytrauma affects how this trauma is treated. Treatments of TBIs focus on many aspects, including preventing secondary injuries and promoting neuroplasticity (Su et al., 2015). There are several rehabilitation techniques for TBI (Kaurani et al., 2024).

Research by Han et al. (2018) examined the effects of cognitive training in patients with chronic TBI. Even nine years after the accident, patients who took selective attention, abstract reasoning, and other thinking strategies courses improved their cognitive control abilities.

According to research, cognitive rehabilitation improves attention, memory, and communicative and pragmatic abilities (Galetto & Sacco, 2017). Neurological music therapy has been proven to induce neuroplasticity, and its enhancing effect was shown in executive function after TBI (Siponkoski et al., 2020).

4.3. Stem Cell Therapy

Stem cells can renew, multiply, and turn into various cells. This makes them potential treatments for damaged tissues. Many studies explore stem cells for TBI recovery (Tian et al., 2023). Different stem cell types have shown benefits in animals. They reduce inflammation, boost cognitive function, and promote new blood and nerve cells (Zhou, 2019). More research is needed to confirm safety and effectiveness. So far, results are promising with no severe side effects (Zhang et al., 2020).

4.4. Pharmacological Interventions

4.4.1. Neurotrophic Factors

Neurotrophic factors (NTF) are proteins that are critical for cell survival. NTFs are crucial in cell proliferation, migration, differentiation, and regeneration (Liu, 2018). NFTs have been shown to reduce inflammation and increase neuroplasticity after TBI (Rejdak et al., 2023). Also, NFTs are associated with reducing cognitive impairment and decreasing the risk of dementia in the long term (Corrigan et al., 2017).

4.4.2. Anti-Inflammatory Drugs

TBIs trigger an inflammatory response in the brain which, slows down or reduces neuroplasticity. Anti-inflammatory drugs manage this inflammation and create a more favorable environment for healing. In animal models, anti-inflammatory drugs show promising results, but these outcomes are not replicated in human clinical trials. More research is needed on this front (Kalra et al., 2022).

4.5. Negative Neuroplasticity After TBI

The brain has a remarkable potential to recover from TBI. This recovery largely depends on neuroplasticity. However, neuroplasticity doesn't always have beneficial outcomes. Especially following an injury, a lack of social interactions and an environment deprived of stimuli can start a downward spiral of negative neuronal connections (Tomaszczyk et al., 2014). Quality of life after the accident and neurorehabilitation are important to prevent negative neuroplasticity.

5. Neuroplasticity After Traumatic Brain Injury (TBI) in Children

Between birth and 18, the brain undergoes rapid transformations characterized by distinct milestones and developmental stages. This period is marked by significant changes, which can be likened to a dynamic construction site, where new connections are constantly being formed and old ones refined. Children's brains are incredibly malleable, and even minor alterations can have lasting effects on their developmental trajectory.

However, childhood injuries can be a major setback, disrupting the delicate balance of brain development and leading to heterogeneous outcomes. For instance, a study by Maria et al. (2019) highlighted the brain's remarkable ability to repair itself after an injury but also underscored the risk of this repair process deviating from the child's normal development. Fortunately, researchers have found that behavioral and pharmacological treatments can help steer this process back on track, ensuring a smoother recovery.

One critical distinction between children's and adults' brains is their response to injury. In children, the body's initial response to injury is characterized by a stronger, longer-lasting pro-inflammatory response, accompanied by atypical pruning and dendritic branching. In contrast, adults tend to exhibit more robust metabolic dysfunctions (Serpa et al., 2021). This disparity has significant implications for treatment strategies, as what works for adults may not be effective for children.

Studies have yielded inconsistent results regarding the impact of brain development on recovery (Kolb, 2014; Levin et al., 2001). While being younger is associated with faster recovery, it also increases the risk of disruptions to development. A review of the literature revealed that children are often caught in a double bind, where their rapid recovery is tempered by the long-term consequences of their injury. Reviewing literature, children were found to be more vulnerable to the lasting impacts of TBI rather than adults (Blackwell & Grell, 2023).

6. Conclusions

TBIs are a major global health concern. They lead to significant disability and death. TBIs disrupt brain functions due to external factors. Common causes include sports, traffic, and work injuries. These injuries can lead to cognitive, physical, and psychological issues (Khellaf et al., 2019). The Glasgow Coma Scale assesses TBI severity. Higher scores indicate milder injuries (Teasdale & Jennett, 1974).

TBIs progress through acute and chronic stages. Primary injuries occur at impact, causing mechanical and chemical changes. These trigger secondary injuries, disrupting brain balance and causing further damage (Li et al., 2022). Even years later, TBIs can lead to neurodegenerative diseases (Brett et al., 2022).

Depending on the injury severity and type there are several interventions including surgical and pharmacological. The goal in the acute stage is to stabilize the patient and limit further damage. In chronic stages, the aim is to boost recovery. Early treatment improves outcomes (Königs et al., 2018).

The brain's ability to change itself is called neuroplasticity. This change can happen due to intrinsic or extrinsic stimuli (Puderbaugh & Emmady, 2023). Secreted neurotransmitters and changes in synapse strength offset the neuroplasticity. The brain unmasks and recruits new and secondary pathways to keep functioning efficiently. Inhibitory networks start to shift to excitatory. Neuronal proliferation and synaptogenesis begin. Damaged cells start to be replaced, gliotic scar tissue begins to alleviate and revascularization starts. After a while remodeling and cortical changes happen by the upregulation of synaptic markers and axonal sprouting (Puderbaugh & Emmady, 2023). The brain activates new pathways and enhances existing ones. Damaged cells are replaced, and blood flow improves.

Several factors affect neuroplasticity and TBI outcomes. These include injury severity and location, cognitive reserve, age, and timing of treatment. Rehabilitation and social support are also crucial (Huang et al., 2024; Herbet et al., 2016; Venkatesan et al., 2022; Levin, 2003; Königs et al., 2018; Zotey et al., 2023; Appelbaum et al., 2023; Cortes & Pera, 2021).

Research shows promise in non-invasive brain stimulation and stem cell therapy for enhancing neuroplasticity (Zhang et al., 2020). Various rehabilitation techniques also aid recovery (Su et al. 2015; Kaurani et al., 2024; Siponkoski et al., 2020). Studies are exploring the effects of NFTs and anti-inflammatory drugs on neuroplasticity (Corrigan et al., 2017; Kalra et al., 2022).

Neuroplasticity sometimes has maladaptive outcomes. Negative neuroplasticity can delay or disrupt recovery. Quality of life after the injury is very important to prevent negative neuroplasticity (Tomaszczyk et al., 2014).

Recovery from TBIs differs between adults and children. Childhood is marked by critical development stages. Disruptions can have lasting effects (Blackwell & Grell, 2023). However, children's brains are more adaptable than adults' (Maria et al., 2019).

7. Takeaways

1. Pathologies or alterations in brain functions caused by an external force are called traumatic brain injuries.
2. Depending on the skull intactness, TBIs can be closed or penetrating.
3. Different forces, including penetration, deceleration, acceleration-deceleration, rotation, or compression can cause injury.
4. The Glasgow Coma Scale (GCS) is used to assess the severity of TBI, with scores ranging from 1 (severe) to 15 (mild).
5. TBIs are associated with neurodegenerative diseases in the long term.
6. The injuries that occur at the time of the impact are called primary injuries, while the ones that develop over time are secondary injuries.
7. TBI causes chemical (Glutamate release, calcium influx, production of reactive oxygen species, and increased inflammatory response) and mechanical changes (disruption of the blood-brain barrier, cell deaths, diffuse axonal injury) in the brain.
8. Neuroplasticity is the capacity of the brain to change to intrinsic or extrinsic stimuli.
9. Neurotransmitter release and synapse strength changes set off a recovery in the brain.
10. Mild trauma recoveries are swifter and easier compared to moderate to severe traumas.
11. Neurotransmitter release and change in the synapse strength start the initial neuroplasticity process and recovery. Further recovery includes neuronal proliferation, synaptogenesis, and revascularization.

12. The impact of TBI varies based on the brain region affected, with cortical areas typically showing higher plasticity than connective tracts.
13. Higher pre-injury cognitive reserve is associated with greater cognitive recovery.
14. Research shows the conflicted effects of age on recovery.
15. Earlier interventions are crucial for optimizing recovery and maximizing neuroplasticity.
16. Individual differences (genetic factors, social support, pre-existing health) affect neuroplasticity after TBI.
17. Rehabilitation and social support help the recovery process.
18. Non-invasive brain stimulation and stem cell therapies show promising results in TBI therapy.
19. Cognitive therapies enhance neuroplasticity and help with the recovery.
20. Neurotrophic factors and anti-inflammatory drugs are being studied for their roles in improving neuroplasticity and recovery.
21. Neuroplasticity can have negative outcomes depending on the life quality and lack of rehabilitation options after the injury.
22. TBI in children presents unique challenges due to rapid brain development, with treatments focusing on minimizing disruption to developmental milestones.
23. Children's recovery has unique challenges; while high plasticity positively affects recovery, missing out on developmental periods has lasting effects.

Abbreviations

TBI: Traumatic Brain Injury
TBIs: Traumatic Brain Injuries
ICP: Intracranial Pressure
GSC: Glasgow Coma Scale
ROS: Reactive Oxygen Species
DAI: Diffuse Axonal Injury
VNS: Vagus Nerve Stimulation
TMS: Transcranial Magnetic Stimulation
tDCS: Transcranial Direct Current Stimulation
NFT: Neurotrophic Factors

Bibliography

1. Agrawal, S. N. (2021). The Pathophysiology of Traumatic Brain Injury (TBI): A Review. Asian Journal of Research and Reports in Neurology, 4(1), 1–10.
2. Al-Dahhak, R., Khoury, R., Qazi, E., & Grossberg, G. T. (2018). Traumatic brain injury, chronic traumatic encephalopathy, and Alzheimer's disease. Clinics in geriatric medicine, 34(4), 617–635. 10.1016/j.cger.2018.06.008 External Link
3. Altman J, Das G. D. Autoradiographic and histological evidence of postnatal hippocampal neurogenesis in rats. J Comp Neurol. 1965;124(3):319–335. https://doi.org/10.1002/cne.901240303
4. Appelbaum, L. G., Shenasa, M. A., Stolz, L., & Daskalakis, Z. (2023). Synaptic plasticity and mental health: methods, challenges, and opportunities. Neuropsychopharmacology, 48(1), 113–120. https://doi.org/10.1038/s41386-022-01370-w
5. Blackwell, L. S., & Grell, R. (2023). Pediatric traumatic brain injury: impact on the developing brain. Pediatric neurology. https://doi.org/10.1016/j.pediatrneurol.2023.06.019
6. Brett, B. L., Gardner, R. C., Godbout, J., Dams-O'Connor, K., & Keene, C. D. (2022). Traumatic brain injury and risk of neurodegenerative disorder. Biological psychiatry, 91(5), 498–507. https://doi.org/10.1016/j.biopsych.2021.05.025
7. Cash, A., & Theus, M. H. (2020). Mechanisms of blood-brain barrier dysfunction in traumatic brain injury. International journal of molecular sciences, 21(9), 3344. https://doi.org/10.3390/ijms21093344
8. Conde, V., & Siebner, H. R. (2020). Brain damage by trauma. Handbook of clinical neurology, 168, 39–49. https://doi.org/10.1016/B978-0-444-63934-9.00005-6
9. Corrigan, F., Arulsamy, A., Teng, J., & Collins-Praino, L. E. (2017). Pumping the brakes: neurotrophic factors for the prevention of cognitive impairment and dementia after traumatic brain injury. Journal of neurotrauma, 34(5), 971–986. https://doi.org/10.1089/neu.2016.4589
10. Cortes, D., & Pera, M. F. (2021). The genetic basis of inter-individual variation in recovery from traumatic brain injury. NPJ regenerative medicine, 6(1), 5. https://doi.org/10.1038/s41536-020-00114-y
11. Elliott, D. (2020). The legacy of the Kennard principle. Journal of Undergraduate Neuroscience Education, 19(1), R11. PMCID: PMC8040834 PMID: 33880106

12. Fesharaki-Zadeh, A. (2022). Oxidative stress in traumatic brain injury. International journal of molecular sciences, 23(21), 13000. https://doi.org/10.3390/ijms232113000
13. Finan, J. D., Vogt, T. E., & Samei, Y. (2024). Cavitation in blunt impact traumatic brain injury. Experiments in fluids, 65(8), 114. https://doi.org/10.1007/s00348-024-03853-6
14. Fraser C. et al. Driving plasticity in human adult motor cortex is associated with improved motor function after brain injury. Neuron. 2002;34(5):831–840. 10.1016/S0896-6273(02)00705-5
15. Fraser, E. E., Downing, M. G., Biernacki, K., McKenzie, D. P., & Ponsford, J. L. (2019). Cognitive reserve and age predict cognitive recovery after mild to severe traumatic brain injury. Journal of neurotrauma, 36(19), 2753–2761.
16. Galetto, V., & Sacco, K. (2017). Neuroplastic changes induced by cognitive rehabilitation in traumatic brain injury: a review. Neurorehabilitation and Neural Repair, 31(9), 800–813. https://doi.org/10.1177/1545968317723748
17. Han, K., Chapman, S. B., & Krawczyk, D. C. (2018). Neuroplasticity of cognitive control networks following cognitive training for chronic traumatic brain injury. NeuroImage: Clinical, 18, 262–278. https://doi.org/10.1016/j.nicl.2018.01.030
18. Hara, T., Shanmugalingam, A., McIntyre, A., & Burhan, A. M. (2021). The effect of non-invasive brain stimulation (NIBS) on executive functioning, attention and memory in rehabilitation patients with traumatic brain injury: a systematic review. Diagnostics, 11(4), 627. https://doi.org/10.3390/diagnostics11040627
19. Hawryluk, G. W., Selph, S., Lumba-Brown, A., Totten, A. M., Ghajar, J., Aarabi, B., … & Bell, R. (2022). Rationale and methods for updated guidelines for the management of penetrating traumatic brain injury. Neurotrauma reports, 3(1), 240–247. https://doi.org/10.1089/neur.2022.0008
20. Herbet, G., Maheu, M., Costi, E., Lafargue, G., & Duffau, H. (2016). Mapping neuroplastic potential in brain-damaged patients. Brain, 139(3), 829–844. https://doi.org/10.1093/brain/awv394
21. Hsu, C. W., Chou, P. H., Brunoni, A. R., Hung, K. C., Tseng, P. T., Liang, C. S., … & Li, C. T. (2024). Comparing different non-invasive brain stimulation interventions for bipolar depression treatment: A network meta-analysis of randomized controlled trials. Neuroscience & Biobehavioral Reviews, 156, 105483. https://doi.org/10.1016/j.neubiorev.2023.105483

22. Huang, S., Han, J., Zheng, H., Li, M., Huang, C., Kui, X., & Liu, J. (2024). Structural and functional connectivity of the whole brain and subnetworks in individuals with mild traumatic brain injury: predictors of patient prognosis. Neural Regeneration Research, 19(7), 1553–1558. https://doi.org/ 10.4103/1673-5374.387971
23. Jha, R. M., Kochanek, P. M., & Simard, J. M. (2019). Pathophysiology and treatment of cerebral edema in traumatic brain injury. Neuropharmacology, 145, 230–246. https://doi.org/10.1016/j.neuropharm.2018.08.004
24. Kaurani, P., de Marchi Apolaro, A. V. M., Kunchala, K., Maini, S., Rges, H. A., Isaac, A., … & Nazir, Z. (2024). Advances in Neurorehabilitation: Strategies and Outcomes for Traumatic Brain Injury Recovery. Cureus, 16(6).
25. Khellaf, A., Khan, D. Z., & Helmy, A. (2019). Recent advances in traumatic brain injury. Journal of neurology, 266(11), 2878–2889. https://doi.org/10.1007/s00415-019-09541-4
26. Kleim, J. A., & Jones, T. A. (2008). Principles of experience-dependent neural plasticity: implications for rehabilitation after brain damage. https://doi.org/10.1044/1092-4388(2008/018)
27. Kolb B. Brain plasticity and behavior during development. In. In: Uzzell B. P, editor. Recovery after Traumatic Brain Injury. Psychology Press.; 2014. pp. 199–212.
28. Königs, M., Beurskens, E. A., Snoep, L., Scherder, E. J., & Oosterlaan, J. (2018). Effects of timing and intensity of neurorehabilitation on functional outcome after traumatic brain injury: a systematic review and meta-analysis. Archives of physical medicine and rehabilitation, 99(6), 1149–1159. https://doi.org/10.1016/j.apmr.2018.01.013
29. Levin H. S. et al. Word fluency in relation to severity of closed head injury, associated frontal brain lesions, and age at injury in children. Neuropsychologia. 2001;39(2):122–131. https://doi.org/10.1016/S0028-3932(00)00111-1
30. Levin, H. S. (2003). Neuroplasticity following non-penetrating traumatic brain injury. Brain Injury, 17(8), 665–674. https://doi.org/10.1080/0269905031000107151
31. Li, Y. F., Ren, X., Zhang, L., Wang, Y. H., & Chen, T. (2022). Microglial polarization in TBI: Signaling pathways and influencing pharmaceuticals. Frontiers in aging neuroscience, 14, 901117. https://doi.org/10.3389/fnagi.2022.901117

32. Liu, S. (2018). Neurotrophic factors in enteric physiology and pathophysiology. Neurogastroenterology & Motility, 30(10), e13446. https://doi.org/10.1111/nmo.13446
33. Maas, A. I., Menon, D. K., Manley, G. T., Abrams, M., Åkerlund, C., Andelic, N., … & Zemek, R. (2022). Traumatic brain injury: progress and challenges in prevention, clinical care, and research. The Lancet Neurology, 21(11), 1004–1060. https://doi.org/10.1016/S1474-4422(22)00309-X
34. Maria, N. S. S., Sargolzaei, S., Prins, M. L., Dennis, E. L., Asarnow, R. F., Hovda, D. A., … & Giza, C. C. (2019). Bridging the gap: mechanisms of plasticity and repair after pediatric TBI. Experimental neurology, 318, 78–91. https://doi.org/10.1016/j.expneurol.2019.04.016
35. McCrea, M. A., Giacino, J. T., Barber, J., Temkin, N. R., Nelson, L. D., Levin, H. S., … & TRACK-TBI Investigators. (2021). Functional outcomes over the first year after moderate to severe traumatic brain injury in the prospective, longitudinal TRACK-TBI study. JAMA neurology, 78(8), 982–992. 10.1001/jamaneurol.2021.2043
36. Mckee, A. C., & Daneshvar, D. H. (2015). The neuropathology of traumatic brain injury. Handbook of clinical neurology, 127, 45–66. https://doi.org/10.1016/B978-0-444-52892-6.00004-0
37. Merzenich, M. M., Van Vleet, T. M., & Nahum, M. (2014). Brain plasticity-based therapeutics. Frontiers in human neuroscience, 8, 385. https://doi.org/10.3389/fnhum.2014.00385
38. Meythaler, J. M., Peduzzi, J. D., Eleftheriou, E., & Novack, T. A. (2001). Current concepts: diffuse axonal injury–associated traumatic brain injury. Archives of physical medicine and rehabilitation, 82(10), 1461–1471. https://doi.org/10.1053/apmr.2001.25137
39. Mira, R. G., Lira, M., & Cerpa, W. (2021). Traumatic brain injury: mechanisms of glial response. Frontiers in physiology, 12, https://doi.org/10.3389/fphys.2021.740939
40. Pease, M., Gupta, K., Moshé, S. L., Correa, D. J., Galanopoulou, A. S., Okonkwo, D. O., … & Castellano, J. F. (2024). Insights into epileptogenesis from post-traumatic epilepsy. Nature Reviews Neurology, 20(5), 298–312. https://doi.org/10.1038/s41582-024-00954-y
41. Puderbaugh, M., & Emmady, P. D. (2023). Neuroplasticity. In StatPearls [Internet]. StatPearls Publishing.
42. Rejdak, K., Sienkiewicz-Jarosz, H., Bienkowski, P., & Alvarez, A. (2023). Modulation of neurotrophic factors in treating dementia, stroke and TBI: Effects of Cerebrolysin. Medicinal Research Reviews, 43(5), 1668–1700. https://doi.org/10.1002/med.21960

43. Rubiano, A. M., Carney, N., Chesnut, R., & Puyana, J. C. (2015). Global neurotrauma research challenges and opportunities. Nature, 527(7578), S193-S197. https://doi.org/10.1038/nature16035
44. Schlaug G, Marchina S, Norton A. Evidence for plasticity in white-matter tracts of patients with chronic Broca's aphasia undergoing intense intonation-based speech therapy. Ann N Y Acad Sci. 2009;1169:385–394. https://doi.org/10.1111/j.1749-6632.2009.04587.x
45. Scholten, A. C., Haagsma, J. A., Andriessen, T. M. J. C., Vos, P. E., Steyerberg, E. W., Van Beeck, E. F., & Polinder, S. (2015). Health-related quality of life after mild, moderate and severe traumatic brain injury: patterns and predictors of suboptimal functioning during the first year after injury. Injury, 46(4), 616–624. https://doi.org/10.1016/j.injury.2014.10.064
46. Schuett S, Zihl J. Does age matter? Age and rehabilitation of visual field disorders after brain injury. Cortex. 2013;49(4):1001–1012. https://doi.org/10.1016/j.cortex.2012.04.008
47. Schwertfeger, J. L., Beyer, C., Hung, P., Ung, N., Madigan, C., Cortes, A. R., … & Madhavan, S. (2023). A map of evidence using transcranial direct current stimulation (tDCS) to improve cognition in adults with traumatic brain injury (TBI). Frontiers in neuroergonomics, 4, https://doi.org/10.3389/fnrgo.2023.1170473
48. Serpa, R. O., Ferguson, L., Larson, C., Bailard, J., Cooke, S., Greco, T., & Prins, M. L. (2021). Pathophysiology of pediatric traumatic brain injury. Frontiers in neurology, 12, 696510. https://doi.org/10.3389/fneur.2021.696510
49. Siponkoski, S. T., Martínez-Molina, N., Kuusela, L., Laitinen, S., Holma, M., Ahlfors, M., … & Särkämö, T. (2020). Music therapy enhances executive functions and prefrontal structural neuroplasticity after traumatic brain injury: evidence from a randomized controlled trial. Journal of neurotrauma, 37(4), 618–634. https://doi.org/10.1089/neu.2019.6413
50. Sowers, J. L., Sowers, M. L., Shavkunov, A. S., Hawkins, B. E., Wu, P., DeWitt, D. S., … & Zhang, K. (2021). Traumatic brain injury induces region-specific glutamate metabolism changes as measured by multiple mass spectrometry methods. Iscience, 24(10). https://doi.org/10.1016/j.isci.2021.103108
51. Stocchetti, N., Carbonara, M., Citerio, G., Ercole, A., Skrifvars, M. B., Smielewski, P., … & Menon, D. K. (2017). Severe traumatic brain injury: targeted management in the intensive care unit. The Lancet Neurology, 16(6), 452–464.

52. Su, S., Veeravagu, A., & Grant, G. (2015). Neuroplasticity after traumatic brain injury..
53. Surendrakumar, S., Rabelo, T. K., Campos, A. C. P., Mollica, A., Abrahao, A., Lipsman, N., … & Hamani, C. (2023). Neuromodulation therapies in pre-clinical models of traumatic brain injury: systematic review and translational applications. Journal of Neurotrauma, 40(5–6), 435–448.
54. Taupin P, Gage F. H. Adult neurogenesis and neural stem cells of the central nervous system in mammals. J Neurosci Res. 2002; 69(6): 745–749. https://doi.org/10.1002/jnr.10378
55. Teasdale G, Jennett B (1974) Assessment of coma and impaired consciousness: a practical scale. Lancet 304(7872) :81–84 https://doi.org/10.1016/S0140-6736(74)91639-0
56. Tian, Z., Yu, T., Liu, J., Wang, T., & Higuchi, A. (2023). Introduction to stem cells. Progress in Molecular Biology and Translational Science, 199, 3–32. https://doi.org/10.1016/bs.pmbts.2023.02.012
57. Tomaszczyk, J. C., Green, N. L., Frasca, D., Colella, B., Turner, G. R., Christensen, B. K., & Green, R. E. (2014). Negative neuroplasticity in chronic traumatic brain injury and implications for neurorehabilitation. Neuropsychology Review, 24, 409–427. https://doi.org/10.1007/s11065-014-9273-6
58. Venkatesan, U. M., Rabinowitz, A. R., Bernier, R. A., & Hillary, F. G. (2022). Cognitive reserve in individuals aging with traumatic brain injury: Independent and interactive effects on cognitive functioning. The Journal of Head Trauma Rehabilitation, 37(3), E196-E205. https://doi.org/10.1097/HTR.0000000000000697
59. Voss, P., Thomas, M. E., Cisneros-Franco, J. M., & de Villers-Sidani, É. (2017). Dynamic brains and the changing rules of neuroplasticity: implications for learning and recovery. Frontiers in psychology, 8, 274878. https://doi.org/10.3389/fpsyg.2017.01657
60. Xiong, H. Y., Zheng, J. J., & Wang, X. Q. (2022). Non-invasive brain stimulation for chronic pain: state of the art and future directions. Frontiers in molecular neuroscience, 15, 888716. https://doi.org/10.3389/fnmol.2022.888716
61. Yi, J. H., & Hazell, A. S. (2006). Excitotoxic mechanisms and the role of astrocytic glutamate transporters in traumatic brain injury. Neurochemistry international, 48(5), 394–403. https://doi.org/10.1016/j.neuint.2005.12.001

62. Zhang, S., Lachance, B. B., Moiz, B., & Jia, X. (2020). Optimizing stem cell therapy after ischemic brain injury. Journal of Stroke, 22(3), 286. https://doi.org/10.5853/jos.2019.03048
63. Zhou, Y., Shao, A., Xu, W., Wu, H., & Deng, Y. (2019). Advance of stem cell treatment for traumatic brain injury. Frontiers in cellular neuroscience, 13, 301.
64. Ziesel, D., Nowakowska, M., Scheruebel, S., Kornmueller, K., Schaefer, U., Schindl, R., … & Rienmüller, T. (2023). Electrical stimulation methods and protocols for the treatment of traumatic brain injury: a critical review of preclinical research. Journal of NeuroEngineering and Rehabilitation, 20(1), 51.
65. Zotey, V., Andhale, A., Shegekar, T., & Juganavar, A. (2023). Adaptive neuroplasticity in brain injury recovery: Strategies and insights. Cureus, 15(9). https://doi.org/ 10.7759/cureus.45873

Stroke Rehabilitation: Harnessing Neuroplasticity for Functional Recovery

Mehmet Kaan Altunok and Aygül Köseoğlu Kurt

1. Introduction

This book section explores neuroplasticity's role in stroke recovery. It reviews treatments that enhance movement, thinking, and emotional health. After a stroke, neuroplasticity helps the brain recover by reorganizing itself. Therefore, promoting it is crucial in rehabilitation. The section lists interventions such as electrical stimulation, robotics, virtual reality, music therapy, and exercise, examining their impact on neuroplasticity and movement. Additionally, it highlights advanced treatments like brain–computer interfaces and virtual reality, focusing on their effects on neuroplasticity.

2. Stroke

Stroke is a major global issue, leading to disability and death. The American Heart Association and the American Stroke Association define it as a sudden neurological deficit due to blood vessel issues in the brain. These issues include cerebral infarction, intracerebral hemorrhage, and subarachnoid hemorrhage (Sacco et al., 2013). Previously, stroke was the second leading cause of death in the United States. However, it has now dropped to fifth place, thanks to early diagnosis and treatment. However, it remains the second leading cause of death worldwide and the top cause of long-term disability (Herpich & Rincon, 2020). The global cost of stroke is around $891 billion. This burden is higher in low- and middle-income countries (Feigin et al., 2022; Owolabi et al., 2022).

Strokes are mainly ischemic (85%) or hemorrhagic (15%). Ischemic strokes include various types, such as cardioembolic and lacunar strokes. Hemorrhagic strokes involve intracerebral and subarachnoid hemorrhages. The severity differentiates transient ischemic attacks from ischemic strokes (Musuka et al., 2015).

Stroke risk factors are unmodifiable (age, gender, ethnicity, genetics) or modifiable (hypertension, diabetes, atrial fibrillation, high cholesterol, smoking, alcohol, drug use, obesity, physical inactivity, inflammation) (Kuriakose & Xiao, 2020; Potter et al., 2022).

Stroke symptoms vary by affected area. Common symptoms include paralysis, weakness, spasticity, vision problems, difficulty speaking, lack of attention, coordination issues, and swallowing problems (Murphy & Werring, 2020). These symptoms can lead to serious complications. For instance, swallowing problems can cause pneumonia (Martino et al., 2005). Motor disorders can severely limit daily activities.

Another important consequence of stroke is cognitive decline, which is as significant as motor impairment. About 30% of stroke survivors show cognitive disorders within three months. Key risk factors include age, ethnicity, diabetes, atrial fibrillation, aphasia, depression, past strokes, and the current stroke's severity and location (Kalaria and Ballard, 2001). Common cognitive declines include attention, language, memory, visual perception, abstract reasoning, and executive functions (Leśniak et al., 2008; Nys et al., 2007). Post-stroke cognitive decline is significant, especially in the elderly and those with recurrent strokes. In the first year, 22% of patients develop cognitive impairments, severely affecting their quality of life. This highlights the need for cognitive rehabilitation (Douiri et al., 2013).

Psychological issues post-stroke are equally crucial. Common problems include depression, anxiety, and stress disorder. Depression affects 30% to 50% of stroke survivors, hindering recovery. These psychological issues make physical rehabilitation harder (Hackett & Pickles, 2014).

Research shows strokes lead to various complications beyond physical, cognitive, and psychological effects. These complexities demand a team approach in rehabilitation. Treatments that boost recovery and support brain healing are vital.

Shahid et al. (2023) outlined key focus areas for stroke rehabilitation.

1. *Gait training:* The aim is to improve walking abilities using treadmills, task-focused locomotor training, and body weight-supported treadmill exercises.
2. *Postural control exercises:* Proprioceptive Neuromuscular Facilitation (PNF), parallel bars, balance boards, and weight transfer exercises improve body posture control.

3. *Exercises to reduce hypertonus:* Continuous stretching, rhythmic rotations, muscle energy techniques (MET), and PNF can reduce excessive muscle tension.
4. *Exercises targeting sensory losses:* The aim is to improve sensory functions with mirror therapy, sensory discrimination exercises, electrical stimulation, compression techniques, and laser applications.
5. *Prevention of neglect syndrome:* Active visual scanning strategies, participation in daily life activities, focus marking, optimizing visual stimuli, PNF, active movements, speech, and proprioceptive stimuli direct patients to the side they have lost attention.
6. *Preservation of flexibility and joint integrity:* Active and passive range of motion (ROM) exercises, stretching exercises, soft tissue mobilizations, positioning strategies, resting splints, robot-assisted treatment, and traditional physiotherapy aim to keep joints healthy.
7. *Exercises to prevent muscle weakness:* Strengthening exercises, progressive resistance training, hydrotherapy, and aquatic exercise.

3. Neuroplasticity in Stroke

Santiago Ramón y Cajal, known as the "father of neuroscience," coined "neuronal plasticity" to describe structural changes in the adult brain. This concept was controversial. Many believed adult brains could not replace dead cells. They thought the number of neurons was fixed. Neuroplasticity, however, allows the brain to adapt its structure and functions. This occurs in response to changing conditions. It includes forming new neurons, restructuring networks, and biochemical changes. In developing brains, processes like neurogenesis and connection formation are key. In adults, neuroplasticity responds to internal and external stimuli. It occurs at molecular, cellular, and systems levels. Adaptive changes improve function. Maladaptive changes worsen function or increase damage. Adaptive plasticity is different from compensatory behaviors in healthy networks (Cramer et al., 2011; Fuchs & Flügge, 2014; Innocenti, 2022).

After a stroke, some nerve connections in the brain can be reactivated or strengthened. This helps improve motor, sensory, and and cognitive functions. The process involves reconnecting axons (axonal sprouting) and forming new synapses (synaptogenesis) (Kossut, 2019). Rehabilitation can boost these reconnections. Small losses often heal on their own. However, significant losses can lead to permanent issues (Robertson & Murre, 1999). Hebbian learning explains recovery after a stroke. Donald Hebb's theory states, "Units that fire together connect." Repeated interaction between two neurons strengthens their connection. This is known as long-term potentiation. If they seldom interact, the connection weakens (long-term depression) (Munakata & Pfaffly, 2004). Thus, Hebbian learning is key to understanding brain plasticity and the neurobiology of learning.

4. Neuroplasticity for Functional Recovery in Stroke

In individuals who have had a stroke, the functions of damaged brain areas can be taken over by the healthy areas he surrounding healthy brain areas. This process contributes to the recovery of motor skills by forming new neural connections. In the cortical remapping process, long-term potentiation is necessary for the recovery of motor skills, and this seems possible with intensive repetitive treatment programs.

4.1. Electrical Stimulation

Research shows mixed results for electrical stimulation in stroke recovery. Some studies found it ineffective in improving upper limb function (Vafadar et al., 2015). However, others noted improvements in daily activities and motor functions (Eraifej et al., 2017; Howlett et al., 2015). Popović et al. (2019) found better outcomes in early treatment phases, especially when combining electrical stimulation with exercise. A study by Kimberley et al. (2004) showed that three weeks of neuromuscular electrical stimulation significantly improved stroke patients' hand function and brain activity. The sham group showed no improvement. Yet, similar benefits were noted when treatments switched groups. Dimyan and Cohen (2011) suggested that non-invasive techniques might boost motor skills and brain adaptability. This could aid in rehabilitation for motor disorders. Overall, early electrical stimulation combined with rehabilitation seems beneficial for improving motor function and brain plasticity in stroke patients.

4.2. Robotic Devices

New research suggests neurorobotic devices can boost neuroplasticity in stroke patients. One study tested a wearable Ekso™ exoskeleton alongside standard treatment for chronic stroke patients with one-sided weakness (hemiparesis). It found the exoskeleton significantly improved walking, brain activity, and muscle function compared to standard training (Calabrò et al., 2018). Another study compared robotic gait training with treadmill training in hemiparetic patients. Both improved walking and motor function. However, the robotic group showed better motor function (Kim et al., 2020). A separate study examined an electromagnetic robotic exoskeleton for wrist and hand rehabilitation in chronic stroke patients. Both the exoskeleton and standard therapy improved motor function. Yet, the exoskeleton group significantly improved movement range and brain activity (Singh et al., 2021). These findings highlight the potential of neurorobotic devices to enhance motor function and brain activity in stroke patients. They suggest combining these devices with traditional treatments could lead to better outcomes.

4.3. Virtual Reality-Based Applications

Our previous research showed that virtual reality treatments can help stroke patients with motor and cognitive issues (Altunok et al., 2024). Another study focused on chronic stroke patients. It tested virtual reality motor training for its effects on inflammation, oxidative stress, neuroplasticity, and motor function. The results were positive, showing improvements in serum biomarkers and motor skills. This highlighted the need for more virtual reality rehabilitation (Huang et al., 2022). In a different study, chronic ischemic stroke patients used virtual reality alongside traditional cognitive rehab. Researchers measured neuroplasticity changes using EEG. They found significant alpha and beta wave power increases in the virtual reality group. However, theta wave power showed no change. These findings suggest that increases in alpha and beta wave power are associated with neuroplastic changes, marking virtual reality as an effective treatment (Arya et al., 2021). These studies suggest that virtual reality applications improve motor and cognitive functions in stroke patients. Thus, they may be preferred over traditional treatments, as they enhance neuroplasticity. This is evident from serum biomarkers and EEG data.

4.4. Exercise

After a stroke, boosting neuroplasticity is crucial. This helps the brain form new connections and regain function. Exercise is key to this. It aids motor recovery through neuroplastic changes. Hill et al. (2023) studied moderate aerobic exercise in stroke patients. They split subjects into two groups: one exercised for 20 minutes, and the other stayed inactive. They measured brain activity using transcranial magnetic stimulation. Results showed that cycling increased neuroplasticity compared to inactivity. Another study looked at low-intensity cycling versus rest. No significant change in brain activity or neuroplasticity was found. This suggests that short-term exercise or brain stimulation might be ineffective (Murdoch et al., 2016). Yet, the long-term effects of intensive aerobic exercise need more study. Comparing intensive training with standard treatment showed different results. The training group showed more brain activity and better motor function (Hubbard et al., 2015). This indicates the need for intensive, repetitive programs for stroke patients. However, traditional physiotherapy and combined treatments showed different outcomes. Both improved balance and function, but neither significantly boosted neuroplasticity (Rahayu et al., 2020). In another study, different exercise methods were tested on stroke patients with cognitive issues. Results showed that each method activated different brain areas (Yeh et al., 2023). This highlights the need for diverse exercise approaches in stroke patients. A study on chronic hemiparetic stroke patients compared treadmill exercises with stretching. The treadmill group showed a 51% increase in walking speed and an 18% boost in fitness. Brain activation also increased significantly (Luft et al., 2008). This supports the link between increased activation and improved walking function.

Studies highlight neuroplasticity's key role in stroke recovery. Different exercises boost neuroplasticity, enhancing brain activity and recovery. Intensive exercises, like cycling and specific tasks, significantly improve brain activation. However, the short-term effects of aerobic exercise on neuroplasticity are debated. Long-term, regular aerobic exercise is believed to be beneficial. In the recovery of arm movement and walking, focused and rhythmic exercises enhance brain activity and function. These findings suggest stroke rehab should be personalized and varied. Incorporating different exercises could activate more brain areas. Thus, testing various methods to boost neuroplasticity in trials is essential.

4.5. Music Therapy

Language disorders can occur after a stroke. In one study, stroke patients listened to vocal music, instrumental music, or audiobooks for three months. Researchers then evaluated their language and verbal memory. They found that both music groups showed better connections in the language network and improved verbal memory. The authors noted this method is simple, low-cost, and supports language disorder treatment (Sihvonen et al., 2021).

Another study focused on stroke patients with chronic aphasia. One group received spoken language therapy plus music therapy, while the other only received spoken therapy. The music therapy group showed better spontaneous speech and health scores, suggesting that music therapy enhances the effects of speech therapy (Raglio et al., 2016).

In a different study, stroke patients were trained with instruments for 20 sessions to improve motor skills. Researchers measured motor functions, cognitive status, and brain changes. They found significant improvements in motor functions, reduced brain activity in certain areas, and better cognitive and mood scores (Ripollés et al., 2016).

Another study involved chronic stroke patients with hand paralysis. They used a MIDI piano and an electronic drum pad for 20 weeks. The aim was to improve fine and gross motor skills. The results showed that music therapy significantly improved hand motor functions and brain neuroplasticity (Amengual et al., 2013).

In a study comparing music therapy with traditional treatments, both groups improved. However, there was no significant difference between them (Grau-Sánchez et al., 2018). Yet, a recent study showed different results. Patients received either music therapy or standard home rehabilitation. Those in the music therapy group had better upper limb functions and quality of life (Segura et al., 2024).

Overall, music therapy boosts neuroplasticity and improves language and motor functions after a stroke. Listening to music enhances language and verbal memory. Music therapy combined with speech therapy improves speaking skills in aphasia patients. Instrument training also aids motor skills and induces brain changes. While some studies found no difference between music therapy and traditional methods, those with remote follow-ups showed music therapy to be superior. Thus, music therapy and instrument training seem promising for stroke rehabilitation.

4.6. Brain–Computer Interface

Brain–computer interfaces are now being used for stroke rehabilitation. They connect the brain to devices, but their use for leg recovery is limited in research. A study compared a brain-controlled robot with traditional therapy in stroke patients. Both groups showed improved thinking and leg movement. However, the group using the robot did better in leg movement. This group also had higher levels of brain-supporting factors. Thus, brain-computer interfaces seem promising for stroke recovery, enhancing movement and brain adaptability (Zhao et al., 2022). Research shows brain-computer interfaces are often used for the upper limbs. After a stroke, these interfaces can boost recovery. They enhance visual–spatial skills and promote brain adaptability (Mrachacz-Kersting et al., 2016).

In a study on stroke patients, researchers compared a brain–computer interface (BCI) combined with a Haptic Knob robot to standard arm therapy. All groups improved motor functions, but the BCI and robot group did best. The Haptic Knob group, however, matched the standard therapy group. This showed that the BCI and robot treatment aid motor recovery after a stroke (Ang et al., 2014). In another study, BCI rehab outperformed traditional methods for upper limb function in ischemic stroke patients (Wang et al., 2024). Similarly, adding BCI training to standard rehab improved both upper limb function and attention more than traditional rehab alone (Liu et al., 2023). Another study found that BCI-based training was better than task-focused training for hand functions (Fu et al., 2023). Also, combining action observation training with BCI and electrical stimulation significantly boosted motor functions and brain activity compared to traditional methods (Lee et al., 2022).

These studies suggest BCIs are promising for stroke rehab. They improve motor functions and boost brain adaptability. Adding BCIs to traditional treatments enhances cognitive and motor skills. BCIs raise neurotrophic factors, aiding brain plasticity. They also improve visual–spatial skills and upper limb function. Thus, BCI technology is effective for stroke patients.

5. Conclusion

Approaches that enhance neuroplasticity in stroke rehabilitation are crucial for patients' functional recovery. The various interventions outlined in this section demonstrate how neuroplasticity aids in restoring motor and cognitive functions following a stroke. Techniques such as electrical stimulation, neurorobotic devices, virtual reality applications, exercise programs, and music therapy all play a role in reorganizing neural connections and improving functionality. Furthermore, emerging technologies like brain–computer interfaces boost recovery of motor functions by fostering neuroplasticity in stroke rehabilitation. Notably, intensive and repetitive treatment programs that promote neuroplastic changes and increase brain activity enhance the effectiveness of rehabilitation efforts. In summary, these findings indicate that personalized, multidisciplinary, and neuroplasticity-focused rehabilitation methods should be prioritized for stroke patients.

6. Takeaways

1. Post-stroke motor disorders negatively affect individuals' daily activities and reduce functional capacity. Cognitive disorders negatively affect the quality of life. Depression and anxiety disorders are among the psychological problems frequently seen in the post-stroke recovery process.
2. Treatment methods that support neuroplasticity can help patients regain their independence by increasing functional recovery in post-stroke rehabilitation.
3. Hebbian learning plays an important role in regaining motor skills by supporting neuroplasticity by strengthening synaptic connections between neurons.
4. Functional electrical stimulation can effectively improve motor functions, and its early application can contribute positively to rehabilitation.
5. Neurorobotic devices promote neuroplasticity and improve lower and upper extremity motor functions.
6. Virtual reality-based rehabilitation methods help improve motor and cognitive functions and increase neuroplasticity.
7. Exercise promotes recovery of motor functions by increasing neuroplasticity, and intensive and repetitive exercise programs support neuroplastic changes.
8. Music therapy supports neuroplasticity by contributing to the recovery of language and motor functions after a stroke.

9. The brain–computer interface contributes to the recovery of motor functions by increasing neuroplasticity and improving cognitive functions.
10. Brain–computer interface technology allows for further recovery of motor and cognitive functions when used in addition to traditional rehabilitation.

Abbreviations

EEG: Electroencephalography
MET: Muscle Energy Techniques
PNF: Proprioceptive Neuromuscular Facilitation
ROM: Range of Motion

Bibliography

1. Altunok, M. K., Kurt, A. K., Aydın, S., Göçen, H. B., Albayrak, H. E., Kethüdaoğlu, M. O., & Özmen, S. (2024). The effect of virtual reality based rehabilitation on cognitive functions in stroke patients: A review. Journal of Experimental and Clinical Medicine, 41(2), 390–396. https://doi.org/10.52142/omujecm.41.2.28
2. Amengual, J. L., Rojo, N., Veciana De Las Heras, M., Marco-Pallarés, J., Grau-Sánchez, J., Schneider, S., … & Rodriguez-Fornells, A. (2013). Sensorimotor plasticity after music-supported therapy in chronic stroke patients revealed by transcranial magnetic stimulation. PLoS One, 8(4), e61883. https://doi.org/10.1371/journal.pone.0061883
3. Ang, K. K., Guan, C., Phua, K. S., Wang, C., Zhou, L., Tang, K. Y., … & Chua, K. S. G. (2014). Brain-computer interface-based robotic end effector system for wrist and hand rehabilitation: results of a three-armed randomized controlled trial for chronic stroke. Frontiers in neuro-engineering, 7, 30. https://doi.org/10.3389/fneng.2014.00030
4. Arya, K. N., Pandian, S., Agarwal, G. G., Chaudhary, N., & Joshi, A. K. (2021). Effect of NEuroplasticity-Principles-based SEnsory-Rehabilitation (NEPSER) on Sensori-motor recovery in stroke: study protocol for a randomized controlled trial. Neurological research and practice, 3, 1–9. https://doi.org/10.1186/s42466-021-00108-1
5. Calabrò, R. S., Naro, A., Russo, M., Bramanti, P., Carioti, L., Balletta, T., … & Bramanti, A. (2018). Shaping neuroplasticity by using

powered exoskeletons in patients with stroke: a randomized clinical trial. Journal of neuroengineering and rehabilitation, 15, 1–16. https://doi.org/10.1186/s12984-018-0377-8

6. Cramer, S. C., Sur, M., Dobkin, B. H., O'Brien, C., Sanger, T. D., Trojanowski, J. Q., … & Vinogradov, S. (2011). Harnessing neuroplasticity for clinical applications. Brain, 134(6), 1591–1609. https://doi.org/10.1093/brain/awr039
7. Dimyan, M. A., & Cohen, L. G. (2011). Neuroplasticity in the context of motor rehabilitation after stroke. Nature Reviews Neurology, 7(2), 76–85. https://doi.org/10.1038/nrneurol.2010.200
8. Douiri, A., Rudd, A. G., & Wolfe, C. D. (2013). Prevalence of poststroke cognitive impairment: South London stroke register 1995–2010. Stroke, 44(1), 138–145. https://doi.org/10.1161/STROKEAHA.112.67084
9. Eraifej, J., Clark, W., France, B., Desando, S., & Moore, D. (2017). Effectiveness of upper limb functional electrical stimulation after stroke for the improvement of activities of daily living and motor function: a systematic review and meta-analysis. Systematic reviews, 6, 1–21. https://doi.org/10.1186/s13643-017-0435-5
10. Feigin, V. L., Brainin, M., Norrving, B., Martins, S., Sacco, R. L., Hacke, W., … & Lindsay, P. (2022). World Stroke Organization (WSO): Global stroke fact sheet 2022. International Journal of Stroke, 17(1), 18–29. https://doi.org/10.1177/17474930211065917
11. Fu, J., Chen, S., Shu, X., Lin, Y., Jiang, Z., Wei, D., … & Jia, J. (2023). Functional-oriented, portable brain–computer interface training for hand motor recovery after stroke: a randomized controlled study. Frontiers in neuroscience, 17, 1146146. https://doi.org/10.3389/fnins.2023.1146146
12. Fuchs, E., & Flügge, G. (2014). Adult neuroplasticity: more than 40 years of research. Neural plasticity, 2014(1), 541870. https://doi.org/10.1155/2014/541870
13. Grau-Sánchez, J., Duarte, E., Ramos-Escobar, N., Sierpowska, J., Rueda, N., Redón, S., … & Rodríguez-Fornells, A. (2018). Music-supported therapy in the rehabilitation of subacute stroke patients: a randomized controlled trial. Annals of the New York Academy of Sciences, 1423(1), 318–328. https://doi.org/10.1111/nyas.13590
14. Hackett, M. L., & Pickles, K. (2014). Part I: Frequency of depression after stroke: An updated systematic review and meta-analysis of

observational studies. International Journal of Stroke, 9(8), 1017–1025. https://doi.org/10.1111/ijs.12357
15. Herpich, F., & Rincon, F. (2020). Management of acute ischemic stroke. Critical care medicine, 48(11), 1654–1663.
16. Hill, G., Johnson, F., Uy, J., Serrada, I., Benyamin, B., Van Den Berg, M., & Hordacre, B. (2023). Moderate intensity aerobic exercise may enhance neuroplasticity of the contralesional hemisphere after stroke: a randomised controlled study. Scientific reports, 13(1), 14440. https://doi.org/10.1038/s41598-023-40902-2
17. Howlett, O. A., Lannin, N. A., Ada, L., & McKinstry, C. (2015). Functional electrical stimulation improves activity after stroke: a systematic review with meta-analysis. Archives of physical medicine and rehabilitation, 96(5), 934–943. https://doi.org/10.1016/j.apmr.2015.01.013
18. Huang, C. Y., Chiang, W. C., Yeh, Y. C., Fan, S. C., Yang, W. H., Kuo, H. C., & Li, P. C. (2022). Effects of virtual reality-based motor control training on inflammation, oxidative stress, neuroplasticity and upper limb motor function in patients with chronic stroke: a randomized controlled trial. BMC neurology, 22(1), 21. https://doi.org/10.1186/s12883-021-02547-4
19. Hubbard, I. J., Carey, L. M., Budd, T. W., Levi, C., McElduff, P., Hudson, S., … & Parsons, M. W. (2015). A randomized controlled trial of the effect of early upper-limb training on stroke recovery and brain activation. Neurorehabilitation and neural repair, 29(8), 703–713. https://doi.org/10.1177/1545968314562647
20. Innocenti, G. M. (2022). Defining neuroplasticity. Handbook of Clinical Neurology, 184, 3–18. https://doi.org/10.1016/B978-0-12-819410-2.00001-1
21. Kalaria, R. N., & Ballard, C. (2001). Stroke and cognition. Current atherosclerosis reports, 3(4), 334–339. https://doi.org/10.1007/s11883-001-0028-5
22. Kim, H., Park, G., Shin, J. H., & You, J. H. (2020). Neuroplastic effects of end-effector robotic gait training for hemiparetic stroke: A randomised controlled trial. Scientific reports, 10(1), 12461. https://doi.org/10.1038/s41598-020-69367-3
23. Kimberley, T. J., Lewis, S. M., Auerbach, E. J., Dorsey, L. L., Lojovich, J. M., & Carey, J. R. (2004). Electrical stimulation driving functional improvements and cortical changes in subjects with stroke.

Experimental Brain Research, 154, 450–460. https://doi.org/10.1007/s00221-003-1695-y
24. Kossut, M. (2019). Basic mechanism of neuroplasticity. Neuropsychiatria i Neuropsychologia/Neuropsychiatry and Neuropsychology, 14(1), 1–8. https://doi.org/10.5114/nan.2019.87727
25. Kuriakose, D., & Xiao, Z. (2020). Pathophysiology and treatment of stroke: present status and future perspectives. International journal of molecular sciences, 21(20), 7609. https://doi.org/10.3390/ijms21207609
26. Lee, S. H., Kim, S. S., & Lee, B. H. (2022). Action observation training and brain-computer interface controlled functional electrical stimulation enhance upper extremity performance and cortical activation in patients with stroke: a randomized controlled trial. Physiotherapy theory and practice, 38(9), 1126–1134. https://doi.org/10.1080/09593985.2020.1831114
27. Leśniak, M., Bak, T., Czepiel, W., Seniów, J., & Członkowska, A. (2008). Frequency and prognostic value of cognitive disorders in stroke patients. Dementia and geriatric cognitive disorders, 26(4), 356–363. https://doi.org/10.1159/000162262
28. Liu, X., Zhang, W., Li, W., Zhang, S., Lv, P., & Yin, Y. (2023). Effects of motor imagery based brain-computer interface on upper limb function and attention in stroke patients with hemiplegia: a randomized controlled trial. BMC neurology, 23(1), 136. https://doi.org/10.1186/s12883-023-03150-5
29. Luft, A. R., Macko, R. F., Forrester, L. W., Villagra, F., Ivey, F., Sorkin, J. D., … & Hanley, D. F. (2008). Treadmill exercise activates subcortical neural networks and improves walking after stroke: a randomized controlled trial. Stroke, 39(12), 3341–3350. https://doi.org/10.1161/STROKEAHA.108.527531
30. Martino, R., Foley, N., Bhogal, S., Diamant, N., Speechley, M., & Teasell, R. (2005). Dysphagia after stroke: incidence, diagnosis, and pulmonary complications. stroke, 36(12), 2756–2763. https://doi.org/10.1161/01.STR.0000190056.76543.eb
31. Mrachacz-Kersting, N., Jiang, N., Stevenson, A. J. T., Niazi, I. K., Kostic, V., Pavlovic, A., … & Farina, D. (2016). Efficient neuroplasticity induction in chronic stroke patients by an associative brain-computer interface. Journal of Neurophysiology, 115(3), 1410–1421. https://doi.org/10.1152/jn.00918.2015

32. Munakata, Y., & Pfaffly, J. (2004). Hebbian learning and development. Developmental science, 7(2), 141–148. https://doi.org/10.1111/j.1467-7687.2004.00331.x
33. Murdoch, K., Buckley, J. D., & McDonnell, M. N. (2016). The effect of aerobic exercise on neuroplasticity within the motor cortex following stroke. PloS one, 11(3), e0152377. https://doi.org/10.1371/journal.pone.0152377
34. Murphy, S. J., & Werring, D. J. (2020). Stroke: causes and clinical features. Medicine, 48(9), 561–566. https://doi.org/10.1016/j.mpmed.2020.06.002
35. Musuka, T. D., Wilton, S. B., Traboulsi, M., & Hill, M. D. (2015). Diagnosis and management of acute ischemic stroke: speed is critical. Cmaj, 187(12), 887–893. https://doi.org/10.1503/cmaj.140355
36. Nys, G. M. S., Van Zandvoort, M. J. E., De Kort, P. L. M., Jansen, B. P. W., De Haan, E. H. F., & Kappelle, L. J. (2007). Cognitive disorders in acute stroke: prevalence and clinical determinants. Cerebrovascular Diseases, 23(5–6), 408–416. https://doi.org/10.1159/000101464
37. Owolabi, M. O., Thrift, A. G., Mahal, A., Ishida, M., Martins, S., Johnson, W. D., … & Shobhana, A. (2022). Primary stroke prevention worldwide: translating evidence into action. The Lancet Public Health, 7(1), e74-e85. https://doi.org/ 10.1016/S2468-2667(21)00230-9
38. Popović, D. B., Sinkjaer, T., & Popović, M. B. (2009). Electrical stimulation as a means for achieving recovery of function in stroke patients. NeuroRehabilitation, 25(1), 45–58. https://doi.org/10.3233/NRE-2009-0498
39. Potter, T. B., Tannous, J., & Vahidy, F. S. (2022). A contemporary review of epidemiology, risk factors, etiology, and outcomes of premature stroke. Current Atherosclerosis Reports, 24(12), 939–948. https://doi.org/10.1007/s11883-022-01067-x
40. Raglio, A., Oasi, O., Gianotti, M., Rossi, A., Goulene, K., & Stramba-Badiale, M. (2016). Improvement of spontaneous language in stroke patients with chronic aphasia treated with music therapy: a randomized controlled trial. International Journal of Neuroscience, 126(3), 235–242. https://doi.org/10.3109/00207454.2015.1010647
41. Rahayu, U. B., Wibowo, S., Setyopranoto, I., & Hibatullah Romli, M. (2020). Effectiveness of physiotherapy interventions in brain plasticity, balance and functional ability in stroke survivors: A randomized controlled trial. NeuroRehabilitation, 47(4), 463–470. https://doi.org/10.3233/NRE-203210

42. Ripollés, P., Rojo, N., Grau-Sánchez, J., Amengual, J. L., Càmara, E., Marco-Pallarés, J., … & Rodríguez-Fornells, A. (2016). Music supported therapy promotes motor plasticity in individuals with chronic stroke. Brain imaging and behavior, 10, 1289–1307. https://doi.org/10.1007/s11682-015-9498-x
43. Robertson, I. H., & Murre, J. M. (1999). Rehabilitation of brain damage: brain plasticity and principles of guided recovery. Psychological bulletin, 125(5), 544. https://doi.org/10.1037/0033-2909.125.5.544
44. Sacco, R. L., Kasner, S. E., Broderick, J. P., Caplan, L. R., Connors, J. J., Culebras, A., … & Vinters, H. V. (2013). An updated definition of stroke for the 21st century: a statement for healthcare professionals from the American Heart Association/American Stroke Association. Stroke, 44(7), 2064–2089. https://doi.org/10.1161/STR.0b013e318296aeca
45. Segura, E., Grau-Sánchez, J., Cerda-Company, X., Porto, M. F., De la Cruz-Puebla, M., Sanchez-Pinsach, D., … & Rodríguez-Fornells, A. (2024). Enriched music-supported therapy for individuals with chronic stroke: a randomized controlled trial. Journal of Neurology, 1–12. https://doi.org/10.1007/s00415-024-12570-3
46. Shahid, J., Kashif, A., & Shahid, M. K. (2023). A comprehensive review of physical therapy interventions for stroke rehabilitation: impairment-based approaches and functional goals. Brain Sciences, 13(5), 717. https://doi.org/10.3390/brainsci13050717
47. Sihvonen, A. J., Pitkäniemi, A., Leo, V., Soinila, S., & Särkämö, T. (2021). Resting-state language network neuroplasticity in post-stroke music listening: A randomized controlled trial. European Journal of Neuroscience, 54(11), 7886–7898. https://doi.org/10.1111/ejn.15524
48. Singh, N., Saini, M., Kumar, N., Srivastava, M. P., & Mehndiratta, A. (2021). Evidence of neuroplasticity with robotic hand exoskeleton for post-stroke rehabilitation: a randomized controlled trial. Journal of neuroengineering and rehabilitation, 18(1), 76. https://doi.org/10.1186/s12984-021-00867-7
49. Vafadar, A. K., Côté, J. N., & Archambault, P. S. (2015). Effectiveness of functional electrical stimulation in improving clinical outcomes in the upper arm following stroke: a systematic review and meta-analysis. BioMed research international, 2015(1), 729768. https://doi.org/10.1155/2015/729768

50. Wang, A., Tian, X., Jiang, D., Yang, C., Xu, Q., Zhang, Y., … & Meng, X. (2024). Rehabilitation with brain-computer interface and upper limb motor function in ischemic stroke: A randomized controlled trial. Med, 5(6), 559–569. https://doi.org/10.1016/j.medj.2024.02.014
51. Yeh, T. T., Chang, K. C., Wang, J. J., Lin, W. C., & Wu, C. Y. (2023). Neuroplastic Changes Associated With Hybrid Exercise-Cognitive Training in Stroke Survivors With Mild Cognitive Decline: A Randomized Controlled Trial. Neurorehabilitation and Neural Repair, 37(9), 662–673. https://doi.org/10.1177/15459683231200220
52. Zhao, C. G., Ju, F., Sun, W., Jiang, S., Xi, X., Wang, H., … & Yuan, H. (2022). Effects of training with a brain–computer interface-controlled robot on rehabilitation outcome in patients with subacute stroke: A randomized controlled trial. Neurology and Therapy, 11(2), 679–695. https://doi.org/10.1007/s40120-022-00333-z

The Role of Neuroplasticity in Pain Management

Hande Besna Göçen

1. Introduction

Although pain has been defined in various ways, the most widely recognized and contemporary definition is offered by the International Association for the Study of Pain (IASP), which characterizes it as "an unpleasant sensory and emotional experience associated with actual or potential tissue damage" (Raja et al., 2020).

Neuroplasticity denotes the brain's capacity to undergo both structural and functional reorganization. Traditionally, it was believed that such plastic changes occurred exclusively during specific critical periods of neural development. However, emerging evidence suggests that neuroplasticity persists throughout the lifespan, representing a fundamental and ongoing property of the human brain. This continuous adaptability is integral to the acquisition of new motor skills. In everyday life, individuals incrementally cultivate a repertoire of motor competencies through repetitive practice and environmental interaction—skills such as walking, driving, cycling, and eating serve as prominent examples (Stanisic, 2022). This chapter explores the intricate interplay between neuroplasticity and pain.

2. Definition and History of Pain

Pain is frequently precipitated by physical tissue damage; however, its perception is shaped by a complex interplay of physiological attributes, emotional states, and prior experiences. As a fundamentally subjective phenomenon, pain is modulated by a range of individual variables—including age, gender, and disability status—as well as contextual factors such as sociocultural background. Thus, pain may be conceptualized as an abstract and multifaceted experience that individuals encounter at various stages of life, one that disrupts engagement in daily activities and significantly diminishes overall quality of life (Yılmaz & Atay, 2014).

The recognition that individuals vary significantly in numerous dimensions is not a recent development. Over two millennia ago, Plato remarked, "No two people are born exactly alike, but each differs from the other in his natural abilities." These individual differences are particularly salient in the context of pain and have captured the attention of pain researchers for decades. More than 70 years ago, in justifying the implementation of psychophysical studies to examine pain sensitivity among healthy adults, it was noted that "A common observation is the striking difference in the intensity of pain experienced in diseases with apparently similar lesions." Historically, such inter-individual variability in pain perception was often regarded as a confounding factor rather than a legitimate focus of scientific investigation. However, the advent of the genomic era and the emergence of precision medicine have reinvigorated scholarly interest in understanding and leveraging these individual differences (Fillingim, 2017).

3. Classification of Pain

Pain is classified in different ways and from different sources.

3.1. Axis System Classification

One of the earliest and most established approaches to categorizing pain is the axis-based classification system proposed by the International Association for the Study of Pain (IASP). This framework delineates five distinct axes for the systematic assessment of pain. The first axis identifies the anatomical region in which the pain is localized; the second specifies the physiological systems affected; the third pertains to the temporal onset of the pain; the fourth addresses the intensity and duration of the pain experience; and the fifth axis focuses on the underlying etiology (Harutoğlu et al., 2016).

3.2. Multiaxial Pain Classification

This classification is grounded in the empirical integration of psychosocial and behavioral data obtained from individuals with chronic pain. Utilizing the Multidimensional Pain Inventory as its primary assessment tool, the system has demonstrated both validity and reliability. Within this framework, it is posited that patients exhibit consistent patterns of psychosocial and behavioral responses to chronic pain. These characteristic responses facilitate the categorization of patients into relatively homogeneous subgroups, independent of their underlying medical diagnoses (Harutoğlu et al., 2016).

3.3. Medical Pain Classification

It is divided into four categories of pain according to neurophysiological mechanism, duration, etiology, and location (Harutoglu et al., 2016).

According to the neurophysiological mechanism, pain is divided into two.

1. **Nociceptive pain:** Nociceptive pain emerges from nerve activity triggered by harmful or potentially damaging stimuli and represents the most prevalent form of chronic pain, commonly observed in conditions such as arthritis and back pain (Cohen et al., 2021). This type of pain is generally classified into two categories: somatic and visceral.

   Somatic pain is typically described as sharp, sudden, and well-localized. It originates in body regions innervated by specific peripheral nerves and tends to intensify with movement while subsiding with rest. Its underlying causes often include injuries to the skin, muscles, ligaments, bones, nerves, or blood vessels, but it does not involve internal organs. Pharmacologically, somatic pain tends to respond well to nonsteroidal anti-inflammatory drugs (NSAIDs) and opioids.

   In contrast, visceral pain is more diffuse and often associated with internal organs. Interestingly, it is not always a direct reflection of tissue injury—procedures such as cutting the intestines may not produce pain. Instead, visceral pain is typically linked to processes such as organ distention, ischemia, muscular contractions, or the release of certain chemical mediators. It may also be accompanied by autonomic symptoms such as nausea, vomiting, or reflex muscle contractions (Harutoğlu et al., 2016).
2. **Neuropathic pain:** Nociceptive pain is typically described as dull, aching, or throbbing in nature. In contrast, neuropathic pain is often characterized by sharp, stabbing, or shooting sensations. Accounting for approximately 15–25% of chronic pain cases, neuropathic pain encompasses conditions such as diabetic neuropathy, postherpetic neuralgia, and radiculopathy. Unlike most nociceptive pain or acute nerve injuries—which may subside or adapt over time—chronic neuropathic pain is inherently maladaptive and often persists without resolution.

Interestingly, the correlation between pain intensity and functional disability tends to be weak in neuropathic cases. Nevertheless, this form of pain has a disproportionately negative impact on quality of life, often more profound than nociceptive pain of equivalent severity (Cohen et al., 2021).

3. **Psychogenic pain:** Psychogenic pain refers to a type of pain that originates primarily from psychological factors rather than nociceptive or neuropathic mechanisms. It is typically experienced at an intensity disproportionate to any identifiable physical source. This form of pain is often chronic in nature and necessitates psychological or psychiatric intervention as part of its management. Importantly, psychogenic pain can affect individuals across the lifespan, regardless of age (Harutoğlu et al., 2016).

Duration-related pain is divided into two:

1. **Acute pain** starts suddenly and indicates tissue damage. Its severity matches the damage's extent and location. As healing occurs, pain reduces. It lasts under 3–6 months. Untreated, it may become chronic.
2. **Chronic pain** lasts over 3–6 months. It can be nociceptive, neuropathic, or a mix. This type of pain harms quality of life. It leads to abnormal behaviors and is influenced by psychological factors (Erdine, 2007).

Cancer pain is complex and often debated. It varies based on clinical and pathological factors, as well as treatment response. Although its classification is ancient, modern definitions need to consider pain characteristics and underlying issues (Harutoglu et al., 2016).

Headache is a common regional pain. It's one of the most frequent issues in children and teens visiting clinics and emergency rooms. It often appears in childhood and is more common in adolescence (Ceylan et al., 2014).

4. Physiology of Pain

Pain arises when nociceptors detect noxious stimuli and transmit signals to the brain through specific nerve fibers. These signals are primarily conveyed by myelinated A-delta fibers and unmyelinated C-fibers. A-delta fibers are typically activated by mechanical or thermal stimuli and are responsible for the rapid transmission of sharp, well-localized pain sensations. In contrast, C-fibers respond to more intense stimuli—such as extreme heat or chemical irritants—and mediate slower, diffuse, and longer-lasting pain, often perceived as burning in nature. The process of nociception refers to the mechanism by which nociceptors relay potentially harmful stimuli to the central nervous system (CNS), thereby initiating protective physiological responses (Di Maio et al., 2023).

5. Pathophysiological Pain Processes

Pain management strategies vary depending on the underlying type and mechanism of pain, making a clear understanding of pain pathways essential for effective treatment. Broadly, three primary processes contribute to pain perception. The first is peripheral sensitization, in which nociceptors become hypersensitive, leading to the misinterpretation of normally innocuous stimuli as painful. The second involves hyperexcitability of certain sensory neurons, often occurring in proximity to nerve injury—a phenomenon frequently observed in peripheral neuropathies and one that presents considerable therapeutic challenges. The third process, central sensitization, takes place within the spinal cord and brain, and is mediated in part by the N-methyl-D-aspartate (NMDA) receptor. Pharmacological inhibition of this receptor has shown promise in alleviating centrally mediated pain (Harutoğlu et al., 2016).

6. Mechanisms of Pain

6.1. Peripheral Mechanisms

6.1.1. Primary Afferent Nociceptors

Peripheral nerve endings are capable of detecting a range of noxious stimuli, including thermal, mechanical, and chemical inputs. Nociceptors—specialized sensory receptors for pain—transmit signals to the spinal cord in response to these stimuli. Two primary types of nociceptors are recognized: A-delta mechanothermal nociceptors, which respond rapidly to sharp and localized pain, and C polymodal nociceptors, which respond more slowly and are activated by multiple stimulus modalities. Under normal conditions, high-threshold nociceptors are selectively responsive to intense or damaging stimuli, such as extreme heat, strong pressure, or chemical irritants. However, the nature of pain changes in the context of sustained injury or tissue damage. Inflammatory processes triggered by such damage sensitize nociceptors, leading to an amplified response to stimuli (Van Neerven & Mouraux, 2020).

The painful stimulus follows a path to the higher centers in four stages:

6.1.1.1. Transduction

It is the transformation of one energy into another energy. For a hot stimulus to become painful, it must rise above a specific temperature. Nociceptors remain insensitive to an average temperature but become sensitive to an increased temperature.

6.1.1.2. Transmission

The transmission of pain perceived by nociceptors to higher centers. Myelinated A-delta and unmyelinated C-fibers play an active role in transmission. Myelinated A-delta fibers are fast-conducting, open to sensitization, and stimulated by thermal and mechanical stimuli. On the other hand, Unmyelinated C-fibers are polymodal, sensitive to all kinds of stimuli, and slow-conducting fibers. Other nerve fibers also participate in the transmission of painful stimuli in various ways.

The three types of neurons responsible for pain transmission:

1. Primary afferent neurons: Synapse with the second neuron in the second dorsal horn in the dorsal root ganglion within the vertebral foramen at each spinal cord level.
2. Second neuron axons reach the thalamus through the contralateral spinothalamic tract.
3. Neurons in the second row synapse with those in the third row in the thalamus. These then pass through the corona radiata and internal capsule to reach the postcentral gyrus of the cerebral cortex.

6.1.1.3. Modulation

This is an event that occurs at the level of the spinal cord. The painful stimulus changes at this level and is transmitted to higher centers.

6.1.1.4. Perception

It is pain perception that transmits the stimulus passing through the spinal cord to the higher centers through various output pathways (Di Maio et al., 2023).

6.1.2. Peripheral Sensitization

Tissue damage resulting from a noxious stimulus initiates a cascade of structural and biochemical changes. In the aftermath of such injury, vasodilation leads to swelling and heightened sensitivity in the affected area, prompting the migration of immune cells to the site. This marks the onset of neurogenic inflammation, during which a variety of pro-inflammatory mediators are released—including neurokinin A, substance P, calcitonin gene-related peptide (CGRP), nitric oxide (NO), serotonin, histamine, bradykinin, potassium ions, arachidonic acid, leukotrienes, and prostaglandins. These substances activate high-threshold nociceptors and contribute to peripheral sensitization, a process in which even normally innocuous stimuli evoke pain (hyperalgesia). As a result, sensitivity to heat increases and previously non-painful stimuli begin to provoke discomfort. Pharmacological agents such as aspirin, paracetamol, and other nonsteroidal anti-inflammatory drugs (NSAIDs) can interrupt this inflammatory cascade and mitigate the associated pain response (Marek-Jozefowicz et al., 2023).

6.1.3. Peripheral Effects of Opioids

Opioids usually have central effects. However, endogenous opioids can be used locally by determining their peripheral effects after tissue destruction (Phillips et al., 2017).

6.1.4. Peripheral Nerve Damage

Alterations in biochemical, physiological, and structural processes following peripheral nerve injury can independently give rise to pain. One key mechanism involves increased sensitivity and excitability of nociceptors located in the dorsal root ganglion. Additionally, compromised blood flow to myelinated fibers may lead to demyelination, which, in turn, facilitates the generation of aberrant neural signals. These changes can manifest clinically as sharp, burning, or stabbing pain sensations (Costigan et al., 2009).

6.1.5. Sympathetic Nervous System

Following nerve injury, stimulation of the sympathetic nervous system or administration of noradrenaline can activate alpha-adrenergic receptors on primary afferent fibers, leading to increased excitability. Additionally, sympathetic terminals located within the dorsal root ganglia may contribute to aberrant sympathetic activity, further amplifying nociceptive signaling (Donello et al., 2011).

6.1.6. Silent Receptors

These primary afferent neurons, which lack myelin, typically do not react to extreme mechanical or noxious stimuli under normal conditions. However, when inflammation or chemical sensitization occurs, these neurons become more responsive and can transmit painful sensations (Donello et al., 2011).

6.2. Central Mechanisms

The dorsal horn of the spinal cord serves as the primary termination site for afferent nociceptive fibers. These fibers typically terminate within laminae I, II, and V, where they synapse with second-order neurons. Two main types of second-order neurons are present in the dorsal horn: nociceptive-specific (or high-threshold) neurons and wide dynamic range (WDR) neurons. Nociceptive-specific neurons are primarily located in the superficial laminae and respond exclusively to noxious stimuli. In contrast, WDR neurons reside in deeper laminae and are capable of responding to both noxious and non-noxious inputs. Notably, WDR neurons can become hypersensitive under certain conditions. When the cumulative activity of these neurons surpasses a critical threshold, even non-painful tactile stimuli may be perceived as painful—a phenomenon known as allodynia (Yam et al., 2018).

6.2.1. Neurotransmitters

A variety of neurotransmitters and neuromodulators play critical roles in the processing of pain within the dorsal horn of the spinal cord. Notably, excitatory amino acids such as glutamate and aspartate are key contributors, acting primarily through NMDA and non-NMDA receptor pathways. Additionally, neuropeptides released from primary afferent fibers—such as substance P, neurokinin A, and calcitonin gene-related peptide (CGRP)—are actively involved in nociceptive signaling. Several other receptor systems modulate this process, including opioid receptors (mu and kappa subtypes), gamma-aminobutyric acid (GABA), serotonin (5-HT), and adenosine receptors (Yam et al., 2018).

6.2.2. Central Sensitization

Following trauma, individuals may develop heightened sensitivity to stimuli that are normally non-painful—a phenomenon known as allodynia. Unlike primary hyperalgesia, allodynia typically does not involve altered responses to thermal stimuli. These sensory changes are primarily attributed to neural adaptations occurring in the dorsal horn of the spinal cord, collectively referred to as central sensitization. This concept underscores the capacity of the central nervous system (CNS) to undergo dynamic changes that amplify its responsiveness to pain. Experimental studies have shown that repeated activation of C-fibers by a noxious stimulus can progressively augment neuronal excitability not only within the dorsal horn but also throughout broader areas of the nervous system. A key mechanism underlying this sensitization is the so-called "wind-up" phenomenon, which is mediated by NMDA receptor activation. During wind-up, neurons exhibit increased responsiveness to subsequent inputs, reinforcing central sensitization.

Beyond this, central sensitization is marked by additional neurophysiological alterations in the dorsal horn. One such change involves the expansion of receptive fields, wherein spinal neurons begin to respond not only to stimuli within their original domain but also to adjacent regions. Moreover, both the magnitude and duration of pain responses become exaggerated, and the threshold for pain perception is lowered. These cumulative changes contribute to various acute pain states—such as postoperative pain—as well as the transition to chronic pain conditions (Harutoğlu et al., 2016).

6.2.3. Spinal Level Modulation

Painful stimuli undergo modulation at multiple levels along the neuraxis, including within the dorsal horn of the spinal cord. Upon reaching the dorsal horn, afferent inputs engage various inhibitory mechanisms that act to attenuate the transmission of subsequent nociceptive signals. This modulation is supported by both local inhibitory interneurons and descending inhibitory pathways originating from higher centers. Moreover, the processing of nociceptive input in the dorsal horn is significantly influenced by the activity of several receptor systems—including opioid, alpha-adrenergic, GABAergic, and glycinergic receptors—operating at both pre- and postsynaptic sites. These receptors are modulated by both endogenous compounds and exogenous pharmacological agents (Basbaum et al., 2009).

7. Pathways in Pain Perception

7.1. Spinal Structures

It includes the spinothalamic, spinoreticular, and spinomesencephalic pathways located in the contralateral and anterolateral quadrants of the spinal cord. In addition to these classical pathways, visceral nociceptive pathways are deep in the posterior column.

7.2. Supraspinal Structures

The second row of neurons travels upward in the spinal cord and terminates in various supraspinal structures from the brainstem to the thalamus and cortex.

7.3. Cortical Structures

Positron emission tomography (PET) studies have demonstrated that painful stimuli activate multiple cortical regions, including the premotor, sensorimotor, frontal, parietal, occipital, and anterior cingulate cortices. Beyond neuroanatomical processing, individual pain perception is shaped by a range of psychosocial factors such as cultural background, anxiety levels, suggestibility, and past experiences. These variables influence both the pain threshold and behavioral responses to pain. The parietal cortex is thought to play a central role in the cognitive evaluation of pain, while the frontal cortex is more closely associated with the emotional and affective dimensions of the pain experience (Al-Chalabi et al., 2023).

8. Descending Inhibition in Pain

Melzack and Wall highlighted the brain's capacity to modulate and inhibit pain through various descending control systems. These descending pathways play a crucial role in the regulation of nociceptive input and originate from several distinct regions. Specifically, descending projections extend from the cerebral cortex to structures such as the nucleus gracilis, nucleus cuneatus, reticular formation, and thalamus. Additional key pathways include those originating in the periaqueductal gray matter, which project directly to the dorsal horn of the spinal cord; pathways that descend via the raphe nuclei of the medulla before terminating in the dorsal horn; and those arising from the locus coeruleus, which similarly project to dorsal horn neurons to modulate pain processing (Harutoğlu et al., 2016).

9. Theories of Pain

9.1. Specificity Theory

Pain signals are transmitted through distinct neural pathways that carry information from the periphery to the spinal cord, thalamus, and ultimately to various cortical regions. Within the central nervous system (CNS), these signals are processed in anatomically and functionally specialized areas. Each type of noxious stimulus is detected by specific receptors, which contribute to defining the qualitative nature of the pain experience. For example, a painful stimulus applied to the skin results in a localized pain sensation that is interpreted by the brain based on the type of input received. Historically, an older theory proposed the existence of four distinct sensory receptors—responsive to heat, cold, touch, and pain. However, this theory has since been rendered obsolete in light of more advanced neurophysiological understanding (Yam et al., 2018).

9.2. Pattern Theory

With the exception of those innervating the hair follicles, the terminal endings of all skin fibers are structurally identical, and pain arises from the overstimulation of these fibers. For a pain sensation to manifest, the stimulus must accumulate after being transmitted into the spinal cord. This accumulation of neural activity is what gives rise to the perception of pain. The responses of somatic organs are generated in response to the intensity of stimuli within their dynamic range. Since different sensory organs exhibit varying levels of sensitivity to different intensities of stimuli, each stimulus is processed and evaluated differently within the central nervous system (CNS). However, this particular theory has been discredited in modern pain research (Dubin & Patapoutian, 2010).

9.3. Gate Control Theory

When noxious stimuli are transmitted to the spinal cord via specific nerve fibers, this information is initially processed and evaluated within the spinal cord itself. At this stage, spinal neurons attempt to inhibit the transmission of the painful signal. Interestingly, non-painful stimuli have the capacity to block the transmission of painful stimuli. Thick afferent fibers initially cause an increase in T-cell activity, which is followed by a subsequent reduction. The initial increase occurs due to the direct activation of second-order neurons by primary afferents. The subsequent decrease in activity results from the indirect activation of substantia gelatinosa cells by thick afferents, effectively "closing the gate." On the other hand, thin afferent fibers increase T-cell activity by activating inhibitory interneurons, which reduce the activity of substantia gelatinosa cells and effectively "open the gate." When the balance between the input from thin and thick fibers is disrupted, it reaches a critical threshold, leading to the activation of second-order neurons.

Activation of the ascending pain pathways results in the conscious perception of pain, which is followed by a corresponding behavioral response. The descending control system modulates the gate mechanism that regulates pain sensation at the spinal level, influenced by emotional states and past experiences. When the substantia gelatinosa is activated, the "gate" closes, limiting the sensory input that reaches the T-cells. Conversely, when the substantia gelatinosa is inactive, the gate remains open. The positioning of the gate—whether open or closed—is determined by the dynamic balance between the activity of thick and thin sensory neurons (Yam et al., 2018).

9.4. Pain Memory

The size and intensity of pain memory are influenced by several factors, including the current intensity of pain, the individual's emotional state, their expectations regarding the pain experience, and the intensity of pain experienced in prior instances (Ameringer, 2010).

9.5. Neuromatrix and Pain

A matrix of neurons is shaped by sensory inputs and personal experiences, resulting in the formation of a neuromatrix that is unique to each individual. This neuromatrix is influenced by a combination of genetic factors, sensory inputs, cognitive processes, and stress. Key components of this matrix include tactile and bodily sensations, visual stimuli, neural signals from the brain, and more. Additionally, neural inhibition and the activation of stress-regulation mechanisms contribute to the overall functioning of the neuromatrix (Nirvanie-Persaud & Millis, 2022).

9.6. Unexplained Pain

Pain is processed and evaluated by the brain within the neuromatrix, which does not merely detect stimuli but also generates subjective experiences. This process helps explain phenomena such as phantom pain, where pain is perceived without any corresponding physical injury. Imaging techniques, including PET and functional magnetic resonance imaging (fMRI), provide evidence of the brain's involvement in pain perception. These techniques reveal changes in somatosensory regions, limbic structures, and the frontal and posterior parietal cortices during pain (Erdine, 2007).

10. Assessment of Pain

Pain management assessment involves a comprehensive approach, beginning with a detailed patient history, physical examination, and diagnostic tests. Initially, demographic, social, and occupational information is gathered. Following this, the characteristics of the pain, its impact on daily functioning, and previous treatment interventions are assessed. Psychological factors, such as attentional difficulties and anxiety, are also considered. A physical examination then evaluates neurological function, reflexes, and the musculoskeletal system. This is followed by laboratory tests and imaging procedures, including X-rays and MRIs. Pain intensity and its effect on daily activities are measured using standardized scales and questionnaires. This holistic approach provides a deeper understanding of the pain experience and informs the development of effective treatment strategies (Fenske et al., 2021).

11. Pain Treatment

The IASP working group advocates for pain treatment centers to incorporate a multidisciplinary team of health professionals, in line with the biopsychosocial model of pain. Essential members of this team include at least two general practitioners or psychiatrists, a clinical psychologist, and a physiotherapist, with additional specialists included as needed based on the patient's specific requirements. Regular interdisciplinary meetings are critical to ensuring coordinated care. The team should employ a variety of methods tailored to address individual concerns, including clinical examinations, pharmacological treatments, and psychosocial interventions. Psychosocial therapies may encompass assessments and cognitive behavioral therapy, while physiotherapy could involve manual therapy and structured exercises aimed at recovery. When necessary, patients should be referred to specialized centers for more focused care (Danilov et al., 2020).

11.1. Biopsychosocial Model

The biopsychosocial model emphasizes the complex interplay between brain and body in understanding pain and illness. It provides a comprehensive framework by considering a range of factors, including physiological, psychological, social, and environmental influences. Additionally, it takes into account neurological, immunological, and hormonal variables. Daily life activities, stress levels, communication patterns, family dynamics, and social support systems also play a significant role in shaping pain experiences. Cultural influences, medical history, prior treatment outcomes, and work-related factors further contribute to the overall pain perception. Given the multifaceted nature of pain, personalized assessments are essential for effective management (Kusnanto et al., 2018).

11.2. Cognitive Behavioral Therapy

Cognitive Behavioral Therapy (CBT) is grounded in psychological models and has been shown to effectively treat a variety of psychiatric disorders. In contrast, behavioral therapies focus on utilizing learning principles to address behavioral concerns, emphasizing the modification of harmful behaviors and the development of strategies to manage anxiety (McCarberg & Peppin, 2019).

Research indicates that CBT can lead to changes in brain connectivity in individuals with chronic pain. In a study using functional magnetic resonance imaging (fMRI), chronic pain patients who underwent CBT exhibited reduced connectivity between the anterior Default Mode Network (DMN) and the amygdala/periaqueductal gray regions. This finding contrasted with the control group, which did not receive CBT (Şpaner et al., 2014). These results contribute to the growing body of evidence suggesting that CBT can be a valuable treatment for chronic pain. Additionally, complementary methods such as imaging, mirror therapy, and virtual reality interventions may also play a role in addressing maladaptive brain plasticity associated with chronic pain (McCarberg & Peppin, 2019).

11.3. Exercise

Exercises such as resistance training, flexibility exercises, yoga, and tai chi have been shown to alleviate pain and enhance memory. Research on the relationship between exercise and chronic pain is expanding rapidly, with growing evidence supporting the idea that exercise activates various physiological systems that contribute to pain relief. These systems include opioid, nitrergic, serotonergic, noradrenergic, and endocannabinoid pathways. Additionally, exercise stimulates the release of anti-inflammatory cytokines. Studies indicate that regular physical activity not only helps alleviate chronic pain but can also serve as a preventive measure by modulating the immune system (McCarberg & Peppin, 2019).

12. Neuroplasticity and Pain

Chronic pain has been shown to alter the hippocampus and prefrontal cortex, impacting both memory and cognitive flexibility (McCarberg & Peppin, 2019). Long-term memory formation relies on the generation of new neurons and synapses through processes involving protein synthesis and gene regulation. Synaptogenesis is crucial for learning, memory consolidation, and the development of chronic pain (Nijs et al., 2021). Changes in neuroplasticity play a pivotal role in the persistence and modulation of pain (Khera & Rangasamy, 2021). In this context, neuroplasticity refers to the alterations in pain responses observed in individuals with chronic pain or those exhibiting heightened sensitivity. It is a fundamental process that underpins learning, memory, and the onset and maintenance of chronic pain.

The amygdala plays a crucial role in learning and memory processes. Glucocorticoids, which influence memory function, activate the amygdala. Elevated glucocorticoid levels can negatively affect working memory and hinder information retrieval. Within the limbic system, the amygdala is responsible for processing both the sensory and emotional dimensions of pain, while also modulating these aspects. Recent studies have highlighted the amygdala's central role in pain processing and perception. Chronic pain may arise when limbic-cortical circuits transform acute pain into a persistent condition, involving both emotional and limbic structures. The medial prefrontal cortex, along with its connections to limbic regions, plays a vital role in modulating the amygdala's response to pain (McCarberg & Peppin, 2019).

12.1. Glial Cells and Synaptic Plasticity

Neurons are commonly associated with memory and learning, but glial cells also play a crucial role in these processes. They contribute to synaptic and structural plasticity, which can significantly influence conditions such as chronic pain (Bradesi, 2010). For instance, oligodendrocytes enhance myelin production, a process essential for memory formation (Czopka et al., 2013). Therefore, glial cells are key in linking the mechanisms of learning, memory, and chronic pain.

Chronic pain impacts brain regions involved in learning and memory, with the hippocampus and amygdala being key areas affected. These regions can undergo structural and functional changes in response to pain. Studies in animal models have demonstrated that chronic pain alters behavior associated with the hippocampus, leading to changes in the amygdala and/or hippocampus, including alterations in volume, synaptic plasticity, and neurogenesis (Mutso et al., 2012). Similar variations are observed in the anterior cingulate cortex (ACC) (Yang et al., 2015). There is growing evidence linking chronic pain with impairments in learning and memory functions (Ziv et al., 2010). Additionally, anxiety can exacerbate pain perception. Individuals differ in their sensitivity to pain expectancy, with hippocampal activity playing a role in the exacerbation of pain due to stress. These differences in sensitivity are associated with varying levels of activation in the hippocampus and amygdala (Ziv et al., 2010).

Chronic pain shares several characteristics with learning and memory processes, with neuronal growth factors playing a central role. For example, brain-derived neurotrophic factor (BDNF) is essential for the formation of certain types of memory in animal models (Aarse et nal., 2016). In the context of neuropathic pain, spinal BDNF signaling is implicated in synaptic plasticity following nerve injury. However, in inflammatory pain models, supraspinal BDNF signaling has been shown to amplify pain perception (Guo et al., 2006).

Serotonin signaling is also crucial for brain areas involved in memory. It is widely distributed across the brain and interacts with other neurotransmitters such as dopamine, glutamate, and GABA. These interactions are essential for cognitive functions like learning and memory. Disruptions in serotonin signaling pathways can have a profound impact on the development and persistence of chronic pain (Kayser et al., 2007).

12.2. Pain Pathways and Neural Plasticity

Pain pathways encompass cognitive, emotional, and behavioral dimensions, with key brain regions such as the hippocampus, amygdala, and frontal cortex playing significant roles in both memory and pain processing. Chronic pain shares commonalities with memory, particularly through the involvement of shared neurotransmitters and processes, including central sensitization and long-term potentiation. Imaging studies have demonstrated that chronic pain leads to shifts in brain activity, with functional changes moving from sensory regions to emotional centers. These alterations reflect changes in neural circuits, and the state of these circuits ultimately determines whether pain remains temporary or evolves into a chronic condition. Furthermore, the emotional response to pain is influenced by these circuit changes. Chronic pain may arise from the persistence of pain memories, and certain treatments have the potential to reverse these neural changes (McCarberg & Peppin, 2019).

12.3. Neuropathic Pain and Neuroplasticity

Neuropathic pain can persist even after the underlying tissue damage has healed, suggesting that changes in nerve function are at play. These alterations may arise from direct nerve injury or from the body's adaptive responses to the injury. Recent studies indicate that such adjustments can lead to the development of neuropathic pain, which may occur either near or distant from the site of injury. Microglia in the central nervous system (CNS) are central to this process, as they can directly modify pain pathways or influence them through the release of growth factors.

Understanding how sensory circuits respond to injury could be pivotal in differentiating between pain types, such as nociceptive, neuropathic, and nociplastic pain. Research utilizing targeted methods to block specific neurons is crucial for unraveling the mechanisms behind chronic pain, especially following peripheral or central nervous system injuries. These investigations could ultimately lead to the development of novel treatment strategies. The role of microglia in pain regulation is critical in both the peripheral and central nervous systems, where they not only influence pain pathways directly but also mediate adjustments through the release of growth factors (Hiraga et al., 2022).

12.4. Acute Pain, Chronic Pain, and Neuroplasticity

Rodent studies have demonstrated that both peripheral nociceptor sensitization and synaptic changes within the spinal cord and brain are crucial factors in the development of chronic pain (Zhuo, 2008). Despite the involvement of shared neural pathways, distinguishing the unique networks and mechanisms underlying chronic versus acute pain remains a significant challenge. Understanding these differences is essential for the development of targeted therapeutic strategies.

Human brain imaging studies have identified specific regions involved in the perception of pain. The Anterior Cingulate Cortex (ACC), insular cortex, and various somatosensory cortices are activated by different types of painful stimuli (Bushnell et al., 2013). In healthy individuals, the ACC responds to harmful heat; for example, in the thermal grid illusion, innocuous hot and cold stimuli can induce a false pain sensation. The ACC shows increased activity when both stimuli are presented together, suggesting its role in processing the unpleasantness of acute pain. In chronic conditions, such as neuropathic pain and irritable bowel syndrome, ACC hyperactivity is frequently observed (Apkarian et al., 2005). This region also responds to emotional pain, with heightened activity during experiences of sadness or social rejection (Yoshino et al., 2010). These findings suggest that chronic pain may enhance pain perception through the mediation of the ACC.

Rodent studies have been instrumental in elucidating the mechanisms underlying chronic pain. These studies have revealed synaptic changes in key pain-processing regions, such as the dorsal horn of the spinal cord and the Anterior Cingulate Cortex (ACC) (Zhuo, 2008).

Long-term potentiation (LTP) and long-term depression (LTD) are fundamental processes involved in learning and memory (Bushnell et al., 2013). Chronic pain may represent a form of persistent sensory memory, with a clear connection between LTP and LTD in both the spinal cord and ACC in relation to chronic pain. For instance, LTP at C fiber synapses in the spinal cord can lead to heightened pain sensitivity. However, reversing this LTP has been shown to reduce pain behaviors, suggesting a potential therapeutic approach for alleviating chronic pain (Sandkühler & Gruber-Schoffnegger, 2012).

12.5. Motor Training and Neuroplasticity

Motor training plays a significant role in managing pain and enhancing brain adaptability, a concept referred to as neuroplasticity, which is essential for the development of motor skills. Researchers commonly use transcranial magnetic stimulation (TMS) to assess neuroplasticity in response to motor training. Studies indicate that motor training can enhance neuroplasticity in the motor cortex. However, the presence of pain may impair these beneficial effects. These findings have the potential to transform rehabilitation strategies, particularly in fields such as physiotherapy (Stanisic et al., 2022).

TMS is a valuable tool for evaluating neuroplasticity in motor skill training. Both acute and chronic pain can influence this neuroplasticity, though the results are not always consistent. Some studies report that reduced excitability in the motor cortex during pain, while others indicate increased excitability in chronic pain conditions, such as phantom limb pain, or even during acute pain episodes (Rothwell, 2018).

Pain has been shown to diminish the neuroplastic benefits of training, as evidenced by TMS results. However, measuring all aspects of neuroplasticity using a single method proves challenging. To address this, researchers employ additional techniques such as structural magnetic resonance imaging (sMRI), functional magnetic resonance imaging (fMRI), magnetoencephalography (MEG), and electroencephalography (EEG). These methods are instrumental in identifying neural changes and associated motor behaviors. This comprehensive understanding is crucial for developing more effective pain management strategies and innovative rehabilitation approaches (Stanisic et al., 2022).

13. Conclusion

Neuroplasticity enables the nervous system to adapt and plays a crucial role in the modulation of pain. Chronic pain arises from neuroplastic changes within the nervous system, including synaptic plasticity and neuronal remodeling. Synaptic plasticity alters the perception of pain by modifying the connections between nerve cells. For instance, LTP and LTD, which are types of synaptic plasticity, are essential for processes such as learning, memory, and pain. In the context of chronic pain, these processes may enhance pain signal processing within the central nervous system (CNS), contributing to the persistence of pain.

The connection between pain and neuroplasticity affects both the peripheral and central nervous systems (CNS). Damage to peripheral nerves can lead to alterations in neural pathways, subsequently modifying pain perception within the CNS. Within the CNS, regions such as the hippocampus, amygdala, and prefrontal cortex are critical in shaping pain perception, particularly its emotional and cognitive components. The hippocampus and amygdala, in particular, are essential for processing the emotional and cognitive dimensions of pain. Changes in these areas can significantly influence an individual's pain experience. Therefore, the development of pain management strategies that leverage neuroplasticity could provide effective solutions for addressing both cognitive and emotional aspects of pain. Further research is necessary to deepen our understanding of the intricate relationship between pain and neuroplasticity.

14. Takeaways

1. The IASP recommends that pain centers should involve specialists to address the biological, psychological, and social dimensions of pain.
2. Glial cells are integral in linking chronic pain with learning and memory. Microglia, in particular, influence pain perception and synaptic plasticity.
3. Serotonin interacts with other neurotransmitters such as dopamine, glutamate, and GABA in processes related to learning and memory. Dysregulation of serotonin signaling is associated with the development of chronic pain.
4. Critical brain regions involved in memory processing, such as the prefrontal cortex and hippocampus, are also implicated in pain perception. Chronic pain and memory share common neurotransmitters and underlying mechanisms.
5. Nerve damage can lead to neuropathic pain through neuroplastic changes, which may occur even distant from the injury site.

6. Chronic pain can be conceptualized as a form of persistent sensory memory, with structural and functional changes in specific brain regions contributing to this phenomenon.
7. In the spinal cord, LTP can enhance pain sensitivity, and reversing LTP may help alleviate pain behaviors.
8. Motor training has been shown to enhance neuroplasticity in the motor cortex, but pain can disrupt these neuroplastic benefits.
9. Techniques like TMS, sMRI, fMRI, EEG, and MEG are valuable tools for tracking neural changes and associated motor behaviors.
10. It is essential to develop more effective rehabilitation strategies that not only manage pain but also foster positive neural changes in the brain.

Abbreviations

ACC: Anterior Cingulate Cortex
BDNF: Brain-Derived Neurotrophic Factor
CBT: Cognitive Behavioral Therapy
CNS: Central Nervous System
GABA: Gamma Amino Butyric Acid
IASP: International Association for the Study of Pain
LTD: Long-Term Depression
LTP: Long-Term Potentiation
NMDA: N-methyl-D aspartate
PNS: Peripheral Nervous System
PET: Positron Emission Tomography
sMRI: Structural Magnetic Resonance Imaging
TMS: Transcranial Magnetic Stimulation
WDR: Wide Dynamic Range

Bibliography

1. Aarse, J., Herlitze, S., & Manahan-Vaughan, D. (2016). The requirement of BDNF for hippocampal synaptic plasticity is experience-dependent. Hippocampus, 26(6), 739–751. https://doi.org/10.1002/hipo.22541

2. Al-Chalabi, M., Reddy, V., & Gupta, S. (2023, August 14). Neuroanatomy: Spinothalamic tract. McKinsey & Company; Campbell University. PMID: 29939601
3. Ameringer S. (2010). Barriers to pain management among adolescents with cancer. Pain management nursing: official journal of the American Society of Pain Management Nurses, 11(4), 224–233. https://doi.org/10.1016/j.pmn.2009.05.006
4. Apkarian, A. V., Bushnell, M. C., Treede, R. D., & Zubieta, J. K. (2005). Human brain mechanisms of pain perception and regulation in health and disease. European journal of pain (London, England), 9(4), 463–484. https://doi.org/10.1016/j.ejpain.2004.11.001
5. Barlow, R. L., Alsio, J., Jupp, B., et al. (2015). Markers of serotonergic function in the orbitofrontal cortex and dorsal raphe nucleus predict individual variation in spatial-discrimination serial reversal learning. *Neuropsychopharmacology, 40*(7), 1619–1630. https://doi.org/10.1038/npp.2015.25
6. Basbaum, A. I., Bautista, D. M., Scherrer, G., & Julius, D. (2009). Cellular and molecular mechanisms of pain. Cell, 139(2), 267–284. https://doi.org/10.1016/j.cell.2009.09.028
7. Bradesi, S. (2010). Role of spinal cord glia in the central processing of peripheral pain perception. *Neurogastroenterology & Motility, 22*(5), 499–511. https://doi.org/10.1111/j.1365-2982.2010.01474.x
8. Bushnell, M. C., Ceko, M., & Low, L. A. (2013). Cognitive and emotional control of pain and its disruption in chronic pain. Nature reviews. Neuroscience, 14(7), 502–511. https://doi.org/10.1038/nrn3516
9. Ceylan I, Er A, Akman C, Inecikli MF, Yaslı G, Korkmaz O. Computed Tomography Findings in Childhood and Adolescent Headache, Benign Intracranial Pathologies and Frequencies. Smyrna Medical Journal 2014;4(3):10–5. Smyrna Tıp Dergisi 2014;4(3):10–5.
10. Cohen, S. P., Vase, L., & Hooten, W. M. (2021). Chronic pain: An update on the burden, best practices, and new advances. Lancet (London, England), 397(10289), 2082–2097. https://doi.org/10.1016/S0140-6736(21)00393-7
11. Costigan, M., Scholz, J., & Woolf, C. J. (2009). Neuropathic pain: a maladaptive response of the nervous system to damage. Annual review of neuroscience, 32, 1–32. https://doi.org/10.1146/annurev.neuro.051508.135531
12. Czopka, T., Ffrench-Constant, C., & Lyons, D. A. (2013). Individual oligodendrocytes only have a few hours to generate new myelin

sheaths in vivo. *Developmental Cell, 25*(6), 599–609. https://doi.org/10.1016/j.devcel.2013.05.007

13. Danilov, A., Danilov, A., Barulin, A., Kurushina, O., & Latysheva, N. (2020). Interdisciplinary approach to chronic pain management. Postgraduate medicine, 132(sup3), 5–9. https://doi.org/10.1080/00325481.2020.1757305
14. Di Maio, G., Villano, I., Ilardi, C. R., Messina, A., Monda, V., Iodice, A. C., Porro, C., Panaro, M. A., Chieffi, S., Messina, G., Monda, M., & La Marra, M. (2023). Mechanisms of transmission and processing of pain: A narrative review. International Journal of Environmental Research and Public Health, 20(4), 3064. https://doi.org/10.3390/ijerph20043064
15. Donello, J. E., Guan, Y., Tian, M., Cheevers, C. V., Alcantara, M., Cabrera, S., Raja, S. N., & Gil, D. W. (2011). A peripheral adrenoceptor-mediated sympathetic mechanism can transform stress-induced analgesia into hyperalgesia. Anesthesiology, 114(6), 1403–1416. https://doi.org/10.1097/ALN.0b013e31821c3878
16. Dubin, A. E., & Patapoutian, A. (2010). Nociceptors: the sensors of the pain pathway. The Journal of Clinical Investigation, 120(11), 3760–3772. https://doi.org/10.1172/JCI42843
17. Erdine, S. (2007). Cardiac, vascular, and homeopathic pain. In S. Erdine (Ed.), Pain (3rd ed., pp. 401–412). Nobel Medical Bookstore.
18. Fenske, J. N., Berland, D. W., Chandran, S., Harrison, R. V., Schneiderhan, J., & Hilliard, P. E. (2021, January). Ambulatory clinical guidelines oversight. Michigan Medicine, University of Michigan. Ann Arbor, MI. https://doi.org/10.1037/0000165-000
19. Fillingim R. B. (2017). Individual differences in pain: understanding the mosaic that makes pain personal. Pain, 158 Suppl 1(Suppl 1), S11–S18. https://doi.org/10.1097/j.pain.0000000000000775
20. Guo, W., Robbins, M. T., Wei, F., et al. (2006). Supraspinal brain-derived neurotrophic factor signaling: A novel mechanism for descending pain facilitation. *The Journal of Neuroscience, 26*(1), 126–137. https://doi.org/10.1523/JNEUROSCI.3739-05.2006
21. Harutoglu, H. (Eds.), Oztürk, B., & Zenginler, Y. (Eds.). Heat, light, and hydrotherapy. Hippocrates Bookstore.
22. Hiraga, S. I., Itokazu, T., Nishibe, M., & Yamashita, T. (2022). Neuroplasticity related to chronic pain and its modulation by microglia. Inflammation and regeneration, 42(1), 15. https://doi.org/10.1186/s41232-022-00199-6

23. Kayser, V., Elfassi, I. E., Aubel, B., Melfort, M., Julius, D., Gingrich, J. A., Hamon, M., & Bourgoin, S. (2007). Mechanical, thermal and formalin-induced nociception is differentially altered in 5-HT1A-/-, 5-HT1B-/-, 5-HT2A-/, 5-HT3A/- and 5-HTT-/- knock-out male mice. Pain, 130(3), 235–248. https://doi.org/10.1016/j.pain.2006.11.015
24. Khera, T., & Rangasamy, V. (2021). Cognition and pain: A review. Frontiers in Psychology, 12, 673962. https://doi.org/10.3389/fpsyg.2021.673962
25. Kusnanto, H., Agustian, D., & Hilmanto, D. (2018). Biopsychosocial model of illnesses in primary care: A hermeneutic literature review. Journal of family medicine and primary care, 7(3), 497–500. https://doi.org/10.4103/jfmpc.jfmpc_145_17
26. Marek-Jozefowicz, L., Nedoszytko, B., Grochocka, M., Żmijewski, M. A., Czajkowski, R., Cubała, W. J., & Slominski, A. T. (2023). Molecular Mechanisms of Neurogenic Inflammation of the Skin. International journal of molecular sciences, 24(5), 5001. https://doi.org/10.3390/ijms24055001
27. McCarberg, B., & Peppin, J. (2019). Pain Pathways and Nervous System Plasticity: Learning and Memory in Pain. Pain medicine (Malden, Mass.), 20(12), 2421–2437. https://doi.org/10.1093/pm/pnz017
28. Mutso, A. A., Radzicki, D., Baliki, M. N., & et al. (2012). Abnormalities in hippocampal functioning with persistent pain. The Journal of Neuroscience, 32(17), 5747–5756. https://doi.org/10.1523/JNEUROSCI.5811-11.2012
29. Nijs, J., George, S. Z., Clauw, D. J., & et al. (2021). Central sensitization in chronic pain conditions: Latest discoveries and their potential for precision medicine. The Lancet Rheumatology, 3(5), e383-e392. https://doi.org/10.1016/S2665-9913(21)00023-7
30. Nirvanie-Persaud, L., & Millis, R. M. (2022). Epigenetics and Pain: New Insights to an Old Problem. Cureus, 14(9), e29353. https://doi.org/10.7759/cureus.29353
31. Phillips, J. K., Ford, M. A., & Bonnie, R. J. (Eds.). (2017). Pain management and regulatory strategies to address prescription opioid abuse. National Academies Press. https://doi.org/10.17226/24781
32. Raja, S. N., Carr, D. B., Cohen, M., Finnerup, N. B., Flor, H., Gibson, S., Keefe, F. J., Mogil, J. S., Ringkamp, M., Sluka, K. A., Song, X. J., Stevens, B., Sullivan, M. D., Tutelman, P. R., Ushida, T., & Vader, K. (2020). The revised International Association for the Study of Pain definition of pain: concepts, challenges, and compromises. Pain, 161(9), 1976–1982. https://doi.org/10.1097/j.pain.0000000000001939

33. Rothwell J. (2018). Transcranial brain stimulation: Past and future. Brain and neuroscience advances, 2, 2398212818818070. https://doi.org/10.1177/2398212818818070
34. Sandkühler, J., & Gruber-Schoffnegger, D. (2012). Hyperalgesia by synaptic long-term potentiation (LTP): an update. Current opinion in pharmacology, 12(1), 18–27. https://doi.org/10.1016/j.coph.2011.10.018
35. Stanisic, N., Häggman-Henrikson, B., Kothari, M., Costa, Y. M., Avivi-Arber, L., & Svensson, P. (2022). Pain's Adverse Impact on Training-Induced Performance and Neuroplasticity: A Systematic Review. Brain imaging and behavior, 16(5), 2281–2306. https://doi.org/10.1007/s11682-021-00621-6
36. Van Neerven, S. G. A., & Mouraux, A. (2020). Capsaicin-Induced Skin Desensitization Differentially Affects A-Delta and C-Fiber-Mediated Heat Sensitivity. Frontiers in pharmacology, 11, 615. https://doi.org/10.3389/fphar.2020.00615
37. Yam, M. F., Loh, Y. C., Tan, C. S., Khadijah Adam, S., Abdul Manan, N., & Basir, R. (2018). General Pathways of Pain Sensation and the Major Neurotransmitters Involved in Pain Regulation. International journal of molecular sciences, 19(8), 2164. https://doi.org/10.3390/ijms19082164
38. Yang, J. X., Hua, L., Li, Y. Q., & et al. (2015). Caveolin-1 in the anterior cingulate cortex modulates chronic neuropathic pain via regulation of NMDA receptor 2B subunit. *The Journal of Neuroscience, 35*(1), 36–52. https://doi.org/10.1523/JNEUROSCI.2208-14.2015
39. Yılmaz, F., & Atay, S. (2014). Clinical pain management of nursing students. *J Hacettepe Univ Fac Nurs, 1*(2), 32–41.
40. Yoshino, A., Okamoto, Y., Onoda, K., Yoshimura, S., Kunisato, Y., Demoto, Y., Okada, G., & Yamawaki, S. (2010). Sadness enhances the experience of pain via neural activation in the anterior cingulate cortex and amygdala: an fMRI study. NeuroImage, 50(3), 1194–1201. https://doi.org/10.1016/j.neuroimage.2009.11.079
41. Zhuo, M. (2008). Cortical excitation and chronic pain. Trends in neurosciences, 31(4), 199–207. https://doi.org/10.1016/j.tins.2008.01.003
42. Ziv, M., Tomer, R., Defrin, R., & Hendler, T. (2010). Individual sensitivity to pain expectancy is related to differential activation of the hippocampus and amygdala. Human Brain Mapping, 31(2), 326–338. https://doi.org/10.1002/hbm.20856

Harnessing Neuroplasticity Through Breathing Techniques

Okan Şahin

1. Introduction

Breathing is complex, linking behavior and body processes. It affects and is affected by our emotions and thoughts. This makes it more than just a physical act (Del Negro et al., 2018). Breathing involves exchanging oxygen and carbon dioxide. It boosts health, awareness, and personality. Crucially, it connects the soul, mind, and body. Simply observing your breath with closed eyes can improve well-being. Breathing is said to heal the body, mind, soul, and emotions (Onder, 2019). Neuroplasticity, the key to the nervous system, affects breathing, memory, and thinking (Zhao et al., 2017). The brain constantly adapts to external cues. Different brain areas change in response to stimuli or behaviors, often with surprising results (Doidge, 2007). Specific breathing methods can trigger these changes, affecting both body and behavior (Ma et al., 2017). This section will look at how breathing exercises affect neuroplasticity and cognitive performance, focusing on the brain's adaptability and breathing's neurophysiological aspects.

2. Breathing Control

Neurons in the hindbrain control breathing. It relies on well-developed networks that adapt throughout life (Thoby-Brisson, 2023). The core network regulates breathing patterns, including the preBötzinger Complex (preBötC). The preBötC contains neurons that generate and coordinate breathing cycles, linking them to facial movements and influenced by emotions and thoughts (Del Negro et al., 2018).

The respiratory system develops in stages during pregnancy and after birth. It is more sensitive to stress and adapts to environmental changes (Bavis & MacFarlane, 2017).

Breathing is vital for life. The brain's respiratory network adjusts breathing based on metabolic needs and coordinates it with movement (Krohn et al., 2023). This system is highly adaptable (Courtney, 2009).

Breathing is unique in the autonomic nervous system. Unlike other processes, it can be consciously controlled. The respiratory system sends important signals to the brain about the body's balance (Betka et al., 2022).

3. Functional and Dysfunctional Breathing

Breathing is crucial. It affects how well we breathe and can lead to chronic issues. Good breathing includes nasal breathing, diaphragm awareness, and proper posture. Stress, poor posture, metabolic issues, and diseases can disrupt breathing.

Nasal breathing is more effective than mouth breathing. The nose is the primary airway, even during intense exercise. Mouth breathing might be easier but can cause discomfort when the nose is blocked. The nose also plays a role in lung function through a reflex mechanism, which is vital during intense exercise (Onder, 2019).

Mouth breathing is only advisable for specific cooling exercises and should not be a habit. It can lead to jaw narrowing. Nasal breathing aids in orthodontic treatment success (Onder, 2019).

The diaphragm is key in breathing. It ensures a good oxygen-carbon dioxide exchange. Diaphragmatic breathing brings emotional stability and increases lung use. It also supports posture and aids in digestive processes. This breathing boosts lymphatic drainage, speeding up toxin removal (Onder, 2019).

Good posture is vital for overall health and affects breathing. It should support the spine's natural curves. Poor posture limits lung expansion, making breathing harder (Onder, 2019).

Dysfunctional breathing can stem from various factors. Biomechanical issues include poor posture and chronic mouth breathing. Biochemical causes include diseases and medications. Psychological factors involve stress and anxiety (Onder, 2019).

4. The Importance of Neuroplasticity in Respiration

Studies highlight the growing importance of neuroplasticity in respiratory control. It is vital for the respiratory system and has significant biological and clinical value. As an adaptive process, neuroplasticity adjusts to changes and offers recovery strategies. It aids in aging, stroke, psychiatric issues, spinal cord injury recovery, and managing neurodegenerative symptoms. Thus, neuroplasticity in respiration is crucial. It influences emotions, forms new neural pathways, and is key in respiratory control.

Neuroplasticity is essential for neural control of respiration. It enhances the homeostatic control system's performance. This process aids in rhythm generation, sensory feedback, and motor unit recruitment in breathing control. Notably, there is a strong link between the respiratory system and motor recovery (Fuller et al., 2017). Research has explored phrenic motor plasticity and the effects of intermittent hypoxia on limb and motor functions (Dale et al., 2014).

Neuroplasticity in respiration helps adapt to disorders like respiratory failure, neurodegeneration, and paralysis. It uses cellular, synaptic, and network mechanisms to maintain breathing despite injury or disease (Fuller et al., 2017).

Neuroplasticity offers therapeutic benefits in respiratory rehabilitation. It also aids in treating conditions like spinal cord injuries and sleep apnea (Fuller et al., 2017). Strategies targeting neuroplasticity are vital for coping with strokes, trauma, spinal cord injuries, psychiatric issues, aging, and neurodegenerative symptoms (Cramer et al., 2011).

However, neuroplasticity is not always beneficial. It can harm the respiratory system. For instance, it might permanently impair the hypoxic ventilation response in adults (Fuller et al., 2017). Thus, while neuroplasticity addresses clinical issues, its potential harms should not be ignored.

4.1. Biological Significance of Respiratory Neuroplasticity

The neural elements controlling breathing face many challenges, such as weight changes, childbirth, infections, lung diseases, nerve injuries, and aging. Despite these challenges, they can adapt through neuroplasticity, a key strategy for coping with life changes (Mitchell et al., 1990; Mitchell & Johnson, 2003).

Neuroplasticity is crucial for breathing control. It is a vital trait of the nervous system, ensuring optimal breathing. Furthermore, research also links breathing motor plasticity to neuroplasticity, showing its impact on other motor systems (Dale et al., 2014).

4.2. The Role of Neuroplasticity in Respiratory Rehabilitation

Compensatory plasticity aids respiratory rehabilitation by enhancing motor neuron function. It boosts synaptic inputs, improves transmission, and reallocates effort among motor pools (Mitchell & Johnson, 2003).

Injury or disease can hinder respiratory components. However, compensatory plasticity triggers adaptive changes that sustain function. This is vital in severe conditions like neuromuscular disorders. It ensures alveolar ventilation and gas exchange despite compromised control. The brain's ability to reorganize and adapt is key to survival (Nichols et al., 2013).

When lung issues disrupt gas exchange or mechanics, compensatory processes are vital. These may involve increasing motor output to maintain adequate breathing. Adaptations allow the reorganization of muscles and networks, ensuring ventilation continues (Mahamed & Mitchell, 2007).

5. Breathing Exercises

Breathing exercises vary, including controlled, yogic, and physiological sighing methods. They all focus on slow, deep, connected breathing.

Yogic breathing, or Pranayama, affects the body's biochemical and metabolic processes. It can enhance cognitive abilities and improve autonomic and lung functions (Saoji et al., 2019). These techniques also boost neuroplasticity. Voluntary control during these practices targets specific neurons, reducing anxiety by lowering stimulation in the Locus Coeruleus (LC). Moreover, they extend the breathing cycle and increase nasal airflow. This influence on the olfactory region alters the frontal lobe and limbic structures, leading to long-term plasticity changes (Campanelli et al., 2020).

Pranayama exercises trigger hyperpolarization signals in lung fibroblasts. These signals, combined with inhibitory inputs, sync neural elements. This process influences the nervous system, promotes rhythmic brain activity, and reduces metabolic activity, indicating parasympathetic activation (Jerath et al., 2006).

Slow, deep breathing regulates the Autonomic Nervous System (ANS) through increased lung stretch and hyperpolarizing currents. This process synchronizes heart, lung, limbic system, and cortex elements. Thus, pranayama breathing enhances these signals, promoting parasympathetic dominance (Onder, 2019).

5.1. Alternate Nostril Breathing (Nadi Shodhana)

Nadi Shodhana, or alternate nostril breathing, is a key pranayama technique that balances the body and brain's hemispheres. "Nadi" means energy channels in the body, and "Shodhana" means cleansing. This practice purifies the main energy channels, Ida and Pingala, balancing internal and external tendencies (Onder, 2019).

This technique also affects brain waves. It introduces regular waves into irregular brain activities, regulating brain waves. Studies show it supports body and mind, cleans energy channels, strengthens the nervous system, and boosts self-awareness (Kinabalu, 2005).

To do alternate nostril breathing, sit straight, relax your shoulders, and close your eyes. Start by exhaling through both nostrils. Then, close the right nostril with the right thumb. Inhale deeply through the left nostril. After inhaling, close the left nostril with the right fourth finger or left thumb. Exhale through the right nostril. Now, inhale through the right nostril and then exhale through the left. This completes one cycle (Saoji et al., 2019).

These exercises also benefit the musculoskeletal system. They lower muscle tension, increase blood flow, and reduce blood pressure. This boosts oxygen to the muscles, reducing fatigue and increasing endurance. They also improve cognitive functions, enhancing task concentration (Udaykumar et al., 2021).

5.2. Deep Diaphragmatic Breathing

Deep diaphragmatic breathing, also called abdominal breathing, has been widely studied. It activates the diaphragm and boosts lung oxygen intake. The practice involves slow, controlled, long breaths (Pujari & Parvathisam, 2022).

To do this exercise, sit comfortably. Place one hand on your chest and the other on your abdomen. Only the hand on your abdomen should move. Breathe deeply through your nose for 5 seconds. Then, exhale through your mouth for 5 seconds. Repeat this for ten cycles at six breaths per minute (Rasheed et al., 2021).

5.3. Cyclic Sighing

Cyclic sighing can calm the brain and enhance alertness. During nasal exercises, it boosts activity in the amygdala and hippocampus, key areas for managing emotions. This method also slows breathing (Zelano et al., 2016).

To do cyclic sighing, find a comfortable position. After a 5-minute pause, take a slow breath in. Then, breathe in more deeply. The second breath should be shorter. Exhale slowly. Repeat this for 5 minutes. You can breathe in through the nose and out through the mouth. Alternatively, you can only breathe through the nose (Balban et al., 2023).

5.4. Bhramari Pranayama

Bhramari is a simple pranayama technique named after a black bee in India. Its key feature is exhaling with a humming sound like a bee. This creates vibrations in the throat, sinuses, and nostrils. The humming helps quiet racing thoughts, making it ideal before meditation. It also regulates the autonomic nervous system and is effective against anxiety and fear. Bhramari can be done anywhere, making it convenient (Onder, 2019).

In Bhramari, you inhale through the nose for about 5 seconds. Then, exhale for 15 seconds, humming "mmm." This humming creates vibrations in the throat and nose. A quiet space and closed eyes help focus. After seven breaths, return to normal breathing and keep your eyes closed momentarily (Onder, 2019).

5.5. Circular Breathing

Circular breathing, also known as conscious connected breathing, is central to many breathing practices and involves breathing without pausing between inhalation and exhalation. This breathing technique, like other connected breathing techniques, often uses components such as mindful body awareness, music, and movement to increase the effectiveness of the technique (Pratscher et al., 2023).

6. Relationship Between Breathing Exercises and Cognitive Performance

The brain needs much energy, mainly from oxygen and glucose. Sufficient oxygen is vital for brain health, neuron communication, and thinking. Research shows that shallow or fast breathing can lower brain oxygen levels, which can lead to problems with thinking, attention, and task performance.

However, specific breathing techniques can boost cognitive performance (Zaccaro et al., 2018; Magnon et al., 2021; Jerath et al., 2015).

Breathing is vital for thinking tasks and brain function. Techniques like deep breathing, alternate nostril breathing, and slow breathing enhance cognitive performance. Moreover, they reduce stress and boost well-being. Often, these methods are part of yoga and meditation, supporting brain health (Pujari & Parvathisam, 2022).

6.1. Mechanisms by Which Respiration Affects Cognitive Performance

Neural Activity and Breathing: Breathing affects brain activity in the prefrontal cortex (PFC), which is key for decision-making and control. Studies show slow breathing boosts PFC activity, while fast breathing lowers it, harming cognitive function. These results highlight the need for controlled breathing to keep the PFC working well and to support cognitive performance (Ma et al., 2017; Streeter et al., 2010).

Oxygen and Carbon Dioxide Levels in the Brain: These levels affect brain chemicals, respiration, and cognitive performance. Changes impact blood flow and neuron activity. For instance, breathing slowly increases oxygen intake (Moss & Shaffer, 2017).

Cerebral Blood Flow and Respiration: Cerebral blood flow delivers oxygen and nutrients to the brain, impacting cognitive performance. Breathing patterns can shift blood flow in the brain. Irregular blood flow significantly affects cognitive function (Krucoff et al., 2005).

6.2. Breathing Techniques That Influence Cognitive Performance

Breathing techniques used in yoga and meditation can enhance cognitive performance. Examples include deep diaphragmatic, alternate nostrils, and slow breathing.

These exercises also promote neuroplasticity by causing changes in the body and behavior. They activate brain regions linked to attention, improving focus (Ma et al., 2017).

Yoga and meditation, especially with breathing techniques, boost cognitive performance. They allow control over breathing and body posture.

Regular practice enhances attention, memory, and executive functions. It also increases neuroplasticity, fostering new brain connections, and can raise gray matter in areas linked to cognitive processing (Jiang et al., 2015; Gillberg et al., 1996).

6.2.1. Deep Diaphragmatic Breathing

Diaphragmatic breathing reduces stress and boosts brain power (Pujari & Parvathisam, 2022). Regular deep diaphragmatic exercises are essential. Scientific studies have shown that people who regularly do these exercises have better executive functions, attention, and memory than those with shallow breathing (Jerath et al., 2014; Bhavanani et al., 2014).

6.2.2. Slow Breathing

Slow breathing is generally defined as breathing that occurs at a rate of about six breaths per minute. During this exercise, the respiratory rate is consciously slowed. This controlled breathing pattern stimulates the parasympathetic nervous system, which results in relaxation and fewer stress reactions (Pujari & Parvathisam, 2022).

6.2.3. The Practice of Breathing Through the Opposite Nostril

The opposite nostril breathing technique is also one technique that plays a role in cognitive performance. For this technique, the breath flow through the nostrils must be done alternately. Performing this exercise has a calming effect on the mental state. The opposite nostril breathing technique is thought to regulate the balance in the brain's two hemispheres. Scientific studies show that breathing through the opposite nostril can increase cognitive performance and attention and improve cognitive flexibility (Pujari & Parvathisam, 2022).

Contralateral nostril breathing is a nasal breathing technique. It is known that nasal breathing affects the central nervous system differently than mouth breathing. Nasal breathing synchronizes electrical activity in the olfactory cortex, the amygdala, and the hippocampus. However, mouth breathing does not have a similar effect to nasal breathing (Zelano et al., 2016). Studies have shown that nasal breathing can affect higher-level brain activities. In particular, tasks related to memory processes, visual–spatial tasks, and recalling memories increase during nasal breathing. These effects are seen in the inspiratory phase of nasal breathing, but no such effect is seen during oral breathing (Zelano et al., 2016; Perl et al., 2019).

6.2.4. Bhramari Pranayama

This technique has a calming effect, especially on the prefrontal cortex. Therefore, it is thought to be useful for sleep problems. It is also believed to be an effective technique for sinus infections and thyroid problems. In addition, studies have shown that the Bhramari breathing exercise is associated with high-frequency paroxysmal gamma waves associated with high mental activities (Rajkishor et al., 2006; Vialatte et al., 2009).

7. Breathing Exercise, Stress and Neuroplasticity

The brain is key in handling and adapting to stress. It assesses whether stimuli are threats. Both adult and developing brains can change in response to stress and experiences. These changes involve neuroplastic processes, including replacing neurons, reshaping dendrites, and altering synapses (McEwen, 2017).

The respiratory system is intimately connected to stress regulation. A 2017 study demonstrated that the preBötC, which generates the breathing rhythm, is directly connected to the LC, a brain region involved in alertness, attention, and stress response (Yackle et al., 2017). The preBötC communicates with the LC when the respiratory rate increases, transmitting stress-related signals that are disseminated throughout the brain. This interaction suggests that respiratory centers directly influence higher-order brain functions, modulating the stress-related LC (Onder, 2019).

In the presence of stress, imbalances in the neural circuits related to cognition, decision-making, anxiety, and mood can be observed. This imbalance also affects systemic physiology through neuroendocrine, autonomic, immune, and metabolic mediators. In the long term, it can cause changes in the neural circuit. In this case, pharmacological and behavioral therapies can be used to intervene. The brain's plasticity is a promising feature for treatments that use brain–body interactions (McEwen, 2017).

Research has determined how slow breathing can reduce stress response and change cognitive function mechanisms such as memory and attention. Brainstem neurons affect pulmonary mechanoreceptors with decreased behavioral arousal, and their interactions with central autonomic connections create respiratory-based changes (Jerath et al., 2006).

In addition to affecting the ANS, controlled breathing exercises also have psychological effects. These psychological/behavioral outcomes are reduced arousal, anxiety, depression, anger, and confusion (Zaccaro et al., 2018). Yogic breathing exercises, such as pranayama, produce neurophysiological effects by triggering the release of β-endorphins and altering brain neurotransmitter levels. It reduces stress levels by releasing dopamine and serotonin, mainly affecting emotional changes (Aman, 2014). Diaphragmatic breathing is considered one of the techniques used in stress management. When individuals identify a stress trigger, they can do the diaphragmatic breathing exercise independently. There is no need for a specific environment for this exercise. Diaphragmatic breathing significantly benefits physical and mental health (Ma et al., 2017).

8. Latest Research on Breathing Exercises and Neuroplasticity

Recent studies highlight the benefits of breathing exercises and mindfulness meditation for mood and anxiety. One study found breathing exercises better than meditation for improving mood and arousal. Just 5 minutes a day of either can help. However, cyclic sighing is the best method (Balban et al., 2023).

Research on slow-paced breathing has surged. These studies show improvements in executive functions, heart health, and stress management. The key is breathing at a rate of about 0.1 Hz. This boosts heart rate variability and baroreflex sensitivity (Sevoz-Couche & Laborde, 2022). Slow breathing reduces stress and enhances mental health. It also triggers brain changes due to neuroplasticity (Shalaby et al., 2022).

A trial using MRI scans showed that pranayama exercises for 4 weeks enhanced brain activity linked to emotions. These exercises also lower anxiety and activate brain areas for attention and awareness (Novaes et al., 2020).

Nadishodhana pranayama is a popular relaxation technique that is effective for anxiety and depression. Regular practice improves oxidative stress, regulates functions, and lowers inflammation. It also boosts happiness-related factors in the brain (Udenia et al., 2021).

Bhramari pranayama is known to enhance cognitive performance. It improves reaction time, a key indicator of cognitive function, thus aiding academic success (Prabu Kumar et al., 2020). Another study found it increased reaction time and improved cognitive inhibition (Rajesh et al., 2014).

Deep breathing exercises also benefit children by reducing anxiety and improving math performance. They help during stressful tests (Khng, 2017).

Slow breathing techniques enhance emotional control and well-being. They boost parasympathetic activity, vital for emotional health (Zaccaro et al., 2018).

Diaphragmatic breathing boosts cognitive functions. A study showed its impact on attention and stress responses. Participants did breathing exercises for 8 weeks, averaging four breaths per minute across 20 sessions. After these exercises, sustained attention notably improved (Ma et al., 2017).

9. Conclusion

Breathing exercises significantly benefit the brain by reorganizing neural connections. By controlling our breathing and using specific techniques, we can boost neuron health, improve communication between neurons, and enhance thinking. For example, with its slow, deep breaths, pranayama can enhance focus, memory, and executive function by altering neural pathways.

Additionally, breathing exercises can regulate body functions, lower oxygen use, and reduce inflammation and oxidative stress, leading to brain changes.

Harnessing neuroplasticity is vital in respiratory therapy. It helps maintain breathing control and repair functions in neuromuscular and pulmonary disorders. For instance, pulmonary rehabilitation can benefit from strategies that rebuild neural pathways for better breathing.

Future research will aim to uncover how neuroplasticity works, paving the way for better treatments. As we learn more, we can expect significant progress in respiratory therapy, offering hope to those with breathing disorders.

10. Takeaways

1. Breathing exercises can trigger neuroplastic processes by influencing brain networks.
2. Neuroplasticity mechanisms play a critical role in respiratory control.
3. Conscious breathing control and slow and deep breathing exercises can positively impact neuronal health, synaptic transmission, and overall brain function.
4. Nasal breathing, diaphragmatic awareness, proper posture, and continuous, deep, and slow breathing are key to efficient respiratory function.
5. Breathing exercise techniques such as deep diaphragmatic breathing, slow breathing, opposite nostril breathing practice, and Bhramari pranayama positively affect cognitive performance.
6. Breathing exercises help reduce stress and improve cognitive functions such as attention, executive function, and memory.
7. Cyclic sighing is an effective technique for mood enhancement and reducing respiratory rate.
8. Respiratory rehabilitation incorporates compensatory neuroplasticity strategies to reorganize respiratory muscles and neural networks, compensating for functional losses.

Abbreviations

ANS: Autonomic Nervous System
LC: Locus Coeruleus
PFC: Prefrontal Cortex
The preBötC: The preBötzinger Complex

Bibliography

1. Aman, M. S. (2014). The effects of breathing exercises on IgG, beta-endorphin, and blood glucose secretion. *Asia Pacific Journal of Education, Arts and Sciences,* 1(4), 27–32.
2. Balban, M. Y., Neri, E., Kogon, M. M., Weed, L., Nouriani, B., Jo, B., Holl, G., Zeitzer, J. M., Spiegel, D., & Huberman, A. D. (2023). Brief structured respiration practices enhance mood and reduce physiological arousal. Cell reports. Medicine, 4(1), 100895. https://doi.org/10.1016/j.xcrm.2022.100895

3. Bavis, R. W., & MacFarlane, P. M. (2017). Developmental plasticity in the neural control of breathing. *Experimental neurology, 287(Pt 2), 176–191.* https://doi.org/10.1016/j.expneurol.2016.05.032
4. Betka, S., Adler, D., Similowski, T., & Blanke, O. (2022). Breathing control, brain, and bodily self-consciousness: Toward immersive digiceuticals to alleviate respiratory suffering. *Biological psychology, 171,* 108329. https://doi.org/10.1016/j.biopsycho.2022.108329
5. Bhavanani, A. B., Ramanathan, M., Balaji, R., & Pushpa, D. (2014). Differential effects of uni nostril and alternate nostril pranayamas on cardiovascular parameters and reaction time. *International journal of yoga*, *7*(1), 60–65. https://doi.org/10.4103/0973-6131.123489
6. Campanelli, S., Tort, A. B. L., & Lobão-Soares, B. (2020). Pranayamas and Their Neurophysiological Effects. *International journal of yoga*, *13*(3), 183–192. https://doi.org/10.4103/ijoy.IJOY_91_19
7. Courtney, R. (2009). The functions of breathing and its dysfunctions and their relationship to breathing therapy. *International Journal of Osteopathic Medicine, 12(3), 78–85*. https://doi.org/10.1016/j.ijosm.2009.04.002
8. Cramer, S. C., Sur, M., Dobkin, B. H., O'Brien, C., Sanger, T. D., Trojanowski, J. Q., Rumsey, J. M., Hicks, R., Cameron, J., Chen, D., Chen, W. G., Cohen, L. G., deCharms, C., Duffy, C. J., Eden, G. F., Fetz, E. E., Filart, R., Freund, M., Grant, S. J., Haber, S., … Vinogradov, S. (2011). Harnessing neuroplasticity for clinical applications. *Brain: a journal of neurology*, *134*(Pt 6), 1591–1609. https://doi.org/10.1093/brain/awr039
9. Dale, E. A., Ben Mabrouk, F., & Mitchell, G. S. (2014). Unexpected benefits of intermittent hypoxia: enhanced respiratory and nonrespiratory motor function. *Physiology (Bethesda, Md.)*, *29*(1), 39–48. https://doi.org/10.1152/physiol.00012.2013
10. Del Negro, C. A., Funk, G. D., & Feldman, J. L. (2018). Breathing matters. *Nature reviews. Neuroscience*, *19*(6), 351–367. https://doi.org/10.1038/s41583-018-0003-6
11. Doidge, N. (2007). The brain that changes itself: Stories of personal triumph from the frontiers of brain science. Penguin.
12. Fuller, D. D., & Mitchell, G. S. (2017). Special Issue: Respiratory Neuroplasticity. *Experimental neurology*, *287*(Pt 2), 91–92. https://doi.org/10.1016/j.expneurol.2016.11.004
13. Gillberg, M., Kecklund, G., Axelsson, J., & Akerstedt, T. (1996). The effects of a short daytime nap after a restricted night's sleep. *Sleep*, *19*(7), 570–575. https://doi.org/10.1093/sleep/19.7.570

14. Jerath, R., Barnes, V. A., & Crawford, M. W. (2014). Mind-body response and neurophysiological changes during stress and meditation: central role of homeostasis. *Journal of biological regulators and homeostatic agents*, *28*(4), 545–554.
15. Jerath, R., Crawford, M. W., & Barnes, V. A. (2015). A unified 3D default space consciousness model combining neurological and physiological processes that underlie conscious experience. *Frontiers in Psychology*, *6*, 1204. https://doi.org/10.3389/fpsyg.2015.01204
16. Jerath, R., Edry, J. W., Barnes, V. A., & Jerath, V. (2006). Physiology of long pranayamic breathing: Neural respiratory elements may provide a mechanism that explains how slow deep breathing shifts the autonomic nervous system. *Medical hypotheses*, *67*(3), 566–571. https://doi.org/10.1016/j.mehy.2006.02.042
17. Jiang, T., Soussignan, R., Schaal, B., & Royet, J. P. (2015). Reward for food odors: an fMRI study of liking and wanting as a function of metabolic state and BMI. *Social cognitive and affective neuroscience*, *10*(4), 561–568. https://doi.org/10.1093/scan/nsu086
18. Khng K. H. (2017). A better state of mind: Deep breathing reduces state anxiety and enhances test performance by regulating test cognitions in children. *Cognition & emotion*, *31*(7), 1502–1510. https://doi.org/10.1080/02699931.2016.1233095
19. Kinabalu, K. (2005). The immediate effect of 'nadi-shodhana pranayama' on some selected parameters of the brain's cardiovascular, pulmonary, and higher functions. *Thai journal of physiological sciences*, *18*(2), 10–16. ISSN 0857 – 5754
20. Krohn, F., Novello, M., van der Giessen, R. S., De Zeeuw, C. I., Pel, J. J. M., & Bosman, L. W. J. (2023). The integrated brain network that controls respiration. *eLife*, *12*, e83654. https://doi.org/10.7554/eLife.83654
21. Krucoff, M. W., Crater, S. W., Gallup, D., Blankenship, J. C., Cuffe, M., Guarneri, M., Krieger, R. A., Kshettry, V. R., Morris, K., Oz, M., Pichard, A., Sketch, M. H., Jr., Koenig, H. G., Mark, D., & Lee, K. L. (2005). Music, imagery, touch, and prayer as adjuncts to interventional cardiac care: the Monitoring and Actualisation of Noetic Trainings (MANTRA) II randomized study. *Lancet (London, England)*, *366*(9481), 211–217. https://doi.org/10.1016/S0140-6736(05)66910-3
22. Ma, X., Yue, Z. Q., Gong, Z. Q., Zhang, H., Duan, N. Y., Shi, Y. T., Wei, G. X., & Li, Y. F. (2017). The Effect of Diaphragmatic Breathing

on Attention, Negative Affect and Stress in Healthy Adults. *Frontiers in Psychology*, *8*, 874. https://doi.org/10.3389/fpsyg.2017.00874
23. Magnon, V., Dutheil, F., & Vallet, G. T. (2021). Benefits from one session of deep and slow breathing on vagal tone and anxiety in young and older adults. *Scientific reports*, *11*(1), 19267. https://doi.org/10.1038/s41598-021-98736-9
24. Mahamed, S., & Mitchell, G. S. (2007). Is there a link between intermittent hypoxia-induced respiratory plasticity and obstructive sleep apnoea? *Experimental physiology*, *92*(1), 27–37. https://doi.org/10.1113/expphysiol.2006.033720
25. McEwen, B. S. (2017). Neurobiological and Systemic Effects of Chronic Stress. Chronic stress (Thousand Oaks, Calif.), 1, 2470547017692328. https://doi.org/10.1177/2470547017692328
26. Mitchell, G. S., Douse, M. A., & Foley, K. T. (1990). Receptor interactions in modulating ventilatory activity. *The American journal of physiology*, *259*(5 Pt 2), R911–R920. https://doi.org/10.1152/ajpregu.1990.259.5.R911
27. Mitchell, G. S., & Johnson, S. M. (2003). Neuroplasticity in respiratory motor control. *Journal of Applied Physiology (Bethesda, Md.: 1985)*, *94*(1), 358–374. https://doi.org/10.1152/japplphysiol.00523.2002
28. Moss, D., & Shaffer, F. (2017). The application of heart rate variability biofeedback to medical and mental health disorders. *Biofeedback*, *45*(1), 2–8. https://doi.org/10.5298/1081-5937-45.1.03
29. Nichols, N. L., Van Dyke, J., Nashold, L., Satriotomo, I., Suzuki, M., & Mitchell, G. S. (2013). Ventilatory control in ALS. *Respiratory physiology & neurobiology*, *189*(2), 429–437. https://doi.org/10.1016/j.resp.2013.05.016
30. Novaes, M. M., Palhano-Fontes, F., Onias, H., Andrade, K. C., Lobão-Soares, B., Arruda-Sanchez, T., Kozasa, E. H., Santaella, D. F., & de Araujo, D. B. (2020). Effects of Yoga Respiratory Practice (*Bhastrika pranayama*) on Anxiety, Affect, and Brain Functional Connectivity and Activity: A Randomized Controlled Trial. *Frontiers in psychiatry*, *11*, 467. https://doi.org/10.3389/fpsyt.2020.00467
31. Onder, O. (2019). From breath to breath. Kuraldışı Publishing Education Consulting Ltd.
32. Perl, O., Ravia, A., Rubinson, M., Eisen, A., Soroka, T., Mor, N., Secundo, L., & Sobel, N. (2019). Human non-olfactory cognition phase-locked with inhalation. *Nature Human Behavior*, *3*(5), 501–512. https://doi.org/10.1038/s41562-019-0556-z

33. Prabu Kumar, A., Omprakash, A., Kuppusamy, M., K N, M., B W C, S., P V, V., & Ramaswamy, P. (2020). How does cognitive function measured by reaction time and critical flicker fusion frequency correlate with students' academic performance? *BMC Medical Education*, *20*(1), 507. https://doi.org/10.1186/s12909-020-02416-7
34. Pratscher, S. D., Sibille, K. T., & Fillingim, R. B. (2023). Conscious connected breathing with breath retention intervention in adults with chronic low back pain: protocol for a randomized controlled pilot study. *Pilot and Feasibility Studies*, *9*(1), 15. https://doi.org/10.1186/s40814-023-01247-9
35. Pujari, V., & Parvathisam, S. (2022). Breathing Your Way To Better Brain Function: The Role Of Respiration In Cognitive Performance. *Journal of Pharmaceutical Negative Results*, 8214–8219.
36. Rajesh, S. K., Ilavarasu, J. V., & Srinivasan, T. M. (2014). Effect of Bhramari Pranayama on response inhibition: Evidence from the stop signal task. *International journal of yoga*, *7*(2), 138–141. https://doi.org/10.4103/0973-6131.133896
37. Rajkishor, P., Fumitoshi, M., Bakardjia, H., Vialatte, F., & Cichocki, A. (2006). EEG changes after Bhramari Pranayama. In *SCIS & ISIS SCIS & ISIS 2006* (pp. 390–395). Japan Society for Fuzzy Theory and Intelligent Informatics.
38. Rasheed, H., Ahmad, I., Javed, M. A., Rashid, J., & Javeed, R. S. (2021). Effects of diaphragmatic breathing and abdominal drawing-in maneuvers on stroke patients' trunk stability. *Physical & Occupational Therapy In Geriatrics*, *39*(1), 1–10. https://doi.org/10.1080/02703181.2020.1770395
39. Saoji, A. A., Raghavendra, B. R., & Manjunath, N. K. (2019). Effects of yogic breath regulation: A narrative review of scientific evidence. *Journal of Ayurveda and Integrative Medicine*, *10*(1), 50–58. https://doi.org/10.1016/j.jaim.2017.07.008
40. Sevoz-Couche, C., & Laborde, S. (2022). Heart rate variability and slow-paced breathing: when coherence meets resonance. *Neuroscience and biobehavioral reviews*, *135*, 104576. https://doi.org/10.1016/j.neubiorev.2022.104576
41. Shalaby, W. S., Ahmed, O. M., Waisbourd, M., & Katz, L. J. (2022). A review of potential novel glaucoma therapeutic options independent of intraocular pressure. *Survey of ophthalmology*, *67*(4), 1062–1080. https://doi.org/10.1016/j.survophthal.2021.12.003

42. Streeter, C. C., Whitfield, T. H., Owen, L., Rein, T., Karri, S. K., Yakhkind, A., Perlmutter, R., Prescot, A., Renshaw, P. F., Ciraulo, D. A., & Jensen, J. E. (2010). Effects of yoga versus walking on mood, anxiety, and brain GABA levels: a randomized controlled MRS study. *Journal of alternative and complementary medicine (New York, N.Y.)*, *16*(11), 1145–1152. https://doi.org/10.1089/acm.2010.0007
43. Thoby-Brisson, M. (2023). Central respiratory command and microglia: An early-life partnership. *Current opinion in neurobiology*, *82*, 102756. https://doi.org/10.1016/j.conb.2023.102756
44. Udaykumar, K. P., Ukkirapandian, K., Selvaraj, S., & Kannan, D. (2021). Effect of alternate nostril breathing on cardiorespiratory parameters and muscle strength among rotating shift workers. *Bioinformation*, *17*(2), 320–325. https://doi.org/10.6026/97320630017320
45. Udenia, H., Mittal, S., Agrawal, A., Singh, A., Singh, A., & Mittal, S. K. (2021). Yogic pranayama and diaphragmatic breathing: Adjunct therapy for intraocular pressure in patients with primary open-angle glaucoma: A randomized controlled trial. Journal of Glaucoma, 30(2), 115–123. https://doi.org/10.1097/IJG.0000000000001697
46. Vialatte, F. B., Bakardjian, H., Prasad, R., & Cichocki, A. (2009). EEG paroxysmal gamma waves during Bhramari Pranayama: a yoga breathing technique. *Consciousness and Cognition*, *18*(4), 977–988. https://doi.org/10.1016/j.concog.2008.01.004
47. Yackle, K., Schwarz, L. A., Kam, K., Sorokin, J. M., Huguenard, J. R., Feldman, J. L., Luo, L., & Krasnow, M. A. (2017). Breathing control center neurons that promote arousal in mice. *Science (New York, N.Y.)*, *355*(6332), 1411–1415. https://doi.org/10.1126/science.aai7984
48. Zaccaro, A., Piarulli, A., Laurino, M., Garbella, E., Menicucci, D., Neri, B., & Gemignani, A. (2018). How Breath-Control Can Change Your Life: A Systematic Review on Psycho-Physiological Correlates of Slow Breathing. *Frontiers in human neuroscience*, *12*, 353. https://doi.org/10.3389/fnhum.2018.00353
49. Zelano, C., Jiang, H., Zhou, G., Arora, N., Schuele, S., Rosenow, J., & Gottfried, J. A. (2016). Nasal Respiration Entrains Human Limbic Oscillations and Modulates Cognitive Function. *The Journal of neuroscience: the official journal of the Society for Neuroscience*, *36*(49), 12448–12467. https://doi.org/10.1523/JNEUROSCI.2586-16.2016
50. Zhao, F., Yang, J., & Cui, R. (2017). Effect of Hypoxic Injury in Mood Disorder. *Neural plasticity*, *2017*, 6986983. https://doi.org/10.1155/2017/6986983

Neuromodulation Techniques and Their Impact on Brain Plasticity

Berkay Eren Pehlivanoğlu

1. Introduction

Research into the brain and nervous system is thriving. It has caught the scientific community's attention. Two key concepts are neuroplasticity and neuromodulation. Neuroplasticity allows the brain to change its structure and function. Meanwhile, neuromodulation controls nerve cell communication and activity.

Neuroplasticity involves changes in brain connections and structure. It can strengthen or weaken connections, create new neurons, and reorganize circuits. These changes are vital for learning, memory, and brain health. They are crucial for recovery and lifelong learning.

Neuromodulation adjusts nerve cell communication. It uses chemicals or electrical signals to influence brain activity. This approach is key in treating neurological conditions and managing nervous system disorders.

The link between neuroplasticity and neuromodulation shows how the brain adapts and regulates itself. Research suggests combining these methods could lead to better, personalized treatments. This work offers insights into their interaction and future potential.

2. Neuromodulation

Neuromodulation uses electrical stimulation to treat the central nervous system or specific nerves. It employs both surface and implanted devices. The first known case of this was in 1874. Dr. Roberts Bartholow stimulated a patient with severe scalp infections using his device. This happened at Good Samaritan Hospital in Cincinnati. Remarkably, it showed that the brain's cortex could respond to electrical stimulation, as it caused muscle spasms in a patient who didn't respond to other methods.

In 1884, Sir Victor Horsley, the pioneer of functional brain surgery, conducted the first cortical stimulation during surgery. He applied electrical stimuli to a brain hernia. The twentieth century then saw the rise of battery-powered pain relief devices. This led to safer, more modern devices for both diagnosis and treatment, boosting the field of transcutaneous stimulation (Gildenberg, 2009).

For scientists, neuromodulation covers a broad range of treatments. It involves methods that interact with the nervous system to improve its function for health benefits. This approach offers more diagnostic treatment options than other methods.

2.1. Evolution of Neuromodulation Over Time

Ancient texts show early use of electrical stimulation in medicine. For example, in 3000 B.C., Egyptians described using Nile catfish for electrical therapy. Fishermen observed this too. The practice continued until the late nineteenth century. Then, the rise of drugs and the pharmaceutical industry caused a decline. Many health professionals turned away from electrotherapy, especially with the industry's growth in the early twentieth century (Stillings, 1974; Rossi, 2003; Shealy et al., 1967).

Despite the decline, electrical stimulation has evolved, thanks to better understanding of electromagnetism. The 1967 introduction of dorsal column stimulators, initially criticized, was a key milestone. It established the biophysical principles of this therapy. Collaboration between the pharmaceutical industry and doctors has driven advancement. Improvements in materials, electrode design, energy management, and waveforms have refined the therapy (Shealy et al., 1967; Rossi, 2003). Rechargeable stimulators and closed-circuit systems have transformed electrostimulation today. Now, companies offer spinal cord stimulation systems with recorded alarms and combined action potentials. Additionally, devices for epilepsy treatment have emerged. They detect seizures and provide alarms to stop them (Gildenberg, 2009; Skarpaas et al., 2009; Fountas et al., 2007).

These advancements show how electrostimulation has evolved and its future potential. The link between electrotherapy, neuromodulation, and medical devices is vital. It leads to new, effective treatments. When discussing neuromodulation, we should acknowledge the methods used now and in the past, categorizing them accordingly.

2.1.1. Deep Brain Stimulation (DBS)

Deep brain stimulation (DBS) is a surgical treatment for neurological and psychiatric disorders. It involves placing thin electrodes in the brain to send electrical impulses that regulate activity. DBS is effective for movement disorders like Parkinson's disease, tremors, dystonia, and obsessive-compulsive disorder (OCD). It can also help with depression in some cases. The electrodes are usually implanted in areas like the thalamus, globus pallidus, or subthalamic nucleus. This method reduces or controls symptoms by altering brain cell activity (Benabid, 2003).

2.1.2. Transcranial Magnetic Stimulation (TMS)

Transcranial Magnetic Stimulation (TMS) is a non-invasive method that alters brain activity. It uses an external electromagnet to create safe, brief magnetic fields. These fields temporarily change the electrical activity in brain nerve cells. TMS treats conditions like depression, migraines, OCD, and epilepsy. During treatment, a coil on the patient's head delivers short magnetic pulses to the brain. These pulses help reduce or control symptoms.

TMS is a preferred option for those who don't respond to medication or want to avoid drug side effects. It's non-invasive. Treatment length and frequency depend on the goals and usually involve several sessions (George, 2011).

2.1.3. Vagus Nerve Stimulation (VNS)

Vagus Nerve Stimulation (VNS) is a method that targets the vagus nerve, part of the autonomic nervous system. This nerve links the brain to organs like the heart and lungs. VNS works by sending electrical signals through a device in the neck. These signals help manage symptoms of various disorders.

VNS is mainly used for epilepsy. Researchers are also looking into its effects on depression, migraines, and Parkinson's disease. The strength and frequency of VNS can be adjusted based on the patient's reaction and tolerance. It's a good option for those who don't respond to standard treatments or can't handle the side effects of medications. Since everyone's response to VNS is different, a tailored treatment plan is crucial (Howland, 2014; Demircioglu, 2024)

2.1.4. Spinal Cord Stimulation (SCS)

Spinal Cord Stimulation (SCS) is a method to manage chronic pain. It sends electrical impulses through electrodes to the spinal cord. These impulses block pain signals. SCS is for patients with chronic pain in the spine, lower back, legs, or arms. It helps those with conditions like failed back surgery syndrome, complex regional pain syndrome (CRPS), and neuropathic pain, especially when medication fails.

The SCS process starts with a trial using temporary electrodes. If the patient finds relief, permanent electrodes are added. The device, powered by a battery, is adjustable by the patient. Despite being invasive, SCS significantly reduces pain and enhances life quality (Kumar, 1996).

2.1.5. Transcranial Direct Current Stimulation (tDCS)

Transcranial Direct Current Stimulation (tDCS) is a non-invasive technique that modifies brain electrical activity. It uses low-intensity direct current through electrodes on the scalp. tDCS changes neuron polarization in specific brain areas. This alters neuron activity levels. The method usually requires multiple sessions.

tDCS aims to boost cognitive functions, treat depression, manage chronic pain, and enhance motor skills. It addresses various neurological and psychiatric conditions. Yet, more research is needed to confirm its effectiveness and safety. We also need to understand its long-term effects better (Nitsche, 2000).

2.1.6. Optogenetics

Optogenetics combines genetic engineering and light to control neuron activity. Scientists introduce specific genes or channels into the neurons of mice and other models. This allows them to activate or deactivate these neurons with light. The method is key for studying neuron functions, brain circuits, and potential treatments for neurological disorders (Montagni, 2019).

2.1.7. Pharmacological Neuromodulation

Pharmacological neuromodulation uses drugs to change brain and nervous system functions. These drugs bind to neuron receptors, modifying neural communication. They can increase or decrease brain activity, affecting thinking, emotions, or behavior. This method treats many neurological and psychiatric disorders. It's key in pharmacotherapy for depression, anxiety, and epilepsy (Conn, 2009).

2.1.8. Transcranial Alternating Current Stimulation (tACS)

Transcranial Alternating Current Stimulation (tACS) is a safe, non-invasive method that alters brain activity. It uses electrodes on the scalp to deliver alternating current (AC). tACS applies AC at specific frequencies to influence brainwaves. This can synchronize activity in certain brain areas or change it. For example, a specific tACS frequency might boost thinking or ease symptoms of neurological disorders. tACS can improve thinking, enhance movement, and relieve pain. It also helps with various neurological and psychiatric issues (Lee, 2022).

2.1.9. Transcranial Pulsed Ultrasound (tPU)

Transcranial Pulsed Ultrasound is a non-invasive method that uses ultrasound waves to target specific brain areas. It induces changes in neuronal activity. This technique can alter the electrical and chemical activities of brain nerve cells. It's being explored for treating neurological and psychiatric conditions. For example, it shows potential in treating Alzheimer's disease and managing neuropathic pain. Being non-invasive, it may have fewer side effects and be a safer treatment option (Tsui, 2004).

2.1.10 Epidural Electrical Stimulation (EES)

Epidural Electrical Stimulation (EES) involves placing electrodes in the epidural space to stimulate the spinal cord. It's used for chronic pain management and certain movement disorders. The electrodes can block or modify pain signals. This reduces the sensation of pain. EES treats conditions like post-spinal cord injury pain, complex regional pain syndrome (CRPS), and neuropathic pain. Treatment is tailored to each patient, starting with a trial. If successful, a permanent device is implanted. The process might need surgery, and responses vary by individual (Angeli, 2018).

2.1.11. Peripheral Nerve Stimulation (PNS)

Peripheral Nerve Stimulation (PNS) helps manage pain and improve functions by stimulating peripheral nerves. Typically, electrodes are placed under the skin to achieve this. PNS is effective for chronic pain, neuropathic pain, and migraines. It sends electrical impulses to block pain signals. This method is preferred for patients who don't respond to drugs or face side effects. PNS is usually non-invasive or minimally invasive. Electrodes are implanted under the skin and controlled externally or internally. The treatment is personalized based on the patient's condition. PNS is a viable alternative in pain management (Slavin, 2008).

2.1.12. Cranial Electrical Stimulation (CES)

Cranial Electrical Stimulation refers to methods that apply electrical stimulation to the brain. These techniques alter brain activity by targeting specific regions directly or indirectly (Kirsch, 2013).

2.1.13 Intracranial Electrotherapy Stimulation (iET)

Intracranial Electrotherapy Stimulation (ICES) is a technique that sends electrical signals to the brain. This is done through surgery, placing electrodes in the brain. ICES aims to change brain activity. It's used for conditions like epilepsy and sometimes for obsessive-compulsive disorder. Low-frequency currents from the electrodes adjust nerve cell activity. The decision to use ICES depends on the patient's condition and goals. Doctors weigh the benefits and risks. ICES is considered when drugs or other treatments fail (Yang, 2019).

2.1.14. Magnetic Resonance-guided Focused Ultrasound (MRgFUS)

Magnetic Resonance-guided Focused Ultrasound (MRgFUS) is a non-invasive treatment. It uses high-intensity ultrasound waves, guided by MRI, to target tissues. This method can either heat or mechanically affect tissues, making it useful for destroying tumors or harmful tissues. MRgFUS is especially effective for treating tumors, including benign ones like uterine fibroids. It is also being studied for neurological disorders such as Parkinson's disease. This technique offers a less invasive option compared to traditional surgery, benefiting patient care (Lipsman, 2013).

2.1.15. Permanent Implantable Neuromodulation Devices

Permanent implantable neuromodulation devices are electronic devices implanted in the body for long-term use. Powered by batteries, they directly influence the nervous system. They're often used to treat chronic pain and neurological disorders. Examples include spinal cord and intracranial stimulation. Doctors can tailor the stimulation based on patient responses and needs. Implantation requires surgery and is usually done after a trial period. Patients can adjust settings with an external device. These devices are key for those who don't respond to drugs or suffer side effects (Van Buyten, 2003).2.1.16 Dorsal Root Ganglion (DRG) Stimulation

Dorsal Root Ganglion (DRG) Stimulation is a method for chronic pain relief. It directly targets pain signals in the spinal nerve. The DRG, located near the spinal cord, gathers nerve fibers. This technique is suitable for conditions like chronic pain, complex regional pain syndrome (CRPS), and neuropathic pain, all linked to nerve fiber issues. Electrodes are placed near the DRG, either under the skin or in the epidural space. These electrodes change nerve activity, which can block or modify pain signals. After a successful trial, patients may upgrade to a permanent device. This device, usually battery-powered, can be adjusted externally by the patient. DRG stimulation stands out as an effective alternative to other methods (Deer, 2017).

2.1.17. Transcutaneous Electrical Nerve Stimulation (TENS)

Transcutaneous Electrical Nerve Stimulation (TENS) uses electrical impulses via skin electrodes for pain relief. It's non-invasive, portable, and battery-operated. Electrodes deliver low-voltage currents at specific frequencies, blocking or reducing pain signals. TENS works by modulating nerve activity with these currents. It's effective for both acute and chronic pain, including musculoskeletal pain, neuropathic pain, and migraines. TENS is easy to use, fitting smoothly into daily routines (Johnson, 2010).

2.1.18. Cellular Neuromodulation

Cellular neuromodulation changes how neurons work by targeting their activity. It can adjust neurotransmitter receptors, ion channels, or signaling pathways. This affects nerve transmission and neuron connections. Usually, it involves drugs or genetic methods. It's key to understanding neurological disorders and developing new treatments (Kauer, 2007).

2.1.19. Photobiomodulation (PBM)

Photobiomodulation uses low-level laser or LED light to affect cells and molecules. It targets specific tissues with particular light wavelengths, impacting cell functions. Benefits include improved mitochondrial function, reduced inflammation, and faster healing. Researchers are exploring it for various medical conditions and to boost athletic performance (Hamblin, 2016).

2.1.20. Galvanic Vestibular Stimulation (GVS)

Galvanic Vestibular Stimulation (GVS) is a technique that uses low electrical currents to influence balance and movement. Electrodes, usually placed behind the ears, deliver these currents. GVS works by stimulating the inner ear's vestibular system, sending signals to the brain's balance centers. This can temporarily alter a person's sense of balance or create illusions of movement. Researchers are exploring GVS for treating balance disorders, inner ear diseases, and vertigo. It's also useful for assessing balance functions (Fitzpatrick, 2004).

2.1.21. Targeted Radiofrequency Ablation (RF ablation)

Targeted Radiofrequency Ablation (RF ablation) is a non-surgical, minimally invasive procedure. It uses high-frequency radio waves to damage specific tissues. Doctors often use medical imaging for precise targeting. This technique is commonly applied in treating cancers, such as liver and renal cancer, and bone metastases. It serves as a local treatment option. Moreover, RF ablation is helpful in pain management. It disrupts nerve roots or fibers to block pain signals. Generally, it is deemed low-risk with minimal complications (Cosman, 2005).

2.1.22. Acoustic Neuromodulation

Acoustic neuromodulation uses special sound waves to affect brain or nervous system activity. It targets specific frequencies to alter neural activity. Sound waves are transmitted to brain tissue, influencing neurons directly or indirectly. Typically, this method is applied through the ear canal. Researchers are exploring its potential to treat disorders like depression, anxiety, hearing loss, and memory issues (Sassaroli, 2016).

2.1.23 Cochlear Implants

Cochlear implants are devices for people with hearing loss. Surgeons place them in the cochlea, part of the inner ear. These implants stimulate the auditory nerve, enhancing hearing. Unlike hearing aids, cochlear implants convert sounds into electrical signals. They then send these signals through the auditory nerve to the brain. This allows the brain to process sounds. The implantation process has several steps. First, the internal part of the implant is placed in the ear. Then, a sound processor is fitted behind the ear. This processor captures and processes sounds, sending them to the inner ear electrodes. These electrodes then stimulate the auditory nerve. Cochlear implants are especially beneficial for those with severe hearing loss, significantly improving their ability to understand speech (Wilson, 2008).

2.1.24. Retinal Implants

Retinal implants are devices for people with vision loss. They are placed on the retina and contain electrodes to capture light. The retina detects light and sends visual information to the brain. These implants are used when retinal cells are damaged, such as in advanced retinitis pigmentosa or macular degeneration. They work by catching light and sending signals to healthy retinal cells. Users may regain some basic vision, like seeing light and shapes, which aids in daily activities. The process requires surgery, and users need to learn how to use the device. Research continues to improve these implants and explore their use (Zrenner, 2002).

3. Neuroplasticity and Neuromodulation

Until recently, people thought the human brain only changed in early development. However, studies now show significant changes can occur in adults too, under certain conditions. These changes happen in neural pathways and synapses, influenced by behavior, experiences, needs, or the environment. Even physical trauma can trigger these changes. This ability of the brain to adapt is called neuroplasticity.

Neuroplasticity includes two types: synaptic and non-synaptic plasticity. Synaptic plasticity involves the creation or removal of synapses, based on the strength and activity of connections. Non-synaptic plasticity, however, refers to changes in axons, dendrites, and ion channels in neurons. This type affects how neurons are stimulated and how they function. When combined with synaptic plasticity, these changes can impact higher brain functions (Pagano, 2016; Costandi, 2016; Innocenti, 2022; Cosentino, 2022).

Recent advancements in imaging have improved our understanding of the human nervous system. We've seen neuroplastic changes that help the body adapt after injuries. These changes aid recovery. However, they can also worsen or prolong diseases (Cosentino, 2022; Costandi, 2016; Innocenti, 2022).

Neuromodulation includes various methods to change brain activity or sensitivity. Research shows that short-term changes in brain activity can lead to lasting effects. This suggests that neuromodulation can improve brain function. It might reverse harmful changes, prevent them, or boost recovery after central nervous system damage (Hillery, 2024; Arle, 2022).

Neuroplasticity reorganizes brain networks for memory, thinking, emotions, and behavior. It involves mechanisms like synaptic stimulation, which boosts connections, and reduced activity, which weakens them. These processes affect memory, thinking, behaviors, and emotions. They depend on sensory input and the changes in synaptic connections across the brain (Higa, 2024).

In glutamatergic synapses, specific activity patterns activate NMDA and metabotropic glutamate receptors. This triggers cascades involving protein kinases like PKA, PKC, CaMKII, and ERK. These kinases adjust synaptic strength by moving AMPA receptors to the postsynaptic membrane or changing glutamate release at the presynaptic terminal. They also affect protein synthesis, influencing changes related to plasticity (Higa, 2024).

Hebbian mechanisms constantly reorganize synapses in neural circuits. Neurotransmitter systems drive widespread effects. Neuromodulatory systems, like noradrenergic, serotonergic, dopaminergic, and cholinergic fibers from cerebellar nuclei, oversee synaptic plasticity. They base this on attention, concentration, mood, alertness, and motivation. These neuromodulators activate G-protein-coupled receptors, adjusting the strength of neural connections (Higa, 2024).

Serotonin (5-HT) is vital for brain functions like sensory processing, thinking, emotions, and sleep patterns. It originates from neurons in the brainstem's raphe nuclei and spreads throughout the brain. Different serotonin receptors, such as 5HT1AR and 5HT2AR, have unique effects. For example, 5HT1ARs, found in certain synapses, reduce cAMP production, thus calming neurons. Meanwhile, 5HT2ARs, located in dendritic fields, activate PLC signaling, leading to cellular responses (Albert, 2019; Meunier, 2017; Teixeira, 2018; Higa, 2024).

The hippocampus processes sensory and spatial information. It also forms complex brain connections. Researchers study it for 5-HT's effects on synaptic plasticity. Hippocampal circuits are vital for learning and memory. They experience dynamic changes in synaptic strength and dendritic shape. These changes are influenced by signaling pathways and plasticity mechanisms involving 5-HT receptors (Albert, 2019; Meunier, 2017; Teixeira, 2018; Higa, 2024).

Serotonin (5-HT) and other neuromodulators are vital for synaptic plasticity. They adjust synaptic connections in neural circuits based on factors like attention, mood, and motivation (Higa, 2024). Serotonin plays a key role in sensory processing, cognition, emotions, and circadian rhythms. Different serotonin receptors (e.g., 5HT1AR, 5HT2AR) have unique effects in the brain (Albert, 2019; Meunier, 2017; Teixeira, 2018; Higa, 2024). The hippocampus, crucial for learning and memory, processes sensory and spatial information. Here, synaptic strength and structure change dynamically. Research shows how 5-HT affects hippocampal synaptic plasticity, highlighting its importance in neural functions. Understanding serotonin's role in neural circuit plasticity is key to grasping neurological functions (Albert, 2019; Meunier, 2017; Teixeira, 2018; Higa, 2024).

4. Conclusion

Neuroplasticity and neuromodulation are key areas in neuroscience. They help us understand the brain's adaptability and nerve cell communication. Neuroplasticity allows the brain to reorganize itself for learning, memory, and recovery from injuries. Neuromodulation, on the other hand, uses electrical or chemical methods to alter nerve activity. This is promising for treating chronic pain, epilepsy, and depression.

The link between neuroplasticity and neuromodulation is vital for new treatments. Neuromodulation can spark neuroplastic changes, boosting learning and memory. This collaboration has led to personalized therapies, especially for rehabilitation and neurological disorders.

Recent advances in imaging have improved our understanding of these processes and treatment strategies. As research grows, the combination of neuroplasticity and neuromodulation could transform rehabilitation and patient outcomes.

In summary, these fields are crucial for both research and clinical practice. They pave the way for better brain health and treatment outcomes.

5. Takeaways

1. Neuroplasticity refers to the brain's capacity for structural and functional changes.
2. Neuromodulation encompasses methods that regulate communication between nerve cells through chemical, electrical, or other means.
3. Neuroplasticity involves strengthening or weakening connections between nerve cells, enhancing the brain's capacity for learning and memory.
4. Neuromodulation techniques play a critical role in treating neurological disorders and rehabilitation processes.
5. Historically, neuromodulation practices date back to the late nineteenth century and have seen significant advancements.
6. Electrical stimulation techniques are a key component of neuromodulation, and they are used to treat conditions such as chronic pain, epilepsy, and depression.
7. The interaction between neuroplasticity and neuromodulation reflects the dynamic relationship between the brain's adaptability and the regulation of neural functions.
8. Neuromodulation can contribute to nervous system healing processes by inducing neuroplastic changes.
9. Neuroplasticity supports brain health by reorganizing synaptic connections and forming new nerve cells.
10. Advances in pharmaceutical therapies and electrical stimulation devices in the twentieth century shaped the historical development of neuromodulation.
11. Advanced neuroimaging techniques have provided more profound insights into the complexity of neuroplasticity and the brain's responses to various stimuli.
12. Research in neuroplasticity and neuromodulation enables the development of innovative and effective methods for treating neurological disorders.

13. Neuroplasticity includes changes in neural pathways and synapses influenced by behavior, experiences, physiological needs, or environmental influences.
14. Advances such as rechargeable stimulators and closed-loop systems have propelled the development of neuromodulation.
15. Neuroplasticity and neuromodulation offer revolutionary approaches in brain research and clinical applications, opening new horizons for treating nervous system diseases.

Abbreviations

AC: Alternating Current
CaMKII: Ca^{2+}-calcium/calmodulin-dependent kinase II
CES: Cranial Electrical Stimulation
CRPS: Complex Regional Pain Syndrome
CRPS: Complex Regional Pain Syndrome
DBS: Deep brain stimulation
DC: Direct Current
DRG: Dorsal Root Ganglion
EES: Epidural Electrical Stimulation
ERK: Extracellular Signal-Regulated Kinase
GVS: Galvanic Vestibular Stimulation
iET: Intracranial Electrotherapy Stimulation
mGluRs: Metabotropic Glutamate Receptors
MRgFUS: Magnetic Resonance-guided Focused Ultrasound
MRI: Magnetic Resonance Imaging
NMDA: N-metil-D-aspartat
OCD: Obsessive-Compulsive Disorder
PKA: Protein Kinase A
PKC: Protein Kinase C
PNS: Peripheral Nerve Stimulation
RF: Radiofrequency
SCS: Spinal Cord Stimulation
tACS: Transcranial Alternating Current Stimulation
tDCS: Transcranial Direct Current Stimulation
TENS: Transcutaneous Electrical Nerve Stimulation
TMS: Transcranial Magnetic Stimulation

tPU: Transcranial Pulsed Ultrasound
VNS: Vagus Nerve Stimulation

Bibliography

1. Albert, P. R., & Vahid-Ansari, F. (2019). The 5-HT1A receptor: Signaling to behavior. Biochimie, 161, 34–45. https://doi.org/10.1016/j.biochi.2018.10.015
2. Angeli, C. A., Boakye, M., Morton, R. A., Vogt, J., Benton, K., Chen, Y., … & Harkema, S. J. (2018). Recovery of over-ground walking after chronic motor complete spinal cord injury. New England Journal of Medicine, 379(13), 1244-1250. https://doi.org/10.1056/NEJMoa1803588
3. Arle, J. E. (2022). The neuromodulation approach. In Essential Neuromodulation (pp. 3–17). Elsevier. https://doi.org/10.1016/b978-0-12-817000-7.00001-6
4. Benabid, A. L. (2003). Deep brain stimulation for Parkinson's disease. Current opinion in neurobiology, 13(6), 696–706. https://doi.org/10.1016/j.conb.2003.11.001
5. Cosentino, G., Todisco, M., & Blandini, F. (2022). Noninvasive neuromodulation in Parkinson's disease: Neuroplasticity implication and therapeutic perspectives. In Handbook of Clinical Neurology (pp. 185–198). Elsevier. https://doi.org/10.1016/b978-0-12-819410-2.00010-2
6. Cosman Jr., E. R., & Cosman Sr, E. R. (2005). Electric and thermal field effects in tissue around radiofrequency electrodes. Pain Medicine, 6(6), 405–424. https://doi.org/10.1111/j.1526-4637.2005.00076.x
7. Costandi, M. (2016). Neuroplasticity. MIt Press.
8. Deer, T. R., Levy, R. M., Kramer, J., Poree, L., Amirdelfan, K., Grigsby, E., … & Mekhail, N. (2017). Dorsal root ganglion stimulation yielded higher treatment success rate for complex regional pain syndrome and causalgia at 3 and 12 months: a randomized comparative trial. Pain, 158(4), 669–681. https://doi.org/10.1097/j.pain.0000000000000814
9. Demircioglu, G., Ozden, A. V., & Genc, H. (2024). Comparison of The Efficacy of Auricular Vagus Nerve Stimulation and Conventional Low Back Rehabilitation in Patients with Chronic Low Back Pain. Complementary Therapies in Clinical Practice, 101862. https://doi.org/10.1016/j.ctcp.2024.101862
10. Fitzpatrick, R. C., & Day, B. L. (2004). Probing the human vestibular system with galvanic stimulation. Journal of Applied Physiology, 96(6), 2301–2316. https://doi.org/10.1152/japplphysiol.00008.2004

11. Fountas, K. N., & Smith, J. R. (2007). A novel closed-loop stimulation system in the control of focal, medically refractory epilepsy. Operative Neuromodulation: Volume 2: Neural Networks Surgery, 357–362.
12. George, M. S., & Post, R. M. (2011). Daily left prefrontal repetitive transcranial magnetic stimulation for acute treatment of medication-resistant depression. American Journal of Psychiatry, 168(4), 356–364. https://doi.org/10.1176/appi.ajp.2010.10060864
13. Gildenberg, P. L. (2009). Neuromodulation. In Neuromodulation (pp. 9–20). Elsevier. https://doi.org/10.1016/b978-0-12-374248-3.00003-3
14. Hamblin, M. R. (2016). Shining light on the head: photobiomodulation for brain disorders. BBA clinical, 6, 113–124. https://doi.org/10.1016/j.bbacli.2016.09.002
15. Higa, G. S. V., Viana, F. J. C., Francis-Oliveira, J., Cruvinel, E., Franchin, T. S., Marcourakis, T., Ulrich, H., & De Pasquale, R. (2024). Serotonergic neuromodulation of synaptic plasticity. In Neuropharmacology (Vol. 257, p. 110036). Elsevier BV. https://doi.org/10.1016/j.neuropharm.2024.110036
16. Hillery, T., Mata, N., & Kim, C. (2024). History of neuromodulation. In Neuromodulation Techniques for the Spine (pp. 1–10). Elsevier. https://doi.org/10.1016/b978-0-323-87584-4.00010-3
17. Howland, R. H. (2014). Vagus nerve stimulation. Current Behavioral Neuroscience Reports, 1, 64–73. https://doi.org/10.1007/s40473-014-0010-5
18. Innocenti, G. M. (2022). Defining neuroplasticity. In Handbook of Clinical Neurology (pp. 3–18). Elsevier. https://doi.org/10.1016/b978-0-12-819410-2.00001-1
19. Jeffrey Conn, P., Christopoulos, A., & Lindsley, C. W. (2009). Allosteric modulators of GPCRs: a novel approach for treating CNS disorders. Nature reviews Drug discovery, 8(1), 41–54. https://doi.org/10.1038/nrd2760
20. Johnson, M. I., & Walsh, D. M. (2010). Continued uncertainty of TENS' effectiveness for pain relief. Nature Reviews Rheumatology, 6(6), 314–316. https://doi.org/10.1038/nrrheum.2010.77
21. Kauer, J. A., & Malenka, R. C. (2007). Synaptic plasticity and addiction. Nature Reviews Neuroscience, 8(11), 844–858. https://doi.org/10.1038/nrn2234
22. Kirsch, D. L., & Nichols, F. (2013). Cranial electrotherapy stimulation for treatment of anxiety, depression, and insomnia. Psychiatric Clinics, 36(1), 169–176.

23. Kumar, K., Toth, C., & Nath, R. K. (1996). Spinal cord stimulation for chronic pain in peripheral neuropathy. Surgical neurology, 46(4), 363–369. https://doi.org/10.1016/S0090-3019(96)00191-7
24. Lee, H. J., Jung, D. H., Jung, Y. J., Shin, H. K., & Choi, B. T. (2022). Transcranial alternating current stimulation rescues motor deficits in a mouse model of Parkinson's disease via glial cell line-derived neurotrophic factor production. Brain Stimulation, 15(3), 645–653. https://doi.org/10.1016/j.brs.2022.04.002
25. Lipsman, N., Schwartz, M. L., Huang, Y., Lee, L., Sankar, T., Chapman, M., … & Lozano, A. M. (2013). MR-guided focused ultrasound thalamotomy for essential tremor: a proof-of-concept study. The Lancet Neurology, 12(5), 462–468. https://doi.org/10.1016/S1474-4422(13)70048-6
26. Meunier, C. N., Chameau, P., Fossier, P., & Meunier, C. (2017). Modulation of synaptic plasticity in the cortex needs to understand all the players. Frontiers in Synaptic Neuroscience, 9, 2. https://doi.org/10.3389/fnsyn.2017.00002
27. Montagni, E., Resta, F., Mascaro, A. L. A., & Pavone, F. S. (2019, August). Optogenetics in brain research: from a strategy to investigate physiological function to a therapeutic tool. In Photonics (Vol. 6, No. 3, p. 92). https://doi.org/10.3390/photonics6030092
28. Nitsche, M. A., & Paulus, W. (2000). Excitability changes induced in the human motor cortex by weak transcranial direct current stimulation. The Journal of Physiology, 527(Pt 3), 633. https://doi.org/10.1111/j.1469-7793.2000.t01-1-00633.x
29. Pagano, G., Niccolini, F., & Politis, M. (2016). Imaging in Parkinson's disease. Clinical medicine (London, England), 16(4), 371–375. https://doi.org/10.7861/clinmedicine.16-4-371
30. Rossi U. (2003). The history of electrical stimulation and the relief of pain B.A. Simpson (Ed.), Electrical stimulation and the relief of pain, Vol. 15, Elsevier Science, New York, pp. 5–16
31. Sassaroli, E., & Vykhodtseva, N. (2016). Acoustic neuromodulation from a basic science prospective. Journal of therapeutic ultrasound, 4, 1–14. https://doi.org/10.1186/s40349-016-0061-z
32. Shealy, C. N., Mortimer, J. T., & Reswick, J. B. (1967). Electrical inhibition of pain by stimulation of the dorsal columns: preliminary clinical report. Anesthesia & Analgesia, 46(4), 489–491.
33. Skarpaas, T. L., & Morrell, M. J. (2009). Intracranial stimulation therapy for epilepsy. Neurotherapeutics, 6, 238–243. https://doi.org/10.1016/j.nurt.2009.01.022

34. Slavin, K. V. (2008). Peripheral nerve stimulation for neuropathic pain. Neurotherapeutics, 5(1), 100–106. https://doi.org/10.1016/j.nurt.2007.11.005
35. Stillings, D. (1974). The first observation of electrical stimulation. Medical instrumentation, 8(5), 313.
36. Teixeira, C. M., Pomedli, S. R., Maei, H. R., Kee, N., & Frankland, P. W. (2018). Involvement of the anterior cingulate cortex in the expression of remote spatial memory. Journal of Neuroscience, 28(10), 2317–2326. https://doi.org/10.1523/JNEUROSCI.1068-06.2006
37. Tsui, P. H., & Wang, S. H. (2004). The effect of transducer characteristics on the estimation of Nakagami paramater as a function of scatterer concentration. Ultrasound in medicine & biology, 30(10), 134
38. Van Buyten, J. P., & Itrel 3 Study Group. (2003). The performance and safety of an implantable spinal cord stimulation system in patients with chronic pain: a 5-year study. Neuromodulation: Technology at the Neural Interface, 6(2), 79–87. https://doi.org/10.1046/j.1525-1403.2003.03012.x
39. Wilson, B. S., & Dorman, M. F. (2008). Cochlear implants: a remarkable past and a brilliant future. Hearing research, 242(1–2), 3–21. https://doi.org/10.1016/j.heares.2008.06.005
40. Yang, F., Fang, X., Tang, W., Hui, L., Chen, Y., Zhang, C., & Tian, X. (2019). Effects and potential mechanisms of transcranial direct current stimulation (tDCS) on auditory hallucinations: A meta-analysis. Psychiatry Research, 273, 343–349. https://doi.org/10.1016/j.psychres.2019.01.059
41. Zrenner, E. (2002). Will retinal implants restore vision?. Science, 295(5557), 1022–1025. https://doi.org/10.1126/science.1067996

Computer Interfaces Leveraging Neuroplasticity for Neural Engineering

Ömer Faruk Candal and Adnan Kuşman

1. Introduction to Brain-Computer Interface

Over the past three decades, the rapid march of technology has significantly heightened scientific interest in leveraging biomedical data for a myriad of applications, from communication to motor control, neuro-rehabilitation, and beyond. An expansion in interest is closely related to the evolution of cognitive sciences as an interdisciplinary domain. Among cognitive sciences, the conventional sciences of philosophy, linguistics, cognitive psychology, neuroscience, and anthropology, as well as the relatively new concepts in the literature, like artificial intelligence and robotics, may be related to this rapidly growing interest. The diverse nature of these fields makes studying Brain-Computer Interface (BCI) systems engaging and dynamic.

Moreover, the emergence of Brain-Inspired Cognitive Systems (BCS) is a significant development in bio-cybernetics and cognitive science. Biomedical signal analysis is a substantial diagnostic instrument in numerous medical domains (Al-Ani et al.,2010). An essential biomedical data analysis method is developing an ideal and intuitive Human–Machine Interface (HMI). This data analysis method is a leading issue of interest for many researchers (Abdulkader et al., 2015). BCI plays a vital role in this, particularly in systems where signals are the Primary interface.

Modern HMI systems enable direct communication between a human and an external device, eliminating the need for control equipment like a mouse or keyboard (Soares et al., 2020). These systems have outstanding benefits, mainly for disabled users, as they allow them to interact directly with their environment. Some HMI systems have used distinct biomedical signs for control purposes. These methods include electromyography (EMG) to record muscle activity, electroencephalography (EEG) to record brain activity, electrocorticography (ECoG) to record neural activity directly from the brain surface, and electrooculography (EOG) to measure eye activity.

BCI detects human decisions via electromagnetic currents coming through straight from the brain. The brain is generally considered the most sophisticated organ. It is often compared to a powerful and intricate computer. Its complexity generates the idea that to simulate the brain is impossible. This awe-inspiring complexity has recently led to rapid developments, especially in information technologies and medical sciences. These developments ushered in the epoch of Brain–Computer Interfaces (BCIs) contingent on EEG, predominantly non-invasive versions (Kübler, 2020).

There are two kinds of systems in BCI systems: non-invasive and invasive (Martinez-Marquez et al., 2021). Typically, information is processed and exchanged in real time between the brain and the computer (here, BCI systems). To this end, all brain-related activity, whether direct or indirect, must be measured. Direct connection refers to assessing electrical activity originating from the neurons (e.g., EEG). In contrast, indirect connections can be made by measuring blood oxygen levels via magnetoencephalography (MEG), functional magnetic resonance imaging (fMRI), and functional near-infrared spectroscopy (fNIRS) (Kübler, 2020).

The refinement of BCI systems enabled the evolution from superficial EEG recordings to advanced technologies that optimize brain–computer communication. BCIs analyze signals from the brain and transform them into specific commands, allowing them to replace or supplement environmental devices such as computer keyboards, mice, or joysticks. Terminologically, a BCI system refers to procedures that assess and evaluate the central nervous system (CNS) signals. The systems mentioned earlier are activated by voice or muscle movements and cannot be classified as BCI. Likewise, a standalone EEG device cannot be regarded as a BCI, as it merely measures signals, and EEG cannot give feedback to the employer.

Additionally, considering BCIs as mind-reading instruments is disappointing because extracting information from an individual's mind is impossible. These systems allow individuals to interact with their environment, utilizing brain activity without muscle movements. In this procedure, brain signals are first amplified and digitized. This approach facilitates the extraction of specific features or patterns from brain signals. This information is then converted into commands. The result defines the user's response and forms an essential part of the feedback loop.

2. Basic Components of Brain–Computer Interfaces

BCI systems consist of three basic elements: recording the signal, processing the recorded signal, and interacting with the BCI devices. These elements are interconnected and inseparable. Brain signals act synchronously to enable the targeted BCI applications to move, operate, and analyze. Various specialized BCI hardware can be sent to the brain to stimulate common functions such as vision, speech, and hearing.

BCI is generally classified into two basic groups: invasive (recording signals directly from the brain's surface) and non-invasive (processing hardware from sensors positioned on the scalp's surface to record signals). In invasive BCI, brain signals are read more clearly, but surgical intervention is required. Non-invasive BCI does not require surgical intervention, but brain signals are not as clear as invasive. In addition, non-invasive BCI can be difficult to implement because it requires expensive hardware and advanced signal processing procedures (Chenet et al., 2022).

A BCI system aims to assess and evaluate the characteristics of nerve signals that imply an individual's goals, all translated into real-time commands that put the individual's intentions into action. In order to achieve this system correctly, it involves four consecutive components: signal acquisition, feature extraction, translation, and output. These components are controlled by a software procedure that completes the operation's start and timing, the signal processing details, the device commands' functions, and performance monitoring. Signal acquisition is the measuring of nerve signals using a unique sensory modality. For example, electrophysiological activity is measured by scalp electrodes, and fMRI assesses metabolic activity. The brain signals are processed to levels suitable for processing (filtering is performed to remove electrical noise, 60 Hz power line interference, and artifacts of undesired signal characteristics). The signals are converted into a mathematical function and transmitted to the computer (Schirrmeister et al., 2020).

Feature extraction refers to analyzing the digital signals process, separating unwanted signals from desired signals by analyzing relevant signal features (related to the action the individual desires to perform), and presenting the separated signals in a form suitable for translation into commands. There must be a strong synchronization accompanying the individual's intention. The majority of the important brain activity may be temporary. In the present-day BCI systems, the most frequently extracted signal features are the amplitude oscillations of EEG or ECoG responses, their latencies, power increases in certain frequency bands, and firing rates of cortical neurons, which are crucial for accurate measurements (Revett et al., 2010).

The consequential signal characteristics are conveyed to translation algorithms, transforming them into suitable issues for the output instrument. For example, a decrease in power in a particular frequency band can be interpreted as an upward displacement of the computer cursor. Likewise, the P300 potential may translate into the individual choice of the letter to be stimulated. The translation algorithm adapts to spontaneous and taught changes in signal characteristics and provides device control of the range of values (Chen et al., 2022).

2.1. Current Electrophysiologic Brain–Computer Interface (BCI) Systems

The interface operation of the connections coming from electrophysiological signals is usually structured as follows: first, the electrodes detect the brain signals, then these detected signals are processed by a BCI microcontroller (usually Esp32 or Arm) to eliminate the noise caused by the hardware state of the device. The resulting signal is analyzed to determine the class of the resulting command. "Artificial Neural Networks" are usually used for this purpose due to their high data processing capacity and algorithmic adaptation capabilities. The detected command is sent to an external device according to the probabilities and predictions determined by the hardware and software support systems designed for BCI (Kübler, 2020).

Neural interfaces are divided into three categories based on invasiveness: invasive, semi-invasive, and non-invasive. Since they are directly implanted into the human brain, invasive interfaces provide the highest data area analysis and application of today's systems. However, invasive interfaces carry risks because they are applied with a surgical procedure. Non-invasive interfaces allow brain activity to be analyzed using EEG, MEG, and fMRI technology to measure brain activity through the skull. Semi-invasive BCIs, ECoG, are a technology that measures brain activity completely non-invasively with electrodes positioned on the brain surface under the skull bone. Invasive and semi-invasive neurointerfaces are used to improve the well-being of individuals. These devices are used as new-generation therapeutic and diagnostic methods to correct or prevent various diseases. Non-invasive neural interfaces have recently gained significant attention in the gaming industry. As neuro-helmets emerge and increase, this sector is about to experience a revolution. Another method is functional near-infrared spectroscopy, magnetoencephalography, and similar BCI devices. These methods measure the neural signals we want to measure differently than EEG. It is very important to increase BCI hardware investments because of their higher temporal resolution and better ability to measure the activities of deep brain structures compared to EEG (Kübler, 2020).

2.2. Electroencephalography (EEG) Platform

The electroencephalogram is one of the most common neurotechnological systems to monitor the brain's neural activity. It can be implemented in the diagnosis and treatment of neurological diseases and in monitoring cognitive and emotional processes. Recently, it has led to the development and widespread use of EEG. Its small sensors show accurate performance, are easy to use, and are cheaper than other methods. These advantages have resulted in the widespread use of EEG-based BCI techniques today.

EEG sensors record the neural electrical activity using electrodes on the scalp. Monitoring the activity detects changes related to sleep, wakefulness, or all other cognitive states. Thanks to EEG readings, it can detect some types of abnormal brain activity. The data recorded and collected by EEG facilitates researchers in the analysis of fast cognitive functions such as learning, memory, and attention span. It can be shown that the human brain has four different rhythms that can be categorized according to their frequencies: delta (0.1–4 Hz), theta (4–7.5 Hz), alpha (7.5–12 Hz), beta (12–30 Hz) and gamma (above 30 Hz) bands are available (Rabiner, 1989).

With the recent developments in EEG sensor technologies, they are becoming cheaper, lighter, and safer, providing higher precision and trustworthiness than conventional approaches. The dry electrode system has been introduced with the development of newer systems, eliminating the conductive gel requirement for the appropriate implementation of EEG systems. A new portable device also lets us record your home-based EEG signals without going to a clinic. It allows the data to be transmitted wirelessly to a computer for analysis using Bluetooth. New signal processing algorithms have led to developments for better detection and analysis. For example, detecting subtle changes in short periods is now possible, as is conducting advanced diagnostic abilities (Chen et al., 2022). Finally, new-generation artificial intelligence techniques are under investigation to help stabilize particular facets of raw data processing and provide a more rapid diagnosis (Kok et al., 2024).

The number of electrodes and their positioning affect the quality level of the EEG signal. Increasing the number of electrodes improves spatial resolution and allows for detailed analysis. However, each system's mantle protocol has disadvantages; for example, applying more electrodes increases equipment costs. Researchers and practitioners should choose the most optimal sensor structure and mounting for all these reasons. Positioning is crucial to obtaining high-quality signals during EEG recordings. Differences in positioning, collecting data on different parts of the brain, and providing different levels of information should be carefully considered when deciding which mounts to use. Individual differences in shape and thickness of the skull and inter-individual impedance may require personalized mounting protocols, designs, and adjustments. Channels are usually placed symmetrically to each other. The most important reason for placing channels symmetrically to each other is that artifacts caused by movement adversely affect the signal quality and thus reduce signal analysis and data quality. The effectiveness of an EEG system depends on the selection, number, arrangement and placement of the sensors used. All factors that can minimize the costs associated with hardware requirements should be considered and selected according to the purpose of the research (Presacco et al., 2011).

2.3. Other Platforms

2.3.1. Functional Near-Infrared Spectroscopy (fNIRS)

This non-invasive brain imaging method employs near-infrared light to assess alterations in hemoglobin in the brain. fNIRS measures regional cortical activity. It provides real-time recording of neural activity. Some methods are combined with BCI applications such as EEG to understand cognitive states such as language processing, emotional state, and motor cortex. The number and positioning of diodes are significant in this method. Using a small number of diodes prevents the collection of sufficient information to measure and evaluate brain activity. If there are too many diodes, the signals interfere with each other, preventing the analysis of healthy and accurate data. Positioning diodes close to one another causes the signal value from one sensor to dominate the other. During the experiment, depending on the type of application, it is necessary to evaluate how many channels should be monitored simultaneously and where these channels should be placed relative to each other. When determining the placement setup, attention should be paid to how many channels should be included to ensure accurate data collection (Picton et al., 2000).

2.3.2. Magnetoencephalography

This non-invasive brain imaging method measures the electromagnetic fields produced inside neurons. It provides high temporal resolution and exceptional spatial accuracy. Signal density is a significant factor affecting signal resolution. Sensor density is correlated to the accurate targeting activity of neurons. Each individual's head is modeled in detail for optimal positioning according to the person's specific anatomy before the electrodes are placed. Optimal placement depends on the application type and data we want to obtain (Pekša & Grabis, 2018).

2.3.3. Electrocorticography (ECoG)

ECoG is an invasive method that records neuronal activity by placing electrodes directly on the brain surface. It provides direct access to the desired brain structures with a high temporal resolution (milliseconds), exceptional signal quality, and very low noise levels. It is a surgical method due to its invasive nature. It has higher accuracy than conventional EEG due to its wider electrode coverage areas. However, ECoG may perform better for some small-scale ECoGs with complex structures, such as visual stimulus identification and multitasking classifications (Marshall et al., 2013).

3. Comparison of Brain–Computer Interface (BCI) Platforms

Comparison of BCI Platforms Based on our analysis of the signal processing methods used in BCI in the previous section, we can make some inferences across platforms.

EEG is a conventional, non-invasive method that records neuronal electrical activity. It has good temporal resolution and detects electrical signals within milliseconds. It is a cheap system. The most basic level methodologies are signal quality, feature extraction, classification, and system security. It is known as the most suitable system among BCI systems. Feature extraction can extract meaningful information from raw EEG data and determine users' thoughts. It is the most widely used method to ensure the implementation of commands. Classification is very important in BCI research, with efficient algorithms designed to classify EEG signals into different categories. There are also difficulties in EEG protocols due to system security, installation requirements, long-term use, and different combinations between users and sessions to avoid performance loss (Lotte et al., 2007).

fNIRS assesses differences in oxy and deoxyhemoglobin levels regarding neural activity. It provides remarkable spatial resolution and the ability to detect numerous cortical areas simultaneously. However, its temporal resolution is quite slow compared to EEG. The principal problems are signal quality, spatial, temporal resolution, and cost. The signal quality is poor, and it is very sensitive to noise. The temporal resolution of fNIRS has a relatively slow response time compared to other BCI structures, such as EEG or MEG. More complex cognitive tasks may not be suitable for this technology. Although fNIRS is increasingly being presented to the market at a reasonable price, it remains more expensive than other non-invasive systems, such as EEG. The fNIRS system is generally known to be reliable, but there is a possibility that the brain may be exposed to excessive light in unconscious use, causing eye damage (Lotte et al., 2007).

MEG has a very high temporal resolution. While detecting neuronal firing patterns, systems such as MEG are expensive and require special equipment not available outside of research laboratories or hospitals. MEG signals are strong and easy to detect reliably. Therefore, the biggest disadvantage of MEG is the high cost of the equipment, which limits its practicality and widespread use. MEG's spatial resolution is limited compared to other imaging technologies, like EEG. Data acquisition times for MEG can be long, causing measuring difficulties concerning dynamic processes (Picton at al., 2000).

Due to its direct touch with neurons, ECoG has exceptional temporal and spatial resolution. The surgical intervention required for the ECoG procedure brings well-known surgical risks, such as bleeding and infection. ECoG signals are relatively low amplitude. Reliable signal acquisition is crucial for successful BCI applications. Because of the complexity of neural activities, accurately interpreting data collected from ECoG recordings is a fundamental challenge (Elshenaway & Guirguis, 2021).

4. Signal Processing

The most important intersection of signal processing in BCI technologies is the synchronization between the methods used to connect users' brain signals with computer systems. Synchronization ensures accurate and reliable matching between two signals, while asynchrony allows for temporal changes. If two events or signals have a temporal relationship, synchronization is often utilized to guarantee that all data collected by one device is instantly moved to another device (Wang et al., 2021).

Compared to asynchronous techniques, simultaneous data collection greatly improves signal detection precision, as temporal differences between the information acquired during analysis can be accounted for (Wang et al., 2021). Synchronization of multiple channels allows the alignment of EEG measurements from different electrode sites, such as event-related potentials (ERPs). On the other hand, asynchronous processing allows for a certain degree of variation in situations that do not require high precision. These methods are often used with buffering techniques to help solve the latency problems related to rapidly conveying considerable portions of data between neuronal circuits. Asynchronous strategies have proven effective in reducing the margin of error in BCI applications. A synchronous methods were adequately implemented in motor imaging systems, P300 spellers, and hybrid systems combining EEG and EMG (Noah et al., 2020).

4.1. Independent Component Analysis (ICA) in Brain–Computer Interface (BCI) Systems

ICA is a robust statistical method for distinguishing self-dependent information from data streams. In BCI systems, ICA increases the breadth of EEG data processing by diverging it from the collected discrete noise. ICA helps generate expressive information, such as ERPs, to interpret and analyze the neuronal activity patterns of desired cells better. It is also valuable for decreasing false determinations emerging from motion artifacts (Batista et al., 2024).

ICA signal processing library models such as Infomax, JADE, FastICA, and RADICAL presume that the signal is non-Gaussian or consists of a solitary Gaussian origin. These attitudes do not consider the signal's temporal structure or remove Gaussian distributed elements. New time series-based software and hardware ICA models such as TDSEP/SOBI and AMUSE have been developed. The new extended Infomax ICA library helps to clear both super-Gaussian artifacts and sub-Gaussian signals and prevents hardware overload (Lotte et al., 2018).

4.2. Wavelet Transforms and Autoregressive Modeling in Brain–Computer Interface (BCI)

Wavelet transforms and autoregressive modeling are applied in BCI systems to recognize neural signal patterns. These techniques facilitate the analysis of different types of frequencies related to cognitive tasks. Autoregressive modeling uses the "Linear Regression" model, a sub-AI application method of Machine Learning, to predict the current value of a signal based on past values. It is often combined with AI algorithms such as "Support Vector Machines" or "Artificial Neural Networks" to notice subtle details and modest differences in EEG signals that may indicate mental states. These techniques provide innovative and powerful tools to improve BCI system performance, accuracy, and responsiveness. Discrete Wavelet Transform (DWT) is a technique used to calculate detailed waveform elements by measuring low and high-frequency signals. DWT reconstructs the remaining signal patterns by equating the decomposed coefficients (Urigüen et al., 2015).

Regression analysis is more straightforward than other AI algorithms, but the correct results depend on the predictions (Picton et al., 2000). Regression algorithms require reference channels for artifact detection that depend on external factors such as electrooculography (EOG) or electrocardiography (ECG). However, it is more difficult to determine the minimum reference signal for myographic and other non-biological methods. While regression techniques can identify optical devices, their capability to decrease two-sided unwanted external factors between EEG and EOG is more limited (Klados et al., 2011). To address this limitation, various solutions are produced by sub-filtering the regression algorithms of the recorded signals with the Bayesian method (Senoussi et al., 2017).

4.3. Support Vector Machines (SVM) in Brain–Computer Interface (BCI) Systems

Support Vector Machines (SVMs) are widely applied in BCI systems for classifying and recognizing EEG signal patterns. SVMs extract features from raw EEG data and classify various cognitive states, such as individuals' mental workload levels, into detailed and distinct classes (Lotte et al., 2007).

Classification is performed with a two-step method, including training and testing stages. First, the data is separated as a training set and a test set, and each is assigned data labels so that we can distinguish certain features. In the first stage, a classifier creates a trained model from the training examples by classifying the data into predefined classes. In the second stage, the trained model is used to classify the EEG patterns into possible classes to determine the level of accuracy, and then the classifier's effectiveness is measured (Lotte et al., 2007).

4.4. Hidden Markov Models (HMMs) for Brain–Computer Interface (BCI)

Hidden Markov Models (HMMs) are applied in BCI systems to determine participant's intentions using their brain signals. HMMs detect signal patterns in brain signals in tasks that require high cognitive skills, such as actions to be performed (cognitive tasks), and classify these patterns into different types of assigned classes. These models are repeatedly trained (done with AI accelerator hardware) to predict user actions from subsequent data and are made usable by transferring the data to computer systems. Based on the Bayesian probability approach, the HMM method has been used to classify time series. This technique is extensively applied for analysis of the signal, classification, modeling, and control purposes (Rabiner et al., 1989).

The problems that arise in the HMM system are solved by employing "Forward Procedures" or "Viterbi algorithms" in the decoding phase (Rabiner et al., 1989). HMMs are integrated into the software as probabilistic methods used to calculate the probability of a given feature vector sequence in BCI systems. Each probability condition generates a distribution of a particular feature vector, and those mentioned earlier "Gaussian Models" are widely employed in BCI applications (Lotte et al., 2007).

Although studies have reported high success rates, HMMs have yet to be found across-the-board applications in BCI development. One of the biggest challenges is examining the pattern synchronization triggered by events reported in different channels when analyzing EEG signals is challenging. Instead, defining the classified action labels as a priori, excluding the background neural activity, facilitates the processing and analysis of the signal (Pekša et al., 2018).

4.5. Neural Network Algorithms for Brain-Computer Interface (BCI) Systems

Machine learning models such as Artificial Neural Networks (ANNs) are used in BCI systems to interpret and classify brain signals (Gu et al., 2021). ANNs employing a BCI system is a new-generation AI algorithm method to identify models in EEG data and mental activity changes (emotions, mood). This method makes it possible for BCI systems to reply more precisely and immediately than older machine learning methods (such as support vectors, decision trees, and linear regression). However, it has higher computational power. Therefore, the hardware costs of the AI accelerator used to train the data are very high. As EEG data reveals new signal patterns over time, artificial neural networks can adapt these patterns with high-fidelity simulation (LeCun et al., 1989). Convolutional Neural Networks (CNNs) are this task's most frequently employed type of neural network. While working with EEG signals, CNNs can reduce the image classification problem by feeding a time-varying signal as input (Gu et al., 2021).

4.6. Genetic Algorithms and Particle Swarm Optimization in Brain-Computer Interface (BCI)

Genetic algorithms (GAs) and particle swarm optimization (PSO) are advanced optimization methods used to tune or optimize the parameters of a BCI system. GAs is an evolutionary algorithm that uses mutation, evolution, and natural selection concepts. GAs solve complex data problems by utilizing genetic principles. These are applied in tasks such as uncovering stable parameters for a machine learning model or determining the briefest distance in a network from one point to another. GAs are gaining popularity daily due to their optimization, low energy consumption, and faster computation of processed data (Naebi et al., 2021).

5. Brain-Computer Interface (BCI) Technologies Research and Development

5.1. Prevention

Systems for determining the level of consciousness and their applications in brain-related investigations have been produced. The effects of smoking and drinking alcohol on brain waves regarding attention have been clarified as follows (Jayalakshmi et al., 2023). The importance of these investigations for medical prevention lies in identifying the likely functional losses and decreased vigilance generated by smoking and alcohol consumption. The authors also examined the brain regions most responsive to alcoholism (Shooshtari et al., 2010).

Traffic accidents are considered a leading cause of loss or severe harm (Peng et al., 2022). Later, analyzing the causes of these accidents became a concern for research in numerous domains. For this reason, the attention and focus skills of individuals with motion sickness, especially drivers, are examined. Motion sickness, caused by conflicting sensory information sent to the brain from the inner ear and eyes, typically occurs during transportation. It can lead to traffic accidents due to the reduced ability of the individual to control themselves. EEG power indicators can contribute to the driver's vigilance system by predicting motion sickness. Five brain regions were examined for EEG signals (Wei et al., 2011). Since the human hearing level is part of the sensory information collection process, auditory evoked potential was measured via a BCI-based system. In another study, a virtual reality-based motion sickness platform was designed with a 32-channel EEG system and a joystick used to report motion sickness in real-time experiments. Monitoring of consciousness levels via brain waves was extended to drivers and patients alone (Huggins et al., 2017).

5.2. Detection and Diagnosis

The mental state monitoring function of BCI systems contributes to the prediction and detection of health tribulations such as abnormal brain structures (such as brain tumors), seizures (such as epilepsy), sleep disorders (such as narcolepsy), and brain swelling (encephalitis). EEG is a cost-effective alternative to MRI and CT scans for detecting a tumor resulting from uncontrolled cell division. EEG-based brain tumor monitoring systems have become the focus of research in Türkiye. In addition, studies on detecting breast cancer by using EEG are of interest.

It affects reading and learning abilities, and early detection helps children overcome confidence and self-esteem issues, allowing them to acquire fundamental knowledge and skills. BCI can detect sleep disorders (**Khalil** et al., 2024). Particular approaches for distributing EEG signals to detect idiopathic rapid eye movement (REM) sleep behavior disorder (iSBD) are presented, as iSBD is a critical early predictor of Parkinson's Disease (PD). A plantar pressure measurement system experimentally verified the association between the gait cycle and EEG signals. This association assists in predicting conditions like dyskinesia, peripheral neuropathy, and musculoskeletal disorders (Wang et al., 2021).

5.3. Rehabilitation and Restoration

Mobility rehabilitation is a method of physical rehabilitation for individuals with movement problems. It helps them regain damaged functional abilities and restore their mobility levels or at least make it easier to adapt to their disabilities. Individuals suffering severe wounds or disabilities like strokes may recover completely (Wei et al., 2011).

When blood flow is abrupt for any reason, a shortage of oxygen emerges for neurons. If the blood flow does not restart in a few minutes, neurons die. The result is called a stroke. Meanwhile, the patient may lose the ability to speak suddenly or experience paralysis on one side of the body, and in the long term, the patient may experience memory problems. Stroke and its result, namely disabilities, are investigated by numerous studies that focus on probably solutions involving neural signals. These studies showed that stroke-related injury regions can reorganize, and disturbances in motor functions can be restored through neuroplasticity (Vidaurre et al., 2018).

Mobile robots can assist locked-in individuals in completing daily activities (Barbosa AO. et al., 2010). For patients who cannot regain their previous levels of mobility or communication, BCI-based prosthetic limbs, also known as neuroprosthetic devices, can be used to restore standard functionality (Dato et al., 2018).

Various reality approaches, such as natural, virtual, and augmented reality, were introduced for BCI-based rehabilitation guidance. The natural rehabilitation methodology uses healthy individuals' neural signals and decoded kinematic parameters (Edelman et al., 2024). It helps stroke patients change their belief patterns to correspond to the recorded signals and retrains healthy brain areas to take control. An alternative method for rehabilitation contains virtual reality, where the movement of an avatar is monitored and controlled by outgoing brain waves (Batista et al.,2024). Augmented reality depicts a third procedure to BCI-based rehabilitation, as seen in the augmented mirror box system, which is viewed as an improvement to Mirror Box Therapy (MBT).

MBT employs neural signals from balanced actions involving damaged and healthy limbs (Regenbrecht et al., 2014).

5.4. Neuroergonomics and Smart Environments

As mentioned earlier, the distribution of neuronal signals is not unique to the medical field. Smart environments like homes, workplaces, and transportation systems can also benefit from BCIs to increase daily life's safety, luxury, and physiological control. Moreover, an alliance with the Internet of Things (IoT) and BCI technologies is expected to emerge (Abdulkader et al., 2015).

A cognitive controller system labeled the Brain–Computer Interface-Based Smart Living Environmental Automatic Adjustment Control System (BSLEACS) was proposed. It scans the user's mental state and adjusts the environmental elements appropriately. The system's functionality is further extended by including a Universal Plug and Play (UPnP) home network (Ou et al., 2012).

On the other hand, environmental contributions were considered in developing BCI-based home applications through context awareness. Additionally, integrating healthcare and smart homes for non-invasive mental health care has been an existing approach in brain–computer applications, as shown below (Lee et al., 2019).

The influence of workload, mental fatigue, and task duration on EEG characteristics was also analyzed (Senoussi et al., 2017). The operating room represents a candidate location for BCI-based smart workplace applications (Maddox et al., 2015). The system measures a surgeon's stress levels and provides warnings based on the type of response.

The cognitive state monitoring function of BCIs was used in smart transportation. Several studies examined the driver's behavior. Studies have demonstrated that interference and exhaustion are two significant sources of inattentiveness, a substantial factor in most traffic accidents (Gu et al., 2021). Various preventive measures contributed to determining the cognitive status of drivers (Liang et al., 2012). Fatigue detection has been extensively studied by using EEG signals. Various models have been examined to differentiate distracted drivers. By analyzing ECG and EEG signals, multimodal context recognition has been proposed for intelligent driving systems to predict concentration and stress levels by linking brain signal concentration values to vehicle speed (Hamdy et al., 2009).

Alcohol-impaired drivers, another factor contributing to traffic accidents, can also be characterized using EEG signals (Liang et al., 2012). A visual-auditory virtual environment was developed to assess and analyze driving responses and associated brain signals. Kawamura and colleagues proposed specific driving tasks for a simulated driving model and explored the neural dynamics generated during these tasks. They also depicted employing multiple stimulation procedures to enhance attention levels when dealing with drowsy drivers (Adewole et al., 2016). Another researcher studied the feasibility of employing driver EEG recordings to monitor emergencies, such as a pedestrian who comes out of nowhere (Soares et al., 2020).

5.5. Neuromarketing and Advertising

BCI investigations have been interested in the marketing sector. Studies have shown that EEG assessments are helpful for television commercials related to business and political areas (Vecchiato et al., 2013). BCI-based assessments measure the attention generated by viewing activity (Yoshimura et al., 2012). On the other hand, investigators have addressed additional cognitive functions in neuromarketing, namely, estimating television commercials' memorability, hence using an alternative approach for advertising evaluation (Vecchiato et al., 2013).

5.6. Education and Self-Regulation

Neurofeedback is a favorable practice for improving cognitive task performance by targeting the modulation of human brain activity. It invades educational systems by using brain electrical signals to determine the clarity level of studied information. Based on the outcome of the experienced response, personalized relations are established for every student. (Jamil et al., 2021). Studies have been conducted on learning self-regulation using non-invasive BCIs. Non-invasive BCIs may serve as a tool for individuals to develop cognitive therapeutic strategies. EEG-based emotional intelligence has also been used to manage the stress accompanying sports competitions (Martinez et al., 2021). Within BCI technology, self-regulation and skill learning have been detailed through neurofeedback via fMRI (Adewole et al., 2016).

5.7. Games and Entertainment

The gaming applications and entertainment sector has become a significant economic market for non-medical brain–computer interfaces. Various games use aircraft to reach any point in a 2D or 3D virtual world (Senoussi et al., 2010). Many studies have examined the combination of brain control capabilities with features of existing games, including games that aim to deliver a multi-brain amusement experience. A video game called Brain Arena allows subjects to participate in a cooperative or competitive soccer game via two distinct BCIs. They can score goals by imagining left—or right-hand movements (Vinod & Thomas, 2018).

However, few EEG games are employed for emotional regulation and neuroprosthetic rehabilitation. These games are either built on new concepts or are based on revised ideas. Tan and Nijholt, as a result of their work, represented the Brainball game, aiming to reduce subjective tension levels. The subjects can only move the ball if they are relaxed. This way, the more relaxed subject has a higher chance of winning and learning to control stress while having fun (Arakawa et al., 2010).

5.8. Security and Authentication

Security systems comprise knowledge-based, object-based, and biometric-based confirmation. These systems are vulnerable to many disadvantages, like uncomplicated passwords with low-profile security, shoulder surfing, identity theft, and cancellable biometrics (Hamdy & Traoré, 2009). Cognitive biometrics or electrophysiology, using modalities where biosignals (such as brain signals) are the sole source of identity, offer a solution to this defenselessness (Švogor & Kišasondi, 2012). The inspiration behind the investigating applicability of electrophysiological methods is that external observers cannot randomly obtain biosignals. Furthermore, they could be highly valuable for users with disabilities or those lacking the associated physical characteristics (Revett K., 2010). Using electrophysiology makes such signals challenging to synthesize, increasing the endurance of biometric systems to fraud attacks. In addition to EEG, electrophysiology as a biometric method can send secret alerts when the authorized user is under external coercion (Elshenaway & Guirguis, 2021). Many studies have considered the verification of EEG signals generated from driving behavior as part of smart driving systems. One study used a simplified driving simulator with a mentally tasked condition to authenticate the driver's identity on demand. Unconscious driver authentication was achieved (Nakanishi et al., 2014).

5.9. Brain–Computer Interfaces (BCI) and Artificial Intelligence (AI)

Tremendous technological advancements have started to narrow the boundaries between humans and machines. What once seemed like science fiction, where the mind controls technology, is now becoming a reality with the help of technological innovations. This new technique is called BCI and Artificial Intelligence (AI). While scientific research on BCIs and AI has often developed independently, new insights into brain research and algorithm development efforts in AI technology have led to a preference for merging these technologies. This method enables precise control of external devices through electrical signals. For individuals with severe disabilities, BCI technologies represent one of the most significant technological breakthroughs of the last decade. BCIs, a new-generation technology, are created to convey information between the central nervous system and sensory organs. They can provide communication unconnected with muscle function in individuals with neurodegenerative diseases. BCIs record brain activity and decode its meaning by manipulating activity in specific areas using electrophysiological techniques. Thus, they are rapidly applied to regulate brain functions (Keutayeva et al., 2024)

The BCI systems development faces some constraints. Even though the amount of information obtained from many extracellular electrodes is substantial, transferring this information remains challenging. Neuroscientists struggle to accurately decipher the background electrical activity originating from an individual's brain to understand their intentions and match that intention to robotic arm movements. This attempt has several restrictions because the neurological correlates of psychological phenomena are not yet fully comprehended or determined. However, recent advances in AI methodologies have shown that AI can better decipher and process neural signals than humans. This success of AI means that it has the potential to become an outstanding assistant in processing signals from the brain before they reach the prosthesis (Bi et al., 2013).

AI refers to various approaches that can model intelligent behaviors with minimal human intervention. Moreover, naturally occurring changes become equivalent or superior to human achievement in distinct tasks. When using AI with BCIs, internal parameters like pulse durations and amplitudes, stimulation frequencies, device energy consumption, stimulation or recording intensities, and the electrical properties of neural tissues are continuously provided to the algorithms. After acquiring this information, AI algorithms can determine meaningful aspects and patterns in the data and simultaneously yield the expected functional outcomes. While these studies are mainly preclinical, continuous advancements may indicate clinically viable innovations in BCIs. On the verge of technological change, combining BCIs with artificial intelligence is also attracting attention. In this article, we review the current condition of BCIs, the function of AI in this field, and the forthcoming tendencies of AI-based BCIs, examining existing applications. Early studies focused on controlling computer mouse cursors for paralyzed patients via BCIs with elevated plausibility (Wallstrom et al., 2004). A cursor-control BCI's essential components contain a detector for recording neural signals, a decoder for interpreting action preferences, and a computer cursor interacting with the outer milieu. Numerous healing techniques have been invented to assist patients with stroke to rescue certain functions concerning influenced limbs.

Nevertheless, about 80% of stroke patients with upper extremity motor deficiencies do not gain satisfactory results using these approaches. Recently, methods for lower extremity rehabilitation employing BCIs have been investigated. The general dogma is to produce afferent feedback activity by locking the loop between cortical activity and movement preference, thereby restoring functional corticospinal and corticocortical connections (Edelman et al., 2024).

Cochlear implants are the prevalent and earliest form of BCI use. United States Food and Drug Administration declared that cochlear implants have restored disordered hearing to more than 200,000 patients worldwide since December 2010. This data is a thriving sample of an afferent interface. Cochlear implants bypass the dysfunctional ear component, convert sound waves into electrical signals, and generate auditory signals by directly stimulating the sensory epithelium of the basilar membrane. When the auditory nerve is damaged, cochlear implants can be placed in the cochlear nerve, pontine junction, cochlear nucleus, or inferior colliculus. More satisfactory resolution of sensory information can be obtained by applying the stimulus directly to the auditory cortex rather than the nerve (Senoussi et al.,2017).

Tetraplegic and anarthrous patients can communicate in real-time using neural markers derived from intracortical neural activity and click-control assistance. With the development of mobile communication tools, researchers have invented a variety of virtual keyboards with more efficient text input qualifications, such as DASHER, which was tested as a BCI communication tool based on gesture-guided one-dimensional EEG control. These devices interpret neural activity when people silently mouth words and use this information to generate synthetic speech sounds (Al-Ani et al., 2010).

Age-related macular degeneration, or Leber's congenital amaurosis, results in retinal degeneration. Dystrophy is observed in photoreceptor cells in the total vision loss that develops due to these conditions. The working principle of visual prostheses can be defined as the prosthesis perceiving the light in the patient's physical environment, converting this light into an artificial stimulus, and transmitting the stimulus to the retina. Primitive vision can be achieved by converting images into binary pulses of electrical signals and transmitting them to the visual cortex (Coffey et al., 2016).

This article highlights the existing studies in AI-based BCIs in the last two decades. Researchers have gained more knowledge about the human brain, and the combination of BCI and AI has contributed to new developments in rehabilitation. It suggests an influential manner of manipulating brain functions by directly supplying understanding and management over the neurons that govern human behavior. One of the most significant advantages of AI algorithms, such as Machine Learning and Artificial Neural Networks, is their ability to provide active real-time modulation and adjustments. Algorithms learn from the data and guide external BCI devices for users based on previous preferences with new data. Adaptive machine learning methods can help users gain control by integrating supervised and unsupervised BCI adaptation. Despite the success in this field, some challenges remain. First, the majority of research has focused on restoring motor ability.

Furthermore, using BCIs in the cognitive training field with AI integration is still in its early stages. Secondly, clinical BCI applications remain very limited. For example, smaller-scale stimulation electrode technologies are needed. Thirdly, machine learning algorithms learn to analyze data by producing rarely predictable and understandable algorithms in the real world.

This leads to unknown process issues between a person's thoughts and the technology acting on their behalf (Cannard et al.,2024).

Ethical and social challenges emerge as technologies integrating the brain directly with computers continue developing. Some forms of these technologies can be expensive, creating cost and accessibility issues for severely disabled individuals to access such assistive technologies. Additionally, BCIs equipped with AI software capable of making autonomous decisions raise concerns about human autonomy. Powered AI accelerators can use digitally stored neural data from the brain to predict our memories, interests, and emotional responses. Furthermore, there are reports of applications like deep brain stimulation, used in Parkinson's disease treatment, leading to side effects such as hypersexuality or impulse control problems in some individuals (Bi et al., 2013).

6. Conclusion

AI-supported BCIs are rapidly expanding in the interdisciplinary integration of fields such as medicine, neuroscience, and engineering. These instruments aim to increase the functionality and enhance the quality of life of individuals with paralysis, individuals with spinal cord injury, amputated individuals, and individuals with acquired blindness, deafness, memory loss, and other neurological disorders. These systems, which also improve normal motor, sensory, or cognitive functions, require regulation and control. BCIs will require more technical development, clinical trials, and regulatory approvals before being widely implemented in clinical applications. The progress of this technology could bring revolutionary changes to the field of medicine (Kok et al., 2024).

7. Takeaways

1. Neuroplasticity provides adaptation to dynamically transforming environment and underlying learning ability.
2. BCIs leverage neuroplasticity to enhance neural engineering and rehabilitation processes.
3. BCIs facilitate communication between the brain and external devices, offering new solutions for neurological disorders.

4. Advances in BCIs are transforming neurorehabilitation, providing personalized treatments for various neurological conditions.
5. Electrical stimulation and other neuromodulation techniques enhance neuroplasticity for improved learning and memory.
6. Understanding and utilizing neuroplasticity is key to developing innovative treatments for diseases like stroke, spinal cord injuries, and neurodegenerative disorders.

Abbreviations

AI: Artificial intelligence
ANNs: Artificial Neural Networks
BCI: Brain-Computer Interface
BCIs: Brain-Computer Interfaces
BCS: Brain-Inspired Cognitive Systems
BSLEACS: Brain-Computer Interface-Based Smart Living Environmental Automatic Adjustment Control System
CNNs: Convolutional Neural Networks
CNS: Central Nervous System
ECoG: Electrocorticography
EEG: Electroencephalography
EMG: Electromyography
EOG: Electrooculography
ERPs: Event-Related Potentials
fMRI: Functional Magnetic Resonance Imaging
fNIRS: Functional Near-Infrared Spectroscopy
Gas: Genetic algorithms
HMI: Human-Machine Interface
HMMs: Hidden Markov Models
IoT: Internet of Things
iSBD: Sleep Behavior Disorder
MEG: Magnetoencephalography
PD: Parkinson's Disease
PSO: Particle Swarm Optimization
REM: Rapid Eye Movement
UPnP: Universal Plug and Play

Bibliography

1. Abdulkader, S. N., Atia, A., & Mostafa, M. S. M. (2015). Brain computer interfacing: Applications and challenges. Egyptian Informatics Journal, 16(2), 213–230. https://doi.org/10.1016/j.eij.2015.06.002
2. Adewole, D. O., Serruya, M. D., Harris, J. P., Burrell, J. C., Petrov, D., Chen, H. I., … & Cullen, D. K. (2016). The evolution of neuroprosthetic interfaces. Critical Reviews™ in Biomedical Engineering, 44(1–2). SSN Online: 1943-619X
3. Al-Ani, T., Trad, D., & Somerset, V. S. (2010). Signal processing and classification approaches for brain-computer interface. Intelligent and Biosensors, 1, 25–66.
4. Arakawa, J., Tsuji, T., & Tan, D. S. (2010, November). Personal identification method for robot with whole-body sensing mechanism. In IECON 2010-36th Annual Conference on IEEE Industrial Electronics Society (pp. 1590–1595). IEEE. https://doi.org/110.1109/IECON.2010.5675445
5. Batista, D., Caetano, G., Fleury, M., Figueiredo, P., &Vourvopoulos, A. (2024). Effect of head-mounted virtual reality and vibrotactile feedback in ERD during motor imagery Brain–computer interface training. Brain-Computer Interfaces, 11(1–2), 11–20. https://doi.org/10.1080/2326263X.2023.2264000
6. Bi, L., Fan, X. A., & Liu, Y. (2013). EEG-based brain-controlled mobile robots: a survey. IEEE transactions on human-machine systems, 43(2), 161–176. https://doi.org/10.1109/TSMCC.2012.2219046
7. Cannard, C., Wahbeh, H., & Delorme, A. (2024). Author Spotlight: Advancing the Study of Brain-Heart Interplay with a Comprehensive EEGLAB Plugin for Multimodal Signal Analysis. JoVE (Journal of Visualized Experiments), (206), e65829.
8. Chen, X., Tao, X., Wang, F. L., & Xie, H. (2022). Global research on artificial intelligence-enhanced human electroencephalogram analysis. Neural Computing and Applications, 1–39. https://doi.org/10.1007/s00521-020-05588-x
9. Coffey, A. L. (2016). Ubiquitous and wearable computing solutions for enhancing motor rehabilitation of the upper extremity post-stroke. National University of Ireland, Maynooth (Ireland).
10. Dato, A., & Crepax, T. (2018). Brain Computer Interface: a Data Protection Perspective. Tilburg Law School.

11. Edelman, B. J., Zhang, S., Schalk, G., Brunner, P., Müller-Putz, G., Guan, C., & He, B. (2024). Non-invasive Brain-Computer Interfaces: State of the Art and Trends. IEEE Reviews in Biomedical Engineering. https://doi.org/10.1109/RBME.2024.3449790
12. Elshenaway, A. R., & Guirguis, S. K. (2021). Adaptive thresholds of EEG brain signals for IoT devices authentication. IEEE Access, 9, 100294–100307. https://doi.org/10.1109/ACCESS.2021.3093391
13. Gu, X., Cao, Z., Jolfaei, A., Xu, P., Wu, D., Jung, T. P., & Lin, C. T. (2021). EEG-based brain-computer interfaces (BCIs): A survey of recent studies on signal sensing technologies and computational intelligence approaches and their applications. IEEE/ACM transactions on computational biology and bioinformatics, 18(5), 1645–1666. https://doi.org/10.1109/TCBB.2021.3052811
14. Hamdy, O., & Traoré, I. (2009, May). Cognitive-based biometrics system for static user authentication. In 2009 Fourth International Conference on Internet Monitoring and Protection (pp. 90–97). IEEE. https://doi.org/10.1109/ICIMP.2009.23
15. Huggins, J. E., Guger, C., Ziat, M., Zander, T. O., Taylor, D., Tangermann, M., … & Aarnoutse, E. J. (2017). Workshops of the sixth international brain–computer interface meeting: brain–computer interfaces past, present, and future. Brain-computer interfaces, 4(1–2), 3–36. https://doi.org/10.1080/2326263X.2016.1275488
16. Jamil, N., Belkacem, A. N., Ouhbi, S., & Guger, C. (2021). Cognitive and affective brain–computer interfaces for improving learning strategies and enhancing student capabilities: A systematic literature review. IEEE Access, 9, 134122–134147. https://doi.org/10.1109/ACCESS.2021.3115263
17. Jayalakshmi, P. K., Manimegalai, P., & Sankar, J. S. (2023, February). An Ample Review of Various Deep Learning Skills for Identifying the Stages of Sleep. In International Conference on Biomedical Engineering Science and Technology (pp. 47–65). Cham: Springer Nature Switzerland. https://doi.org/10.1007/978-3-031-54547-4_5
18. Keutayeva, A., Zollanvari, A., & Abibullaev, B. (2024). Motor-Imagery BCI Systems. Discovering the Frontiers of Human-robot Interaction: Insights and Innovations in Collaboration, Communication, and Control, 233.
18.1 Khalil, R. A., Safelnasr, Z., Yemane, N., Kedir, M., Shafiqurrahman, A., & Saeed, N. (2024). Advanced learning technologies for intelligent transportation systems: Prospects and challenges. IEEE Open Journal of Vehicular Technology. https://doi.org/10.1109/OJVT.2024.3369691

19. Klados, M. A., Papadelis, C., Braun, C., & Bamidis, P. D. (2011). REG-ICA: a hybrid methodology combining blind source separation and regression techniques for the rejection of ocular artifacts. Biomedical Signal Processing and Control, 6(3), 291–300. https://doi.org/10.1016/j.bspc.2011.02.001
20. Kok, C. L., Ho, C. K., Aung, T. H., Koh, Y. Y., & Teo, T. H. (2024). Transfer Learning and Deep Neural Networks for Robust Intersubject Hand Movement Detection from EEG Signals. Applied Sciences (2076–3417), 14(17).
21. Kübler, A. (2020). The history of BCI: From a vision for the future to real support for personhood in people with locked-in syndrome. Neuroethics, 13(2), 163–180. https://doi.org/10.1007/s12152-019-09409-4
22. LeCun, Y., Boser, B., Denker, J. S., Henderson, D., Howard, R. E., Hubbard, W., & Jackel, L. D. (1989). Backpropagation applied to handwritten zip code recognition. Neural computation, 1(4), 541–551. https://doi.org/10.1162/neco.1989.1.4.541
23. Lee, M. B., Kim, Y. H., & Park, K. R. (2019). Conditional generative adversarial network-based data augmentation for enhancement of iris recognition accuracy. IEEE Access, 7, 122134–122152. https://doi.org/10.1109/ACCESS.2019.2937809
24. Liang, S. F., Kuo, C. E., Hu, Y. H., Pan, Y. H., & Wang, Y. H. (2012). Automatic stage scoring of single-channel sleep EEG by using multiscale entropy and autoregressive models. IEEE Transactions on Instrumentation and Measurement, 61(6), 1649–1657. https://doi.org/10.1109/TIM.2012.2187242
25. Lotte, F., Congedo, M., Lécuyer, A., Lamarche, F., &Arnaldi, B. (2007). A review of classification algorithms for EEG-based brain–computer interfaces. Journal of neural engineering, 4(2), R1. https://doi.org/10.1088/1741-2560/4/2/R01
26. Maddox, M. M., Lopez, A., Mandava, S. H., Boonjindasup, A., Viriyasiripong, S., Silberstein, J. L., & Lee, B. R. (2015). Electroencephalographic monitoring of brain wave activity during laparoscopic surgical simulation to measure surgeon concentration and stress: Can the student become the master?. Journal of endourology, 29(12), 1329–1333. https://doi.org/10.1089/end.2015.0239
27. Marshall, D., Coyle, D., Wilson, S., & Callaghan, M. (2013). Games, gameplay, and BCI: the state of the art. IEEE Transactions on Computational Intelligence and AI in Games, 5(2), 82–99. https://doi.org/10.1109/TCIAIG.2013.2263555

28. Martinez-Marquez, D., Pingali, S., Panuwatwanich, K., Stewart, R. A., & Mohamed, S. (2021). Application of eye tracking technology in aviation, maritime, and construction industries: A systematic review. Sensors, 21(13), 4289. https://doi.org/10.3390/s21134289
28.1 Naebi, A., Feng, Z., Hosseinpour, F., & Abdollahi, G. (2021). Dimension reduction using new bond graph algorithm and deep learning pooling on EEG signals for BCI. Applied Sciences, 11(18), 8761. https://doi.org/10.3390/app11188761
29. Nakanishi, Y., Wada, F., Saeki, S., & Hachisuka, K. (2014). Rapid changes in arousal states of healthy volunteers during robot-assisted gait training: a quantitative time-series electroencephalography study. Journal of neuroengineering and rehabilitation, 11, 1–9. https://doi.org/10.1186/1743-0003-11-59
30. Noah, S., Powell, T., Khodayari, N., Olivan, D., Ding, M., & Mangun, G. R. (2020). Neural mechanisms of attentional control for objects: decoding EEG alpha when anticipating faces, scenes, and tools. Journal of Neuroscience, 40(25), 4913–4924. https://doi.org/10.1523/JNEUROSCI.2685-19.2020
31. Ou, C. Z., Lin, B. S., Chang, C. J., & Lin, C. T. (2012, July). Brain computer interface-based smart environmental control system. In 2012 Eighth International Conference on Intelligent Information Hiding and Multimedia Signal Processing (pp. 281–284). IEEE. https://doi.org/10.1109/IIH-MSP.2012.74
32. Peng, Y., Xu, Q., Lin, S., Wang, X., Xiang, G., Huang, S., … & Fan, C. (2022). The application of electroencephalogram in driving safety: current status and future prospects. Frontiers in psychology, 13, 919695. https://doi.org/10.3389/fpsyg.2022.919695
33. Pekša, J., &Grabis, J. (2018). Integration of decision-making components in ERP systems. In Proceedings of the 20th International Conference on Enterprise Information Systems (Vol. 1, pp. 183–189). https://doi.org/10.5220/0006779601830189
34. Picton, T. W., van Roon, P., Armilio, M. L., Berg, P., Ille, N., & Scherg, M. (2000). The correction of ocular artifacts: a topographic perspective. Clinical Neurophysiology, 111(1), 53–65. https://doi.org/10.1016/S1388-2457(99)00227-8
35. Presacco, A., Goodman, R., Forrester, L., & Contreras-Vidal, J. L. (2011). Neural decoding of treadmill walking from noninvasive electroencephalographic signals. Journal of neurophysiology, 106(4), 1875–1887. https://doi.org/10.1152/jn.00104.2011

36. Rabiner, L. R. (1989). A tutorial on hidden Markov models and selected applications in speech recognition. Proceedings of the IEEE, 77(2), 257–286. https://doi.org/10.1109/5.18626
37. Regenbrecht, H., Hoermann, S., Ott, C., Müller, L., & Franz, E. (2014). Manipulating the experience of reality for rehabilitation applications. Proceedings of the IEEE, 102(2), 170–184. https://doi.org/10.1109/JPROC.2013.2294178
38. Revett, K., Deravi, F., & Sirlantzis, K. (2010, September). Biosignals for user authentication-towards cognitive biometrics?. In 2010 International Conference on Emerging Security Technologies (pp. 71–76). IEEE. https://doi.org/10.1109/EST.2010.32
39. Schirrmeister, R. T., Springenberg, J. T., Fiederer, L. D. J., Glasstetter, M., Eggensperger, K., Tangermann, M., … & Ball, T. (2017). Deep learning with convolutional neural networks for EEG decoding and visualization. Human brain mapping, 38(11), 5391–5420. https://doi.org/10.1002/hbm.23730
40. Senoussi, M., Verdiere, K. J., Bovo, A., Chanel, C. P. C., Dehais, F., & Roy, R. N. (2017, October). Pre-stimulus antero-posterior EEG connectivity predicts performance in a UAV monitoring task. In 2017 IEEE International Conference on Systems, Man, and Cybernetics (SMC) (pp. 1167–1172). IEEE. https://doi.org/10.1109/SMC.2017.8122770
41. Shooshtari, M. A., & Setarehdan, S. K. (2010, October). Selection of optimal EEG channels for classification of signals correlated with alcohol abusers. In IEEE 10th international conference on signal processing proceedings (pp. 1–4). IEEE. https://doi.org/ 10.1109/ICOSP.2010.5656482
42. Soares, S., Ferreira, S., & Couto, A. (2020). Driving simulator experiments to study drowsiness: A systematic review. Traffic injury prevention, 21(1), 29–37. https://doi.org/10.1080/15389588.2019.1706088
43. Švogor, I., &Kišasondi, T. (2012). Two factor authentication using EEG augmented passwords. In Proceedings of the ITI 2012 34th International Conference on Information Technology Interfaces (pp. 373–378). IEEE. https://doi.org/10.2498/iti.2012.0441
44. Urigüen, J. A., & Garcia-Zapirain, B. (2015). EEG artifact removal—state-of-the-art and guidelines. Journal of neural engineering, 12(3), 031001. https://doi.org/10.1088/1741-2560/12/3/031001
45. Vecchiato, G., Cherubino, P., Trettel, A., & Babiloni, F. (2013). Neuroelectrical brain imaging tools for the study of the efficacy of TV

advertising stimuli and their application to neuromarketing (Vol. 3). Berlin, Germany: Springer.

46. Vidaurre, C., Irastorza-Landa, N., Sarasola-Sanz, A., Insausti-Delgado, A., Ray, A. M., Bibián, C., … & Ramos-Murguialday, A. (2023). Challenges of neural interfaces for stroke motor rehabilitation. Frontiers in Human Neuroscience, 17, 1070404. https://doi.org/10.3389/fnhum.2023.1070404
47. Vinod, A., & Thomas, K. P. (2018). Neurofeedback games using EEG-based brain–computer interface technology. Signal Processing and Machine Learning for Brain–Machine Interfaces, 114, 301–329.
48. Wallstrom, G. L., Kass, R. E., Miller, A., Cohn, J. F., & Fox, N. A. (2004). Automatic correction of ocular artifacts in the EEG: a comparison of regression-based and component-based methods. International journal of psychophysiology, 53(2), 105–119. https://doi.org/10.1016/j.ijpsycho.2004.03.007
49. Wang, Y., Luo, J., Guo, Y., Du, Q., Cheng, Q., & Wang, H. (2021). Changes in EEG brain connectivity caused by short-term BCI neurofeedback-rehabilitation training: a case study. Frontiers in Human Neuroscience, 15, 627100. https://doi.org/10.3389/fnhum.2021.627100

49.1 Wei, C. S., Ko, L. W., Chuang, S. W., Jung, T. P., & Lin, C. T. (2011, April). EEG-based evaluation system for motion sickness estimation. In 2011 5th International IEEE/EMBS Conference on Neural Engineering (pp. 100–103). IEEE. https://doi.org/ 10.1109/NER.2011.5910498

50. Yoshimura, N., DaSalla, C. S., Hanakawa, T., Sato, M. A., & Koike, Y. (2012). Reconstruction of flexor and extensor muscle activities from electroencephalography cortical currents. Neuroimage, 59(2), 1324–1337. https://doi.org/10.1016/j.neuroimage.2011.08.029

Virtual Reality Applications in Neuroplasticity Research and Therapy

Suzan Aydın

1. Introduction

Virtual reality (VR) creates interactive, computer-generated environments. It mimics real-world stimuli, triggering accurate physiological and psychological responses. VR also eliminates obstacles in virtual environments (Ali et al., 2023; Hamad & Jia, 2022). Neuroplasticity involves changes in brain cells and reflects the brain's adaptability (Puderbaugh & Emmady, 2023). Experiences, learning, and interactions influence it. VR can promote neuroplasticity by offering varied experiences. Engaging in a virtual environment changes brain activity, triggering neuroplasticity (González-Tapia et al., 2020). VR also boosts cognitive function. This improvement relies on specific neuroplastic changes, such as adjusting synaptic transmission, reorganizing connections, or reshaping dendrites. This chapter will explore the link between neuroplasticity and VR, exploring how VR shapes the brain.

2. Virtual Reality

VR is a computer-based modeling and simulation technology that allows individuals to interact with an artificial three-dimensional (3D) visual or sensory environment. The person enters the VE through a VR headset and/or headsets and acquires new experiences.

Telepresence

Sensors create a sense of being present elsewhere. They track user movements and instantly adjust screen images. This happens as users move. So, in a virtual environment (VE), users see different views based on their head movements and steps. For instance, data gloves allow users to touch and move objects in the virtual world. These gloves transmit the sense of touch and feedback. VR places users in a new digital space, linking their physical, mental, and emotional health. It also uses a safe, multi-sensory environment as a powerful rehab tool. VR scenarios display objects and events in 3D, offering visual and auditory feedback. This makes users active participants in their therapy (Al-Asi et al., 2023).

2.1. History of Virtual Reality

The roots of VR lie in artists and performers creating imaginary worlds. These efforts predate VR technology and showed how belief in artificial environments could be fostered. For instance, panoramas from the eighteenth and nineteenth centuries created illusory spaces in homes and public areas. They blended two-dimensional images with three-dimensional environments, offering immersive experiences. This tradition paved the way for advancements in the twentieth century. Innovations included futuristic theater designs, stereoscopes, 3-D films, and IMAX cinemas. Notably, the Cinerama format showcased the importance of immersive vision in artificial settings. This spurred the development of technologies to cover the viewer's entire field of vision (Lowood, 2024).

The flight simulator is one of the first interactive systems built. The flight simulator is also considered an important milestone. It was patented in 1931 by Edward Link. In the 1950s, Morton Heilig tried to create an immersive and multi-sensory experience in VR. Sensorama, developed by Heilig, offers users a variety of sensory experiences while watching a film. Sensoroma was a large device the size of a cupboard and contained components such as stereo speakers, fans, a vibrating seat, and odor generators. In this way, it stimulates several senses at the same time, such as sound, airflow, vibration, and odor. It creates a sensory experience for users. Sensorama represents an early example of VR technology (Lowood, 2024; Paro et al., 2022).

Tactile feedback has become a critical component in VR systems for enhancing user interaction with objects in the virtual world. One of the early examples of such feedback is the Grope-III system, developed in 1967 at the University of North Carolina. This system was designed to simulate intermolecular forces and emerged as a technology for teaching tactile feedback. It laid the groundwork for high-precision VR systems that would later be used in surgical training (Virtual Reality Society, n.d.).

2.2. The Three Fundamental Concepts of Virtual Reality

VR encompasses three fundamental concepts: immersion, presence, and interaction.

2.2.1. Immersion

Immersion refers to sensually being present in a fully computer-generated 3D environment. Users enter into this virtual world with sensory stimuli. They interact with the virtual world in a K-personal way. Thanks to VR technology immersion, users feel like they are in a separate digital world. In this way, users can reshape the virtual and real worlds. They also experience changing these worlds (Holden, 2005).

The quality of the equipment used to feel immersion is essential. As the quality of the sensory stimuli offered to users with VR systems increases, the feeling of perceiving the virtual world as the real world also increases. If the sense of immersion in VR environments is at a high level, the brain may be unable to distinguish between the natural world and the computer-generated environment. In this way, the user is completely immersed in the VR environment, and the boundaries between the real and virtual worlds are blurred.

2.2.2. Presence

The second key feature of VR is the "sense of presence," which refers to the user's perception of being part of the experience. This sense is closely related to the amount and variety of sensory and motor channels used in the VR system. The accuracy and responsiveness of user interactions depend on the quality of the integrated sensory inputs and motor outputs. The sense of presence represents a psychological state in which users feel consciously connected to the VE. When using VR, people's reactions can vary—physiological reactions, emotional reactions, and behavioral changes. Therefore, the sense of presence is crucial in VR experiences. It shapes the way people perceive reality and their engagement with the VE. In order to increase the power of presence, people need to be fully immersed in VR.

2.2.3. Interaction

The third VR characteristic is interaction. This means the system responds to user actions in real time. It detects input like voice commands, eye movements, and brain waves. Then, it processes this information to respond appropriately. For instance, in a VR game, a user's movement or voice command triggers an immediate response. This makes the experience more interactive and realistic. Moreover, it broadens VR use in education, training, entertainment, and therapy (Sokołowska, 2024).

2.3. Types of Virtual Reality

2.3.1. Non-Immersive Virtual Reality

In non-immersive VR, users engage with virtual images on screens. They employ devices like mice, keyboards, or hand controllers. Crucially, they can also perceive and interact with the real world. This keeps them from being fully immersed in the virtual environment (VE). VEs are typically viewed on monitors or large TVs, limiting immersion. Still, users can be aware of the real world, such as a room, and virtual elements, like a game character. PCs, tablets, smartphones, TVs, Nintendo Wii™, and Microsoft Kinect™ enable this. They provide virtual experiences while keeping users linked to the real world, thus balancing both environments (Sokołowska, 2024).

2.3.2. Semi-Immersive Virtual Reality

Visual displays are also used in semi-immersive VR systems. However, these displays are advanced, such as panoramic or curved displays. This increases the sense of immersion. In total immersion technology, users cannot see outside the screen. However, this technology lets users see the real world from both sides of the screen. Semi-immersive technology is often used for education and training purposes. For example, it is used in flight simulators for student pilots to increase the sense of immersion (Sokołowska, 2024; Smith, 2021).

2.3.3. Fully Immersive Virtual Reality

In this type of VR, users are completely immersed in the VE. The sense of presence is enhanced. People use several main pieces of equipment in this genre, such as head-mounted displays, full-body suits, motion-tracking gloves, controllers, vibrating platforms, omnidirectional (allowing running in all directions) treadmills, and immersive rooms. In these technologies, users can see avatars (virtual characters). With avatars, people explore and interact with the virtual environment. The purpose of these devices is to block sensory information from the real world. It then replaces these senses with computer-generated stimuli. Thus, providing illusions to people makes the virtual world seem as accurate as the physical world. In addition, extra equipment, such as sensors used in clothes to monitor the user's movements or EEGs, are attached to the head to monitor brain activity. As a result, the immersion process generates and records a significant amount of experimental data, which can then be analyzed (Smith, 2021).

Embodiment

Embodiment refers to the phenomenon where a user identifies their own body with the movements of an avatar, such as its hands, within a virtual reality environment. This concept involves the user fully immersing themselves in the VE and perceiving the avatar's movements as if they were their own. For a user to experience embodiment to its fullest extent, they must perceive three fundamental sensations:

1. **Presence**: The sensation of being located in a specific time and space.
2. **Agency**: The ability to perform actions or interventions that produce effects.
3. **Body Ownership**: The belief that the avatar is an extension of the self.

These sensations boost the feeling of being in the VR world. They make interaction more realistic. Immersive VR, with haptic feedback, deepens the sense of being, moving, and owning space. This opens new ways to assess and manage musculoskeletal problems (Riva et al., 2019).

3. Augmented Reality (Ar) and Related Technologies

3.1. Augmented Reality

Augmented Reality (AR) is characterized by virtual objects (e.g., tables, carpets) appearing as if they were in the real world. AR allows users to see and interact with computer-generated perceptual information and natural objects. AR is considered a subset of VR systems. The difference is that in VR, users are fully immersed in a computer-generated environment. In contrast, in AR, users see real-world environments and can interact with virtual objects placed in that environment. Although AR does not entirely change the real world, it enriches it. Three key features are essential in AR technologies. These combine real-world and virtual objects, full-time interaction, and 3-D visualization. Electronic devices with cameras, such as smartphones and tablets, can capture real-world images as technology advances. AR applications can enrich The captured images with animations or other digital content (Invitto et al., 2015).

3.2. Mixed Reality (MR)

MR combines elements from the natural world and the virtual world. It is a more technologically advanced form of AR that does not provide interaction between objects in both worlds. Form. Through MR, users can interact with both real and virtual objects. Digital technologies are improving every day. Thanks to this development, it may be possible to project 3-D holograms into real-world spaces and allow users to interact with these digital projections and controllers (Georgiev et al., 2021).

3.3. Extended Reality (XR)

XR is a more inclusive term. It covers all immersive technologies and all possible future developments. AR encompasses VR and MR. XR combines elements of VE and natural environments. This technology effectively blends VR and AR. These advanced technologies detail the sensitivities, responsiveness, and functions to be performed. They work very precisely. They are essential, especially in basic research, clinical neuroscience, and modern medical practice (Georgiev et al., 2021).

4. Virtual Reality and Neuroplasticity

Neuroplasticity reflects the ability of neural networks to change their structure and function in response to experiences, learning, or trauma. At the synaptic level, these changes are seen as changes in the strength of synaptic connections or changes in synaptic activity within a network. This process occurs through mechanisms such as long-term potentiation or long-term depression of synaptic connections. It also involves the actional reorganization of cortical pathways and reorganization of cortical mapping, which can occur due to disease or injury (Pekna et al., 2012).

The treatments must be intensive, repetitive, and engaging to increase experience-dependent neuroplasticity and functional recovery.

Interventions, simulated environments, and feedback technologies such as VR and AR effectively establish synaptic connections. VR technologies provide a more prosperous environment and sensory integration than real-world interventions. Technology stimulates the person's sensory, cognitive, and motor skills. Thus, it supports the rehabilitation process. As a result, VR technology supports critical neuroplasticity components compared to traditional rehabilitation methods. It can also enhance motor recovery (Hao et al., 2022).

4.1. Effects of VR on Neural Circuits

Virtual reality potentially alters neural processes. The role of visual inputs on the brain is critical in this change. Visual sensory inputs support the reorganization of sensorimotor circuits and optimize brain function by providing strong signals. For example, visual errors can affect motor cortex regions during motor learning.

Biofeedback helps reduce movement errors. It makes learning active and rewarding, shaping brain activity in motor areas. Also, watching and doing actions boost Motor Evoked Potentials (MEPs) and affect interactions in these regions (Cheung et al., 2014).

Visual information modulates many neurons in motor, premotor, and parietal regions. Movement can be tuned by visual information. Proprioception, visual feedback, and illusions can aid movement in limbs, even without actual movement. For instance, patients with severe cerebral palsy or poor voluntary control might be asked to imagine movements they cannot yet do. The visual system can override other senses, like proprioception, during sensory conflicts (Carey et al., 2006).

Biofeedback enhances neural reorganization. Clinicians can control aspects that are not possible in natural settings. For instance, they can freeze a virtual limb and assess movements to boost movement quality. Skills learned in VR training transfer to real life. Similar shifts in brain hemisphere use have been seen after real-world therapy. Training an affected limb in VR with ongoing feedback hints at beneficial neuroplastic changes akin to real-world training.

VR's unique benefits make it vital for clinicians to steer neuroplastic changes. The long-term effects of VR training are under study. However, current data show that VR can explore neural functions and serve as a training tool (Cheung et al., 2014).

4.2. Virtual Reality Influence on Neuroplastic Mechanisms

Using VR technology in treatments showed notable neural improvements in primary motor areas, aiding cognitive recovery. EEG analyses indicated that VR significantly boosted alpha power in occipital areas and beta power in frontal areas but not theta. This increase underscores VR rehabilitation's potential to spark brain changes. Moreover, VR's immersive environment can specifically target cognitive areas, promoting cognitive and motor recovery (Gangemi et al., 2023).

4.3. Mirror Neuron, VR and Neuroplasticity

Understanding mirror neurons (MN) is important for understanding VR. Because it allows us to explore how VR affects potential neural mechanisms. In the human brain, the MN system includes regions such as the inferior parietal lobule, ventral premotor cortex, and inferior anterior gyrus. These regions are not segregated, isolated areas. The structures mentioned are between the primary and secondary motor regions. Here, it functions as a functionally distributed network. Neural stimulation can be stimulated by motor observation, imitation, and visualization. As a result of stimulation, motor rehabilitation approaches become more effective. These facts are an essential factor in the design and development of VR technology. In virtual environments, the avatar visually represents the user from the outside. In this way, VR users perform motor tasks, and feedback devices can observe and adjust users' movements by providing visual and proprioceptive feedback (Hao et al., 2022).

5. Digital Games and Neuroplasticity: the Role of Serious Games and Exergames

5.1. Serious Games and Neuroplasticity

In the real world, challenges can sometimes resolve on their own. Some games are made to maintain this balance. This balance encourages players to improve their skills. Games are like controlled environments. Here, the main aim is to enjoy, not mirror reality. Unusual behavior in the real world can lead to long-term problems. However, games offer a safe space for different experiences. Game developers build these unique environments. They also decide how much mental skill can grow by tweaking game interactions. Games blend or adjust various elements. These can include strategy, creativity, quick reactions, or social interaction. So, picking a game that matches a player's interest is key. It is also important to assess the player's motor and cognitive skills to foster improvement (Cheung et al., 2014; Kleschnitzki et al., 2022).

Gaming boosts creativity, cognitive flexibility, and executive functions (EFs), making it a promising therapeutic tool to unlock potential (Kleschnitzki et al., 2022). Serious games (SG) aim to educate, improve health, strategize, or enhance social interaction. They also serve therapeutic purposes. SG fosters creativity, boosts brain plasticity, and aids cognitive growth.

Virtual reality (VR) SG tackles cognitive decline effectively by boosting engagement and cognitive function (Guzmán et al., 2024). These games leverage neuroplasticity and motor learning principles. They do this by ensuring repetition, adjusting difficulty, offering feedback, and maintaining motivation (Levin, 2011). SG also enhances executive functions like planning, focus, and decision-making. Strategy games, in particular, improve decision-making and coping skills. They encourage active participation, moving away from passive content consumption. Games adapt to players' learning processes, accommodating individual speeds and paths (Kleschnitzki et al., 2022).

Popular consoles like Xbox 360 KinectTM, Nintendo Wii, and Eye Toy-Play System use game-based VR (Sin & Lee., 2013).

5.2. Exergames and Neuroplasticity

Exergames blend exercise with video games, promoting physical activity. They show promise in neurorehabilitation, mainly aimed at young adults. Initially for fun, these games now demand full-body workouts, enhancing motor skills, coordination, balance, and energy use (Mura et al., 2018).

Exercise benefits the brain by fostering adaptations. This includes new nerve cells in the hippocampus, increased blood vessels, and higher levels of BDNF. It also boosts synaptic plasticity (Mura et al., 2018).

In summary, therapeutic games can enhance mental abilities, improve executive functions, and reduce cognitive decline.

6. Virtual Reality for Neurorehabilitation

6.1. Motor Rehabilitation

Rehabilitation aims to improve people's daily lives, both in quality and capability. It focuses on early intervention, task training, and repetition to build motor skills. Additionally, using tasks that engage different senses can boost recovery. Thus, maximizing these repetitive tasks is vital in rehabilitation (Sveistrup, 2004).

Virtual reality (VR) is a powerful tool for improving motor skills in rehabilitation. It enhances skills, boosts motivation, and offers real-time feedback.

VR exercise therapy is effective for various rehabilitation disorders. It overcomes the limitations of traditional methods. VR is more effective than conventional approaches. It aids in pain management, improves function, increases range of motion, boosts muscle strength, and enhances motivation (Asadzadeh et al., 2021).

In 2022, Feitosa et al. reviewed the impact of VR rehabilitation on motor skills and brain adaptability. They examined studies from 2005 to 2021 that used fMRI. These studies showed that VR rehabilitation altered brain activity and reorganized the cortex, especially in post-stroke patients. This indicates that VR therapy fosters brain adaptability and aids recovery. Moreover, exercise games notably enhance executive functions and visual-spatial skills. Similar effects were seen in older adults and patients with other neurological issues. Thus, VR therapy is effective for both motor and cognitive rehabilitation (Feitosa et al., 2022).

6.2. Cognitive Rehabilitation

Virtual reality (VR) is becoming a key tool in cognitive rehabilitation. It helps improve cognitive skills, especially for those aged 65 and over. This age group is growing, leading to more people at risk of cognitive decline. Aging can affect memory, decision-making, processing speed, and reasoning. These functions are vital for independent living. Moreover, cognitive impairment puts a heavy strain on healthcare and families.

Spotting cognitive decline early is vital for effective treatment. Yet, many people don't stick to their treatment plans. To improve this, experts suggest combining cognitive exercises, physical activity, socializing, and stress relief.

VR rehabilitation allows people to get therapy at home. It makes sessions more engaging, tailored, and motivating. Therapies adjust challenges based on individual performance. Practicing real-life scenarios in VR helps people apply what they learn in daily life.

In 2024, Tortora et al. reviewed VR-based cognitive rehabilitation (VR-CRT) against traditional methods (CRT) in people over 65 with mild cognitive impairment. They found VR-CRT equally effective as CRT, but with added benefits. VR-CRT offers better ecological and adaptive advantages, especially with high immersion.

Many studies highlight VR exercise programs' positive effects on brain plasticity and recovery. For instance, programs combining exercise and computer games boosted hippocampal volume and BDNF levels in Parkinson's patients, enhancing neuroplasticity. In elderly participants, VR improved cognitive functions and BDNF levels.

Currently, there are no set standards for virtual worlds. Fields like neuroscience and medicine highlight the value of virtual environments (VEs). VEs help assess people's brain and psychological health. They also provide objective data. Many studies have used VR, showing VEs to be engaging and reflective of modern life.

VEs offer motivating experiences that spark exploration of brain capabilities, driven by modern advancements. They are easy to customize, safe for testing cognitive and motor skills. Exercises combining tasks effectively enhance both thinking and movement. Additionally, concepts like metadata can improve current methods and inspire future research (Sokołowska, 2024).

6.3. Psychological Rehabilitation

VR is a promising tool in mental health, especially for treating disorders like anxiety and depression. Its effectiveness ties to neuroplasticity. Research shows VR helps even those with severe psychiatric issues. These treatments are both effective and safe.

The advice to engage in enjoyable activities might seem simple, especially for those with severe depression. Yet, it's backed by solid science. This method, called "behavioral activation," encourages activities like going out, exercising, socializing, volunteering, or doing fun things. Such actions can ease depression symptoms (Williams, 2024).

VR aids cognitive restructuring in depressed individuals. It offers immersive experiences with therapeutic content. Users can create and explore different virtual environments. For instance, they might find themselves by the sea, listening to its calming sounds. Or, they could relax in virtual settings like snowy mountains, peaceful lakes, or blooming gardens (Williams et al., 2024).

Virtual Reality Exposure Therapy (VRET)

Exposure therapy helps treat phobic anxiety disorders effectively. It gradually exposes people to anxiety triggers, improving their thoughts, emotions, and physical reactions. Therapists use live or imagined scenarios for this.

Triggers can be anything from animals and objects to tough situations, scary thoughts, or painful feelings. VRET is a modern take on this therapy, using virtual environments (VEs) to present fears. This method gives therapists better control and customization for treatments. It also preserves privacy and is often less frightening than real-life exposure.

VRET is effective for specific phobias like fear of heights (acrophobia) and spiders (arachnophobia). It also helps with social anxiety, autism, PTSD, and addiction issues. Research shows VR is a promising tool for treating anxiety and depression in various settings (Baghaei et al., 2021).

6.4. Learning and Education

VR boosts learning with interactive, visual experiences. It aids understanding of complex topics and enhances neuroplasticity. For language learners, VR simulates real-life interactions, fostering exploration and knowledge (Wang et al., 2020).

In VR, users engage deeply with subjects. This activity stimulates the brain and forms new connections. Unlike textbooks, VR offers immersive, first-person learning. This method personalizes understanding. Textbooks may fall short. VR's immersion eliminates the need for physical presence. It offers a multicultural experience, aiding language learning without geographical limits (Wang et al., 2017).

7. Latest Research on Virtual Reality and Neuroplasticity

Neuroplasticity is vital for recovery after neurological injuries. It helps the brain heal after strokes, traumatic brain injuries, and neurodegenerative diseases (Kumar et al., 2023).

7.1. Stroke

Stroke is a major cause of long-term disability. However, treatments using neuroplasticity can help. The stroke's location and severity determine brain damage. Patients often face issues with movement, speech, thinking, and sensing. Yet, neuroplasticity allows for recovery. Rehabilitation should consider the brain's healing and damage mechanisms. Treatments include VR, transcranial magnetic stimulation (TMS), physiotherapy, occupational therapy, and drugs.

One review studied VR interventions for stroke rehabilitation. It found key benefits, such as:

1. Balance between brain hemispheres.
2. Stronger brain connections.
3. A larger brain map for affected muscles.
4. Better neuroplasticity linked to improved behavior.
5. More activity in the frontal cortex.
6. Involvement of the mirror neuron system.

These results show VR's positive impact on recovery (Hao et al., 2022).

Another study looked at VR motor training in chronic stroke patients. It split 30 patients into two groups: standard therapy and VR therapy. Both groups had 16 sessions, 60 minutes each, 2 to 3 times a week. The VR group used HTC VIVE goggles. The VR group improved significantly. However, there was no change in BDNF levels, suggesting these levels depend on the type of therapy (Huang et al., 2022).

In another study, EEG measured the effects of VR on neuroplasticity in chronic stroke patients. It randomly assigned 30 patients to two groups. One group had VR cognitive training; the other had standard treatment. After training, both groups' EEG rhythms were analyzed. The VR group showed better results in the primary motor circuit. It also had more alpha and beta power in the brain. However, theta power did not change. This suggests VR rehabilitation can promote neuroplasticity even in chronic cases (Gangemi et al., 2023).

7.2. Traumatic Brain Injury (TBI)

Traumatic brain injury (TBI) is sudden brain damage from external force. It can lead to cognitive, behavioral, and physical issues. Damage to specific brain areas, like the frontal and temporal lobes, affects attention, memory, emotions, behavior, and executive function. This happens alongside motor impairments. The brain's healing and adaptation post-TBI is called neuroplasticity (Kleim et al., 2004; Bonanno et al., 2022).

In 2022, a review studied virtual reality (VR) training for TBI patients' attention and mental health. It analyzed 29 studies. It found VR, computer methods, telerehabilitation, and neuromodulation helpful in cognitive rehab. VR attention training notably improved attention, reduced depression, and enhanced cognitive and behavioral outcomes (Bonanno et al., 2022).

7.3. Other Neurological Diseases

Therapies that enhance neuroplasticity significantly aid in treating Alzheimer's, Parkinson's, and multiple sclerosis. These diseases slowly damage brain tissue, leading to cognitive decline, movement issues, and other symptoms. They often involve neuroplasticity disorders. However, the brain retains some plasticity even in such conditions. Enhancing neuroplasticity can slow disease progression and ease symptoms. This can be done through drugs, cognitive tasks, physical activities, and non-invasive brain stimulation techniques like TMS and tDCS (Kleim et al., 2004).

A study involved 17 Parkinson's patients and 18 healthy controls in a 6-week program combining computer games and physical exercise. MRI scans assessed hippocampal regions and BDNF levels before and after the program. In Parkinson's patients, significant growth was noted in specific hippocampal areas post-exercise. The most notable increase was in the dentate gyrus, which had lower initial volumes than healthy controls. Both groups saw BDNF levels rise. This shows that combining games with exercise boosts hippocampal volume and BDNF in Parkinson's patients (Schaeffer et al., 2022).

Another study looked at the impact of a 6-week exercise game on cognitive function and BDNF in elderly care residents. The tailored program ran twice a week for 40 minutes. Cognitive function was tested before and after. Blood samples measured BDNF levels. No overall cognitive score changes were seen. However, improvements in language were noted, with tendencies in abstract thinking and memory. BDNF levels significantly increased. This rise might be linked to cognitive changes (Monteblanco et al., 2021).

A trial compared exergames and traditional training in older women. Twenty-two women were split into two groups. Blood samples and cognitive tests were conducted before and after the trial. Both methods enhanced cognitive performance and BDNF levels. They also improved inflammatory profiles and induced changes in histones. This suggests that exergames could promote cognitive growth in healthy older women. The responses were linked to histone changes and inflammatory markers, varying with exposure duration (Henrique et al., 2023).

8. Impact of Virtual Reality on Different Age Groups

While young people often use virtual reality, many hardware manufacturers specify age restrictions in their product manuals. According to the Safety Center website, the Meta Quest device is unsuitable for children under 13.

Children (13–18 years): Children aged 13–18 can benefit from VR's many opportunities. VR:

- Provides a fully immersive environment where children can interact with other players, which is unique compared to the real world.
- Helps develop their ability to follow rules and guidelines, improving their self-discipline.
- Demonstrates appropriate activities through technological means.
- Encourages children to become respectful digital citizens.
- Allow them to see things from a different perspective.

Adults (18–65 years): Adults between 18 and 65 years of age derive several benefits from using VR:

- It provides opportunities to socialize and develop a sense of community with other players in a VE.
- Allows users to visit places around the world and in remote locations.
- Practice mindfulness with VR-based relaxation applications.
- It is used to treat various phobias such as fear of spiders (arachnophobia), fear of heights (acrophobia), and fear of enclosed spaces (claustrophobia).
- It is accessible to people with disabilities and can be used for learning, entertainment, and stress reduction.

Elderly (65 years and older): Older people over 65 can benefit from VR in the following ways:

- Reduces feelings of loneliness and social isolation.
- Improves the mental health of the elderly.
- Provides the opportunity to relive memories from the past.
- Bring joy and fun to the elderly.
- Help them learn new skills and keep their minds sharp.
- Reduce or eliminate the use of certain medications for anxiety and other mental health conditions.
- VR and AR technologies can also promote a culture of equality and inclusion by providing opportunities for marginalized and underserved groups.

9. Limitations of Virtual Reality

VR has many benefits but also some drawbacks. Firstly, we need more prominent studies to see how well VR works for sensory-motor rehab across different groups. We also lack knowledge on how VR training impacts real-world activities and what settings are best for improvement. It is still unclear if VR is better than traditional training and why. We need to find out if VR is unique or just more fun. It is vital to see if VR keeps people more engaged than other treatments (Cheung et al., 2014).

10. Conclusions

VR can boost neuroplasticity, changing brain functions through exercises. Research shows that VR makes exercises more appealing and easier to manage. It is a key tool for linking cognitive and motor functions. VR aids in boosting neuroplasticity in healthy people and treating neurological diseases. It enhances cognitive and motor flexibility. Such interventions raise treatment success and improve life quality.

However, more studies are needed to fully grasp VR's impact and clinical use. We need extensive, thorough studies to find VR's effectiveness in various conditions and to set up the best rehab plans. Despite this, VR could revolutionize neuroplasticity therapies and set new standards for neurological rehab.

11. Takeaways

1. Virtual reality can produce neuroplastic effects on the brain by exposing users to different environments and experiences.
2. VR, an immersive and engaging environment, can support motor reorganization.
3. Brain-derived neurotrophic factor plays a critical role in neuroplasticity by supporting the survival and growth of nerve cells. Virtual reality may increase BDNF release.
4. Many clinical studies support the positive effects of VR-based treatment and rehabilitation approaches on neuroplasticity.
5. VR can be effective in rehabilitation of motor skills and cognitive functions, especially in recovery processes after stroke and traumatic brain injuries,
6. Compared to traditional rehabilitation methods, virtual reality technology may be more effective in supporting important components of neuroplasticity and enhancing functional recovery outcomes.
7. Virtual reality can be used in motor rehabilitation for different purposes such as exercise, pain management, improving functional abilities, increasing range of motion, supporting muscle strength, and increasing motivation.
8. Virtual reality can be an interactive visualization that supports cognitive restructuring. It allows users to create and manipulate various environments.
9. Using games for therapeutic purposes can positively affect individuals' mental capacities, improve executive functions, and reduce cognitive impairments.
10. Although VR offers many advantages, there are some limitations and uncertainties about practical applications and the generalizability of effects.
11. Future research in VR and neuroplasticity will enable the development of innovative therapies and training strategies

Abbreviations

3D: Three Dimensional
AR: Augmented Reality

BDNF: Brain-Derived Neurotrophic Factor
EEG: Electroencephalography
fMRI: Functional Magnetic Resonance İmaging
MN: Mirror Neurons
MR: Mixed Reality
TBI: Traumatic Brain Injury
TMS: Transcranial Magnetic Stimulation
VE: Virtual Environment
VR: Virtual Reality
VRET: Virtual Reality Exposure Therapy
XR: Extended Reality

Bibliography

1. Al-Ansi, A., Jaboob, M., Garad, A., & Al-Ansi, A. (2023). Analyzing recent educational developments in augmented reality (AR) and virtual reality (VR). Social Sciences & Humanities Open, 8, 100532. https://doi.org/10.1016/j.ssaho.2023.100532
2. Ali, S. G., Wang, X., Li, P., Jung, Y., Bi, L., Kim, J., Chen, Y., Feng, D. D., Magnenat Thalmann, N., Wang, J., & Sheng, B. (2023). A systematic review: Virtual-reality-based techniques for human exercises and health improvement. Frontiers in public health, 11, 1143947. https://doi.org/10.3389/fpubh.2023.1143947
3. Asadzadeh, A., Samad-Soltani, T., Salahzadeh, Z., & Rezaei-Hachesu, P. (2021). Effectiveness of virtual reality-based exercise therapy in rehabilitation: A scoping review. Informatics in Medicine Unlocked, 24, 100562. https://doi.org/10.1016/j.imu.2021.100562
4. Baghaei, N., Chitale, V., Hlasnik, A., Stemmet, L., Liang, H. N., & Porter, R. (2021). Virtual reality for supporting the treatment of depression and anxiety: A scoping review. JMIR Mental Health, 8(9), e29681. https://doi.org/10.2196/29681
5. Bonanno, M., De Luca, R., De Nunzio, A. M., Quartarone, A., & Calabrò, R. S. (2022). Innovative Technologies in the Neurorehabilitation of Traumatic Brain Injury: A Systematic Review. Brain sciences, 12(12), 1678. https://doi.org/10.3390/brainsci12121678
6. Carey, L. M., Abbott, D. F., Egan, G. F., O'Keefe, G. J., Jackson, G. D., Bernhardt, J., & Donnan, G. A. (2006). Evolution of brain activation with good and poor motor recovery after stroke.

Neurorehabilitation and neural repair, 20(1), 24–41. https://doi.org/10.1177/1545968305283053
7. Cheung, K. L., Tunik, E., Adamovich, S. V., & Boyd, L. A. (2014). Neuroplasticity and Virtual Reality. pp. 5–24. https://doi.org/10.1007/978-1-4939-0968-1_2
8. Feitosa, J. A., Fernandes, C. A., Casseb, R. F., & Castellano, G. (2022). Effects of virtual reality-based motor rehabilitation: a systematic review of fMRI studies. Journal of neural engineering, 19(1), 10.1088/1741-2552/ac456e. https://doi.org/10.1088/1741-2552/ac456e
9. Gangemi, A., De Luca, R., Fabio, R. A., Lauria, P., Rifici, C., Pollicino, P., Marra, A., Olivo, A., Quartarone, A., & Calabrò, R. S. (2023). Effects of Virtual Reality Cognitive Training on Neuroplasticity: A Quasi-Randomized Clinical Trial in Patients with Stroke. Biomedicines, 11(12), 3225. https://doi.org/10.3390/biomedicines11123225
10. Georgiev, D. D., Georgieva, I., Gong, Z., Nanjappan, V., & Georgiev, G. V. (2021). Virtual Reality for Neurorehabilitation and Cognitive Enhancement. Brain sciences, 11(2), 221. https://doi.org/10.3390/brainsci11020221
11. González-Tapia, D., González-Ramírez, M. M., Vázquez-Hernández, N., & González-Burgos, I. (2020). Motor learning induces plastic changes in Purkinje cell dendritic spines in the rat cerebellum. El aprendizaje motor induce cambios plásticos en las espinas dendríticas de las células de Purkinje del cerebelo de ratas. Neurologia, 35(7), 451–457. https://doi.org/10.1016/j.nrl.2017.10.007
12. Guzmán, D. E., Rengifo, C. F., Guzmán, J. D., & et al. (2024). Virtual reality games for cognitive rehabilitation of older adults: A review of adaptive games, domains, and techniques. Virtual Reality, 28(1), 92. https://doi.org/10.1007/s10055-024-00968-3
13. Hamad, A., & Jia, B. (2022). How Virtual Reality Technology Has Changed Our Lives: An Overview of the Current and Potential Applications and Limitations. International journal of environmental research and public health, 19(18), 11278. https://doi.org/10.3390/ijerph191811278
14. Hao, J., Xie, H., Harp, K., Chen, Z., & Siu, K. C. (2022). Effects of Virtual Reality Intervention on Neural Plasticity in Stroke Rehabilitation: A Systematic Review. Archives of physical medicine and rehabilitation, 103(3), 523–541. https://doi.org/10.1016/j.apmr.2021.06.024
15. Henrique, P. P. B., Perez, F. M. P., Dorneles, G., Peres, A., Korb, A., Elsner, V., & De Marchi, A. C. B. (2023). Exergame and/

or conventional training-induced neuroplasticity and cognitive improvement by engaging epigenetic and inflammatory modulation in older women: A randomized clinical trial. Physiology & behavior, 258, 113996. https://doi.org/10.1016/j.physbeh.2022.113996
16. Holden M. K. (2005). Virtual environments for motor rehabilitation: review. Cyberpsychology & behavior: the impact of the Internet, multimedia and virtual reality on behavior and society, 8(3), 187–219. https://doi.org/10.1089/cpb.2005.8.187
17. Huang, CY., Chiang, WC., Yeh, YC. et al. Effects of virtual reality-based motor control training on inflammation, oxidative stress, neuroplasticity and upper limb motor function in patients with chronic stroke: a randomized controlled trial. BMC Neurol 22, 21 (2022). https://doi.org/10.1186/s12883-021-02547-4
18. Invitto, S., Spada, I., & De Paolis, L. T. (2015). Augmented reality, embodied cognition, and learning. In L. T. De Paolis & A. Mongelli (Eds.), Augmented and virtual reality (pp. 125–134). Springer.
19. Kleim, J. A., Hogg, T. M., VandenBerg, P. M., Cooper, N. R., Bruneau, R., & Remple, M. (2004). Cortical synaptogenesis and motor map reorganization occur during the late, but not early, phase of motor skill learning. The Journal of neuroscience: the official journal of the Society for Neuroscience, 24(3), 628–633. https://doi.org/10.1523/JNEUROSCI.3440-03.2004
20. Kleschnitzki, J. M., Beyer, L., Beyer, R., & Großmann, I. (2022). The Effectiveness of a Serious Game (MemoreBox) for Cognitive Functioning Among Seniors in Care Facilities: Field Study. JMIR serious games, 10(2), e33169. https://doi.org/10.2196/33169
21. Levin M. F. (2011). Can virtual reality offer enriched environments for rehabilitation? Expert review of neurotherapeutics, 11(2), 153–155. https://doi.org/10.1586/ern.10.201
22. Lowood, H. E. (2024, August 16). virtual reality. Encyclopedia Britannica. https://www.britannica.com/technology/virtual-reality
23. Monteblanco Cavalcante, M., Fraga, I., Dalbosco, B., De Marchi, P., Iraci, L., Baechtold da Silva, M. E., Dani, C., Dal Bosco, A., & Elsner, V. (2021). Exergame training-induced neuroplasticity and cognitive improvement in institutionalized older adults: A preliminary investigation. Physiology & behavior, 241, 113589. https://doi.org/10.1016/j.physbeh.2021.113589
24. Mura, G., Carta, M. G., Sancassiani, F., Machado, S., & Prosperini, L. (2018). Active exergames to improve cognitive functioning in neurological disabilities: a systematic review and meta-analysis.

European journal of physical and rehabilitation medicine, 54(3), 450–462. https://doi.org/10.23736/S1973-9087.17.04680-9

25. Paro, M. R., Hersh, D. S., & Bulsara, K. R. (2022). History of Virtual Reality and Augmented Reality in Neurosurgical Training. World neurosurgery, 167, 37–43. https://doi.org/10.1016/j.wneu.2022.08.042
26. Pekna, M., Pekny, M., & Nilsson, M. (2012). Modulation of neural plasticity as a basis for stroke rehabilitation. Stroke, 43(10), 2819–2828. https://doi.org/10.1161/STROKEAHA.112.654228
27. Puderbaugh, M., & Emmady, P. D. (2023). Neuroplasticity. In StatPearls. StatPearls Publishing.
28. Riva, G., Wiederhold, B. K., & Mantovani, F. (2019). Neuroscience of Virtual Reality: From Virtual Exposure to Embodied Medicine. Cyberpsychology, behavior and social networking, 22(1), 82–96. https://doi.org/10.1089/cyber.2017.29099.gri
29. Schaeffer, E., Roeben, B., Granert, O., Hanert, A., Liepelt-Scarfone, I., Leks, E., Otterbein, S., Saraykin, P., Busch, J. H., Synofzik, M., Stransky, E., Bartsch, T., & Berg, D. (2022). Effects of exergaming on hippocampal volume and brain-derived neurotrophic factor levels in Parkinson's disease. European journal of neurology, 29(2), 441–449. https://doi.org/10.1111/ene.15165
30. Sin, H., & Lee, G. (2013). Additional virtual reality training using Xbox Kinect in stroke survivors with hemiplegia. American journal of physical medicine & rehabilitation, 92(10), 871–880. https://doi.org/10.1097/PHM.0b013e3182a38e40
31. Smith, J. (2021). Exploring the effectiveness of immersive virtual reality interventions in managing musculoskeletal pain: A state-of-the-art review. Physical Therapy Reviews, 26(1), 1–14. https://doi.org/10.1080/10833196.2021.1903209
32. Sokołowska B. (2024). Being in Virtual Reality and Its Influence on Brain Health Overview of Benefits, Limitations and Prospects. Brain sciences, 14(1), 72. https://doi.org/10.3390/brainsci14010072
33. Sveistrup H. (2004). Motor rehabilitation using virtual reality. Journal of neuroengineering and rehabilitation, 1(1), 10. https://doi.org/10.1186/1743-0003-1-10
34. Tortora, C., Di Crosta, A., La Malva, P., Prete, G., Ceccato, I., Mammarella, N., Di Domenico, A., & Palumbo, R. (2024). Virtual reality and cognitive rehabilitation for older adults with mild cognitive impairment: A systematic review. Aging research reviews, 93, 102146. https://doi.org/10.1016/j.arr.2023.102146

35. Wang, C. P., Lan, Y. J., Tseng, W. T., Lin, Y. T., & Gupta, K. C. (2020). On the effects of 3D virtual worlds in language learning–a meta-analysis. Computer Assisted Language Learning, 33(8), 891–915. https://doi.org/10.1080/09588221.2019.1598444
36. Wang, Y. F., Petrina, S., & Feng, F. (2017). VILLAGE–Virtual immersive language learning and gaming environment: Immersion and presence. British Journal of Educational Technology, 48(2), 431–450. https://doi.org/10.1111/bjet.12388
37. Williams, G., Drini, E., & Riches, S. (2024). Virtual reality relaxation for mental health service users with complex care needs: A pilot study in residential care homes and supported accommodation. Journal of Psychosocial Rehabilitation and Mental Health. https://doi.org/10.1007/s40737-024-00399-y
38. Williams, S. C. P. (2024, April 11). Imagining virtual reality as a simple tool to treat depression. Stanford Medicine. https://scopeblog.stanford.edu/2024/04/11/depression-virtual-reality-vr-headset/

Neuroplasticity in Neurodegenerative Diseases

Öznur Kaya Sağlam, Yusuf Yaşasın, Cansu Demir, and Selen Gür Özmen

1. Introduction

Neurodegenerative diseases are a highly heterogeneous group of disorders that affect the nervous system. Among neurodegenerative diseases are Parkinson's Disease (PD), Alzheimer's Disease (AD), Multiple Sclerosis (MS), Huntington's Disease (HD), Amyotrophic Lateral Sclerosis (ALS) and others. In these diseases, neurons in the central or peripheral nervous system undergo degeneration. Neurodegeneration is the progressive damage and death of neurons resulting from neurodegenerative diseases. Neurodegenerative diseases lead to the selective loss of synapses and neurons, mainly due to abnormal protein build-up in the brain and spinal cord (Kovacs, 2019). The exact causes of neuron degeneration remain unclear but involve multiple factors. Key factors include protein misfolding, aggregation, and accumulation. These are crucial in diseases like PD, AD, MS, HD and ALS. Mitochondrial dysfunction is also significant, often due to genetic mutations, toxins, or aging. This dysfunction leads to the production of harmful reactive oxygen species (ROS) and oxidative stress. Such conditions are linked to AD, PD and HD. Neuroinflammation might also contribute to these diseases (Marzola et al., 2023). For example, in MS, inflammatory lesions can trigger further neuron loss (Mey et al., 2023). Both genetic and environmental factors worsen neurodegeneration and disrupt neuroplasticity. Mutations in the TREM2 gene, found in microglia, increase AD risk. Environmental threats like toxins, infections, and brain injuries also raise the risk of neurodegenerative disorders (Marzola et al., 2023).

Neurodegenerative diseases impair neuroplasticity (Yuan et al., 2020). Neuroplasticity, also known as brain plasticity or neural plasticity, enables the brain to adapt and change. It is vital for learning, memory, recovery from injury, and adapting to new environments (Marzola et al., 2023). Neuroplasticity occurs under physiological and pathological conditions. Under physiological conditions, it supports development, learning, memory, and brain repair, while in pathological conditions, it is observed in brain injury, stroke, epilepsy, and neurodegenerative diseases such as AD, PD, and HD (Dorszewska et al., 2020).

Neurodegeneration impairs both synaptic and non-synaptic neuroplasticity.This raises questions about how neuronal loss affects neuroplasticity. In AD, neuroplasticity is notably disrupted. AD features amyloid-beta (Aβ) plaques, which cause neuronal loss and impair neuroplasticity. There is a strong link between neuronal loss in the cortex and the build-up of hyperphosphorylated tau protein. Both these factors harm neural plasticity and mitochondrial function in neurons. This shows how serious plasticity problems are in AD and other dementias. In PD, the build-up of α-synuclein gene (SNCA) leads to neuronal death. This protein forms Lewy bodies in neurons, especially in the substantia nigra. Patients with PD suffer from severe motor issues due to damaged dopaminergic neurons. These neurons are vital for motor signals. Oxidative stress and genetic mutations in SNCA or PINK1 genes worsen these problems. PD patients with dementia have more tau protein in their blood than those without. This suggests that tau aggregates might link cognitive and memory loss in PD, AD and other disorders (Sasmita et al., 2018). In MS, acute and chronic inflammation harms the CNS. This damage disrupts necessary processes for recovery. However, maintaining functional connectivity is tied to cognitive reserve, even with structural damage. This implies that cognitive reserve may boost the brain's adaptability (Kuhlmann et al., 2023).

Neuroplasticity is influenced by a range of factors, including age, experience, genetics, health, exercise, diet, sleep, and many others, as discussed in other chapters of this book. Age is the most crucial factor. It declines with age. Brain aging and neurodegeneration reduce connectivity. This process varies among individuals. Aging neurons become fragile but can still communicate. As people age, they face more diseases. This strain falls on systems with limited repair capabilities. Post-mitotic neurons are particularly vulnerable. This leads to more neurodegenerative diseases in older populations (Popescu et al., 2024).

Neurodegeneration hampers neuroplasticity, reducing the brain's repair ability. This speeds up disease and worsens symptoms. However, research shows some neuroplasticity remains, even in these diseases. This offers hope for new treatments to slow disease and ease symptoms (Aslam et al., 2024). The idea of boosting neuroplasticity is key in recovery efforts. It shows promise for many neurological and neurodegenerative diseases. Approaches include molecular methods, activity-based strategies, and surgery (Sasmita et al., 2018).

Neuroplasticity varies in each neurodegenerative disease, causing unique challenges. For instance, AD patients show reduced neuroplasticity due to memory and learning region damage. In PD, motor control area degeneration hampers movement learning and adaptation (Dorszewska et al., 2020). MS patients face similar issues from brain lesions and inflammation. This contributes to disease progression. Hence, neuroplasticity is vital for understanding these diseases and developing treatments. Approaches like brain stimulation, exercise, cognitive training, and drugs have shown promise in boosting neuroplasticity and aiding recovery (Hortobágyi et al., 2022; Marzola et al., 2023; Yuan et al., 2020).

In clinical practice, assessing neuroplasticity helps monitor disease progression, evaluate treatment responses, and identify neuroplastic potential. It also aids in personalizing rehabilitation, improving adherence, and finding new therapy targets (Tavazzi et al., 2021). Tailoring approaches to each disease's neuroplasticity is crucial. This section explores the link between neuroplasticity and common neurodegenerative diseases, such as PD, AD and MS. It also reviews methods to assess and treat neuroplasticity in these conditions.

2. Parkinson's Disease and Neuroplasticity

Globally, AD and PD are the most common neurodegenerative diseases (Agnello & Ciaccio, 2022). PD affects about 315 people per 100,000 worldwide. It is a complex disease marked by the loss of specific brain neurons and the presence of Lewy bodies. This loss is not limited to dopaminergic neurons. It also affects noradrenergic neurons in the locus coeruleus and cholinergic neurons in the dorsal motor nucleus (Trempe & Gehring, 2023). Non-motor symptoms like loss of smell, constipation, depression, and sleep disorders can appear years before motor symptoms. However, motor symptoms such as resting tremor (often asymmetric), cogwheel rigidity, and bradykinesia are considered key diagnostic criteria for the disease. For an individual experiencing symptoms such as slowing of movements, weakness in one extremity, and tremors, the onset of these symptoms is often seen as the beginning of the underlying problem. Neurodegeneration begins long before symptoms appear. By the time PD is diagnosed, approximately 90% of neurons in the substantia nigra are lost. Advanced PD brings new non-motor symptoms, such as autonomic issues, pain, and cognitive decline (Al-Chalabi, 2021). While some symptoms can be managed with medication or deep brain stimulation (DBS), no treatments currently exist to stop or slow the disease (Trempe & Gehring, 2023). The brain can adapt, reorganizing itself to handle external and internal challenges. By leveraging endogenous protective mechanisms, the brain can reorganize and modify the neuronal environment, protein expression, and signaling within specific areas and networks. In PD, the loss of neurons leads to increased activity in the remaining neurons, providing temporary compensation (Popescu et al., 2024). Impaired synaptic plasticity in the substantia nigra is linked to motor function disorders. This plays a key role in the cognitive and motor decline seen in PD (Yuan et al., 2020).

Various methods have been developed to assess neuroplasticity, showing how neurons adapt. These methods include brain imaging, studying neurotrophic factors, and taking blood or cerebrospinal fluid samples. Some focus on neurochemistry, while others examine brain functions and structures (Johansson et al., 2020). Recently, both invasive and non-invasive brain stimulation techniques have targeted impaired neurotransmission and plasticity in patients with PD. Non-invasive techniques include transcranial magnetic stimulation (TMS), transcranial direct current stimulation (tDCS), transcranial alternating current stimulation (tACS), and transcranial ultrasound stimulation (TUS). Combining these with motor training can boost neuroplasticity and learning, possibly easing motor symptoms in PD (Horiba et al., 2019; Yuan et al., 2020). DBS is a proven method for managing movement disorders, especially in PD patients. DBS involves implanting electrodes in the basal ganglia to modulate neural activity and alleviate symptoms (Horiba et al., 2019). Specifically, DBS targeting the subthalamic nucleus enhances motor function, improves quality of life, and reduces complications in selected patients with moderate to advanced PD. It significantly alleviates PD symptoms and enhances overall well-being (Schuepbach et al., 2013).

Repetitive transcranial magnetic stimulation (rTMS) influences brain function, though its exact mechanisms remain unclear. It is believed to affect cortical plasticity and network activation. This, in turn, impacts motor and cognitive functions and related issues. rTMS can affect distant brain areas, boost neurotransmitter release, and regulate glial activity. It also increases neurotrophic signaling and triggers the release of glutamate and GABA.

Moreover, it enhances synaptic plasticity by facilitating calcium-mediated signaling. Cortical rTMS specifically boosts striatal dopamine release and alters cortical plasticity (Moretti et al., 2020; Yuan et al., 2020).

Rehabilitation for PD can boost medication effects and prevent further injuries, possibly slowing disease progression. Including physical activity and exercise enhances the benefits of dopamine treatment (Shen et al., 2016). Evidence suggests that exercise promotes neuroplasticity, aiding PD prevention and treatment. It improves dopaminergic neuron function and reduces excessive glutamate activity. Both aerobic and targeted exercises cut oxidative stress and balance neurotransmitters. They also improve synaptic plasticity by affecting non-dopaminergic systems. Moreover, physical activity boosts brain health by increasing beneficial substances. Exercise-induced neuroplasticity improves motor control, learning, and memory in PD (Johansson et al., 2020). Aerobic exercise enhances PD patients' balance, walking, and movement (Shen et al., 2016). A study showed that cycling at 50–80% of maximum heart rate three times a week for six months stabilizes movement and improves cognition. It does this by enhancing plasticity in key brain networks. Thus, aerobic exercise benefits brain functions beyond movement in PD patients (Schootemeijer et al., 2022). Neuroimaging studies highlight exercise's positive impact on mobility and cognition in PD. It suggests exercise boosts prefrontal activity, improving cognition and walking control. Strength training also fosters neuroplasticity by activating key networks for walking (Bonavita, 2020). Eccentric training can reduce tremors after six weeks, thanks to its neuroplastic effects on relevant brain centers (Kadkhodaie et al., 2020).

The mechanisms behind exercise's impact on brain-derived neurotrophic factor (BDNF) in PD are unclear. However, they might involve brain processes that strengthen or weaken connections. BDNF and exercise help neurons in a key brain area and improve movement (Hirsch et al., 2018). BDNF supports neuron growth and health. It also aids brain flexibility in areas like the hippocampus and cortex (Kambestad et al., 2023). Moreover, BDNF protects against brain injuries and promotes new neuron formation in various regions. Low BDNF levels are linked to neurodegenerative diseases. PD patients often have lower BDNF than healthy peers, increasing their risk of brain damage (Kaagman et al., 2024). Exercise might raise BDNF levels in PD patients as a treatment (Palasz et al., 2020). BDNF is linked to cognitive issues in PD (Wang et al., 2016). Researchers believe that exercise boosts BDNF in the hippocampus, improving cognition (Kambestad et al., 2023). Exercise may also reduce inflammation and oxidative stress, benefiting cognitive and motor functions in PD (Wang et al., 2016).

Exercise improves brain flexibility, enhancing cognition. Exergames, which combine exercise and technology, might boost hippocampal volume in PD patients (Schaeffer et al., 2022). Even short, moderate-intensity aerobic workouts can raise BDNF levels in older adults (Kambestad et al., 2023). While exercise is shown to boost BDNF in patients, more research is needed to understand the long-term effects, best exercises, and mechanisms (Kaagman et al., 2024).

PD treatment encompasses a multifaceted approach, combining various therapies and medications to alleviate symptoms. Therapies involve a range of exercises, including physical, occupational, and speech therapies, tailored to individual needs. Medications, too, offer a spectrum of options, from typical drugs to levodopa, which can be used alone or in combination with other medications to achieve optimal results. Notably, aerobic and strength training exercises have been shown to boost neuroplasticity, the brain's ability to adapt and change in response to new experiences. However, more research is needed to determine the ideal type, amount, and frequency of exercise that yields the most benefits for patients. For those struggling with drug-resistant tremors and dyskinesias, DBS and levodopa-carbidopa enteral suspension therapy may offer a viable solution (Sciacca et al., 2023). Despite these advances, finding effective preventive or treatment methods for PD remains an enormous challenge. This is largely due to limited studies, an unclear understanding of neurodegeneration mechanisms, varying pathology, and the lack of suitable animal models that can accurately replicate the complexities of the human brain.

3. Alzheimer's Disease and Neuroplasticity

AD is a neurodegenerative disease that affects the elderly population worldwide and is characterized by a progressive decline in cognitive function. AD is the leading cause of dementia. By 2050, over 100 million people may be affected globally (Monteiro et al., 2023). The risk of dementia rises with age. The prevalence is approximately 5–8% among individuals over 65. However, it jumps to 25–50% for those over 85. Men have a 19–29% lower chance of developing AD than women (Khan et al., 2020). Symptoms include language and memory impairments, functional decline, and behavioral and personality changes. In advanced stages, patients face infections and eating difficulties. These issues lower the quality of life and raise care costs, posing major public health challenges (Ozben & Ozben, 2019).

AD's neurodegenerative features disrupt both synaptic and non-synaptic neuroplasticity. Neuronal loss often leads to the dismissal of neuroplasticity theories in these diseases (Sasmita et al., 2018). However, recent studies suggest that synaptic loss may occur before neuronal degeneration. It is suggested that impaired neuroplasticity may play an important role in the development of AD (Koch & Spampinato, 2022).

Neuroplasticity is impaired in AD patients. Aβ plaques damage neurons and further hinder neuroplasticity. Late-onset Alzheimer's disease (LOAD) patients have more Aβ than early-onset Alzheimer's disease (EOAD) patients. Additionally, hyperphosphorylated tau protein accumulation is linked to neuronal loss in AD brains. The Apolipoprotein E4 (APOE4) gene significantly raises LOAD risk and Aβ levels. This toxic build-up of Aβ and tau disrupts both neuroplasticity and mitochondrial function in neurons. This highlights the serious nature of plasticity loss in AD and other dementias (Sasmita et al., 2018). Research shows that tau and Aβ disrupt neuroplasticity in animal models. Direct assessment of synaptic function in living patients with AD remains unfeasible. Various in vivo imaging techniques such as electroencephalography (EEG), functional magnetic resonance imaging (fMRI) and 18F-fluorodeoxyglucose positron emission tomography (FDG-PET) have been developed to determine synaptic dysfunction in AD through biomarkers. Due to their low temporal resolution, fMRI and FDG-PET only indirectly predict synaptic dysfunction (Koch & Spampinato, 2022). Various invasive and non-invasive brain stimulation techniques have been applied in the last decade to assess impaired neuroplasticity in AD patients. TUS, TMS, tACS and tDCS are used as non-invasive brain stimulation techniques. DBS is effective for PD, but more research is needed for other brain stimulation techniques in diseases like PD and AD (Yuan et al., 2020). TMS can assess changes in brain activity in healthy and diseased states (Koch & Spampinato, 2022). Functional near-infrared spectroscopy (fNIRS) provides a more versatile measurement that is increasingly important for investigating cerebral activity. In patients affected by neurodegenerative disorders, fNIRS will potentially be a valuable tool for monitoring neuroplasticity (Bonilauri et al., 2020).

TMS is a non-invasive neuromodulation technique. It is a painless technique that does not require any preliminary preparation. Its use as a therapeutic tool is rapidly increasing for these reasons. Its therapeutic value can be achieved by applying electromagnetic fields (EMFs) to a predetermined cortical target. It involves the application of time-varying magnetic resonance imaging (MRI) strength magnetic fields near the scalp and superficial layer of the cerebral cortex. When an intense stimulus is applied at the appropriate level, this stimulus generates a strong magnetic field. This strong magnetic field stimulates the brain region. TMS applications can be single-pulse (sTMS), paired-pulse (ppTMS) or repetitive (rTMS). It has shown its potential therapeutic effects on AD patients. Deterioration in neural plasticity occurs due to aging, which may worsen with AD. However, TMS improves neuroplasticity (Bashir et al., 2022).

tDCS is a non-invasive brain stimulation technique that promotes transient polarity-dependent changes in spontaneous neuronal activity. This effect is achieved by applying constant low-amplitude electrical currents with electrodes placed epicranially over a specific brain region of interest. Epicranial application of direct currents stimulates a weak electric field strength and produces neuronal membrane potential changes. These changes occur by spontaneously adjusting neuronal activity using calcium and sodium currents. The localized neuronal excitation or inhibition depends on the polarity of the current. Mainly, anodal currents increase excitability, while cathodal currents show inhibitory effects. There is increasing evidence that tDCS increases neuroplasticity by regulating brain activity in vivo animal model studies. (Aurelian et al., 2023; Cavaleiro et al., 2020). Based on the results of laboratory animal research, tDCS is generally considered a promising technique (Cavaleiro et al., 2020).

Transcranial pulse stimulation (TPS) is a new non-invasive technique based on repeated single ultrashort high-intensity ultrasound pulses. Due to the high spatial resolution of deep focal distances, it can non-invasively modulate neural targets deep in the human cortex. It uses lower focal ultrasound frequencies that can reach deep brain structures. TPS increases BDNF expression and neuroplasticity. However, TPS's exact mechanism of action for AD is limited and requires further research (Chen et al., 2024).

To date, only four drugs are available for the treatment of AD: one N-methyl-D-aspartate (NMDA) receptor antagonist (memantine), three acetylcholinesterase (AChE) enzyme inhibitors (donepezil, rivastigmine, and galantamine). These drugs are symptomatic and cannot prevent neuronal loss. The AD drug development process has reported promising results in improving the neuroplasticity of Levetiracetam, Bryostatin 1, and Masitinib, for which phase 2 and phase 3 clinical trials are currently ongoing (Piscopo et al., 2022).

Due to the ineffectiveness of current drug therapy against AD, interest in light-based therapeutic approaches is increasing. Visual stimulation (VS) and photobiomodulation (PBM) are among the most promising approaches. In AD mouse model studies, both PBM and VS applications have been observed to improve neuroplasticity and increase BDNF expression (Monteiro et al., 2022).

BDNF synthesis/release, which supports neuroplasticity, increases with exercise. The ability of exercise to increase BDNF is emphasized in literature studies due to the high amount of BDNF degraded in AD. Exercise modulates BDNF levels. "Moderate intensity resistance exercise training" is recommended for AD patients, and its benefits have been proven (Aurelian et al., 2023).

Virtual Reality (VR) is another modern therapeutic modality that is not a substitute for classical rehabilitation methods but can be added to them. Short films showing beautiful exterior and interior natural landscapes produced by VR will promote neuroplasticity in AD patients (Aurelian et al., 2023).

In summary, various therapeutic approaches have been applied to improve neuroplasticity impaired by AD. The literature suggests that non-invasive brain stimulation techniques (TUS, TMS, tDCS, TPS), light-based therapeutic approaches (PBM and VS), moderate-intensity resistance exercise training and VR rehabilitation can improve neural plasticity. The drugs Levetiracetam, Bryostatin 1, and Masitinib, which are being developed to treat AD, have shown positive effects in increasing neuroplasticity. It would be beneficial to develop new drugs that improve neuroplasticity for AD treatment. Since there are limited studies on the relationship between AD and neuroplasticity, and the relationship remains theoretical, more research is needed to understand the exact mechanisms. Clarifying these mechanisms will encourage the development and use of neuroplasticity-based treatment approaches.

4. Multiple Sclerosis and Neuroplasticity

MS is a chronic disease affecting the central nervous system (CNS). It damages myelin, oligodendrocytes, and axons. MS mainly affects young to middle-aged adults, with a global prevalence of 5 to 300 cases per 100,000. The disease is more common in people living at higher latitudes. The average onset age is 29, and it is three times more frequent in women than in men (McGinley et al., 2021). MS leads to various symptoms, including muscle weakness, spasticity, ataxia, neuropathic pain, psychiatric issues, and fatigue. These symptoms stem from CNS dysfunction and significantly lower quality of life (Du et al., 2024). Diagnosing MS involves assessing patient history, physical exams, MRIs, and lab tests. Key factors include symptom consistency with demyelination, ruling out other diagnoses, and lesion distribution. The 2017 McDonald Criteria guide the use of MRI and cerebrospinal fluid analysis in diagnosis. Patients not meeting these criteria are monitored regularly (London: National Institute for Health and Care Excellence, 2022).

Neuroplasticity is key for MS patients. It helps track disease, evaluate treatments, and tailor rehab. MRI is the main tool for these assessments. It checks both structural and functional brain changes. Diffusion Tensor Imaging (DTI) focuses on structural issues. It detects axon damage, demyelination, and remyelination by analyzing white matter and water movement. DTI metrics like Mean Diffusivity and Fractional Anisotropy indicate neuroplasticity. For structural changes, volumetric measurements assess brain atrophy or growth. Functional changes are evaluated with fMRI. Task-Based fMRI looks at brain activity during tasks. Resting-State fMRI (rs-fMRI) checks activity and connectivity when at rest. This helps understand brain reorganization and compensatory mechanisms (Tavazzi et al., 2021).

MS patients often take drugs to slow progression, prevent attacks, or regain some function. These drugs control inflammation by adjusting the immune system. There are many Disease-Modifying Therapies (DMTs), including interferons, glatiramer acetate, teriflunomide, sphingosine 1-phosphate receptor modulators, fumarates, cladribine, and monoclonal antibodies. DMTs reduce attacks and MRI lesions effectively (McGinley et al., 2021).

TMS provides a strong treatment option for MS patients. It's non-invasive and boosts nerve function and brain activity using electromagnetic pulses. Studies show TMS helps with spasticity, fatigue, bladder issues, hand movement, walking, and memory problems. It improves quality of life in patients with MS (León Ruiz et al., 2022).

Exercise also benefits MS patients by enhancing brain function and producing neurotrophic factors. It increases neuroplasticity, showing promise for neurodegenerative diseases like MS. Exercise might aid neural repair mechanisms and improve cognitive function and motor skills (de Sousa Fernandes et al., 2020).

Recent research shows exercise boosts aerobic capacity, strength, flexibility, balance, reduces fatigue, and enhances cognition in MS patients. It is now seen as a vital treatment to improve life quality and reduce symptoms (Du et al., 2024). Exercise leads to brain changes that improve life quality in MS patients. This includes increased gray matter, better white matter, and stronger brain connectivity. Such changes enhance cognitive function and motor skills (Sandroff et al., 2020).

Exercise benefits MS patients by enhancing brain flexibility, reducing functional loss, and improving quality of life. It protects and repairs brain connections, vital for neuron communication. MS often disrupts these connections due to inflammation and myelin damage. Exercise can maintain these connections and even create new ones, easing MS-related neurological issues (Azzolini et al., 2024).

Various mechanisms triggered by exercise contribute to this process. Increased production of neurotrophins, suppression of inflammation, and reduction of oxidative stress are among the crucial factors supporting synaptic plasticity. Exercise supports repair processes in the brain by increasing the levels of BDNF molecules, in particular. Furthermore, exercise can improve cognitive functions in MS patients by promoting neurogenesis in brain regions associated with learning and memory, such as the hippocampus. Thus, for MS patients, exercise is a key treatment. It can improve brain functions and slow disease progression by enhancing brain plasticity (Azzolini et al., 2024; Sandroff et al., 2020).

Exercise boosts neuroplasticity, improving brain function and structure. Research shows regular aerobic activity can slow brain degeneration and reduce lesions. It does this by increasing volume in certain brain areas. For instance, it decreases volume in regions linked to motor control and cognition, like the putamen and posterior cingulate. Conversely, it increases volume in the parahippocampal gyrus, vital for learning and memory. These findings highlight exercise's potential to enhance specific brain functions (Savšek et al., 2021).

High-intensity aerobic exercise provides significant benefits for both the brain and overall health in MS patients. It strengthens the cardiovascular system by improving heart and lung function, which leads to more efficient oxygen utilization and enhanced stamina during physical activity. This increased oxygen supply optimizes energy metabolism and allows patients to sustain physical activities for longer durations, which directly impacts their quality of life (Orban et al., 2019).

Exercise also contributes to muscle growth and fat reduction. This improvement in muscle strength and endurance can help counteract the motor skill impairments commonly experienced by MS patients. The strengthened musculature aids in movement, reducing the risk of falls and promoting greater independence in daily activities.

Another critical benefit of high-intensity aerobic exercise is its reduction of fatigue, a common symptom in MS patients. Fatigue in MS is often caused by neurogenic factors, such as impaired energy production due to myelin damage in the nervous system. Aerobic exercise helps regulate energy metabolism, thereby reducing fatigue and allowing patients to participate more actively in daily activities. Additionally, the release of endorphins during exercise enhances mood, which can alleviate psychological symptoms like depression and anxiety.

From a neuroplasticity perspective, high-intensity aerobic exercise strengthens brain connections and improves communication between neurons. The exercise stimulates the production of BDNF, which supports synaptic plasticity and promotes neuron survival. This process is particularly beneficial in areas of the brain like the hippocampus, which is crucial for learning and memory, as it encourages the formation of new nerve cells. Moreover, exercise has been shown to reduce inflammation and oxidative stress, contributing to myelin repair.

In summary, high-intensity aerobic exercise improves both physical and cognitive functions in MS patients. The increase in muscle strength, reduction in fatigue, and enhancement of brain plasticity significantly boost motor control, learning, and overall quality of life. These multifaceted benefits make high-intensity aerobic exercise a key rehabilitation strategy in managing MS (Orban et al., 2019).

Research shows that exercise intensity affects brain adaptability. It's proven in young, healthy adults but unclear for older people with neurological diseases (Hortobágyi et al., 2022). So, tailoring exercise for MS patients is vital. Studies reveal exercise cuts neuronal damage in MS patients. It does this by boosting neurotrophic factors, which protect nerves. This suggests exercise might slow MS progression and aid recovery. Neurotrophic factors are crucial for nerve health. Low levels in MS worsen the disease. Exercise raises these factors, supporting nerve repair (Diechmann et al., 2021).

Most studies on exercise and MS focus on one type, like aerobic or RT. However, multi-component training, which combines different exercises, might be better. It merges the cardiovascular benefits of aerobic exercise with the muscle-strengthening effects of RT. This approach could enhance overall health and quality of life for MS patients (Du et al., 2024; Hortobágyi et al., 2022).

Neuroplasticity is key in treating MS patients. It allows the brain to heal and adapt, potentially offsetting MS-related losses and slowing its progression. Therapies like exercise, cognitive rehab, and brain stimulation enhance neuroplasticity, improving life quality and easing symptoms. Ongoing research will deepen our understanding and pave the way for tailored treatments. Future neuroplasticity-based therapies are expected to significantly aid MS patients.

5. Conclusion

Although there are no definitive results on dosage in preserving and increasing neuroplasticity in PD; aerobic and strengthening exercises, DBS and pharmacological treatments have been found to be effective. In AD, non-invasive brain stimulation techniques (TUS, TMS, tDCS, TPS), light-based therapeutic approaches (PBM and VS), moderate-intensity resistance exercise training and VR rehabilitation may play an essential role in improving quality of life and alleviating symptoms by promoting neuroplasticity. In MS, various therapeutic approaches, including exercise, cognitive rehabilitation, and brain stimulation techniques, may also play a vital role. Future research should prioritize in-depth studies that utilize advanced methodologies to explore the role, mechanisms, and progression of neuroplasticity in neurodegenerative diseases.

6. Takeaways

1. Neuroplasticity, the brain's ability to adapt and reorganize, is crucial in managing MS. It offers potential for functional recovery, slowing disease progression, and improving quality of life.
2. Although neurodegeneration directly affects neuroplasticity, interventions targeting neuroplasticity are of great importance in these diseases, as it is potentially preserved to some extent even in neurodegenerative diseases.
3. Aerobic exercise and strength training contribute to increased neuroplasticity in patients with PD.
4. DBS, one of the brain stimulation techniques, plays a role in neuroplasticity development in PD patients by contributing to increased motor functions and neural networks.
5. Disruption of neuroplasticity mechanisms in AD has been associated with accumulating Aβ and hyperphosphorylated tau protein.
6. The value of fNIRS in evaluating neuroplasticity in neurodegenerative diseases, including PD, AD and MS, is expected to increase in the coming years.
7. PBM and VS, among the light-based therapeutic methods, improve neuroplasticity in AD patients.
8. Moderate-intensity resistance exercise training is recommended for AD patients as it supports the development of neuroplasticity.
9. VR interventions stimulate neuroplasticity in AD.
10. Exercise, particularly multi-component training, has emerged as a valuable tool in MS treatment. It promotes neuroplasticity, improves cognitive and motor functions, and enhances overall well-being.
11. Exercise benefits MS patients by increasing neurotrophic factors, reducing inflammation, and promoting neurogenesis, improving synaptic connectivity and overall brain function.

Abbreviations

AChE: Acetylcholinesterase
AD: Alzheimer's Disease
ALS: Amyotrophic Lateral Sclerosis
APOE4: Apolipoprotein E4
Aβ: Amyloid-Beta

BDNF: Brain-Derived Neurotrophic Factor
CNS: Central Nervous System
DBS: Deep Brain Stimulation
DMTs: Disease-Modifying Therapies
DTI: Diffusion Tensor Imaging
EEG: Electroencephalography
EMFs: Electromagnetic Fields
EOAD: Early-Onset Alzheimer's Disease
FDG-PET:18F-Fluorodeoxyglucose Positron Emission Tomography
fMRI: Functional Magnetic Resonance Imaging
fNIRS: Functional Near-Infrared Spectroscopy
HD: Huntington's Disease
LOAD: Late-Onset Alzheimer's Disease
MRI: Magnetic Resonance Imaging
MS: Multiple Sclerosis
NMDA: N-Methyl-D -Aspartate
PBM: Photobiomodulation
PD: Parkinson's Disease
ppTMS: Paired-Pulse Transcranial Magnetic Stimulation
rTMS: Repetitive Transcranial Magnetic Stimulation
rs-fMRI: Resting-State fMRI
sTMS: Single-Pulse Transcranial Magnetic Stimulation
SNCA: Alpha Synuclein Gene
tACS: Transcranial Alternating Current Stimulation
tDCS: Transcranial Direct Current Stimulation
TMS: Transcranial Magnetic Stimulation
TUS: Transcranial Ultrasound Stimulation
TPS: Transcranial Pulse Stimulation
VR: Virtual Reality
VS: Visual Stimulation

Bibliography

1. Agnello, L., & Ciaccio, M. (2022). Neurodegenerative Diseases: From Molecular Basis to Therapy. *International Journal of Molecular Sciences*, *23*(21), 12854. https://doi.org/10.3390/ijms232112854

2. Al-Chalabi, A. (2021). Preventing neurodegenerative disease. *Brain, 144*(5), 1279–1280. https://doi.org/10.1093/brain/awab151
3. Aslam, R. M. N., Noor-ul-Ain, Ali, H., Muzammil, A., & Iqbal, A. (2024). A Review on the Role of Neuroplasticity in Neurodegenerative Disorders and Its Possible Pharmacological and Non-Pharmacological Interventions. *Asian Journal of Medicine and Health, 22*(1), 72–86. https://doi.org/10.9734/ajmah/2024/v22i1977
4. Aurelian, S., Ciobanu, A., Cărare, R., Stoica, S.-I., Anghelescu, A., Ciobanu, V., Onose, G., Munteanu, C., Popescu, C., Andone, I., Spînu, A., Firan, C., Cazacu, I. S., Trandafir, A.-I., Băilă, M., Postoiu, R.-L., & Zamfirescu, A. (2023). Topical Cellular/Tissue and Molecular Aspects Regarding Nonpharmacological Interventions in Alzheimer's Disease—A Systematic Review. *International Journal of Molecular Sciences, 24*(22). https://doi.org/10.3390/ijms242216533
5. Azzolini, F., Dolcetti, E., Bruno, A., Rovella, V., Centonze, D., & Buttari, F. (2024). Physical exercise and synaptic protection in human and pre-clinical models of multiple sclerosis. *Neural Regeneration Research, 19*(8), 1768–1771. https://doi.org/10.4103/1673-5374.389359
6. Bashir, S., Uzair, M., Abualait, T., Arshad, M., Khallaf, R., Niaz, A., Thani, Z., Yoo, W.-K., Túnez, I., Demirtas-tatlidede, A., & Meo, S. (2022). Effects of transcranial magnetic stimulation on neurobiological changes in Alzheimer's disease (Review). *Molecular Medicine Reports, 25*(4). https://doi.org/10.3892/mmr.2022.12625
7. Bonavita, S. (2020). Exercise and Parkinson's Disease. In *Physical Exercise for Human Health, Advances in Experimental Medicine and Biology* (Vol. 1228). Springer.
8. Bonilauri, A., Sangiuliano Intra, F., Pugnetti, L., Baselli, G., & Baglio, F. (2020). A Systematic Review of Cerebral Functional Near-Infrared Spectroscopy in Chronic Neurological Diseases—Actual Applications and Future Perspectives. *Diagnostics, 10*(8), 581. https://doi.org/10.3390/diagnostics10080581
9. Cavaleiro, C., Martins, J., Gonçalves, J., & Castelo-Branco, M. (2020). Memory and Cognition-Related Neuroplasticity Enhancement by Transcranial Direct Current Stimulation in Rodents: A Systematic Review. Neural Plasticity, 2020, 1–23. https://doi.org/10.1155/2020/4795267
10. Chen, X., You, J., Ma, H., Zhou, M., & Huang, C. (2024). Transcranial pulse stimulation in Alzheimer's disease. *CNS Neuroscience & Therapeutics, 30*(2). https://doi.org/10.1111/cns.14372

11. de Sousa Fernandes, M. S., Ordônio, T. F., Santos, G. C. J., Santos, L. E. R., Calazans, C. T., Gomes, D. A., & Santos, T. M. (2020). Effects of Physical Exercise on Neuroplasticity and Brain Function: A Systematic Review in Human and Animal Studies. *Neural Plasticity, 2020*, 1–21. https://doi.org/10.1155/2020/8856621
12. Diechmann, M. D., Campbell, E., Coulter, E., Paul, L., Dalgas, U., & Hvid, L. G. (2021). Effects of Exercise Training on Neurotrophic Factors and Subsequent Neuroprotection in Persons with Multiple Sclerosis—A Systematic Review and Meta-Analysis. *Brain Sciences, 11*(11). https://doi.org/10.3390/brainsci11111499
13. Dorszewska, J., Kozubski, W., Waleszczyk, W., Zabel, M., & Ong, K. (2020). Neuroplasticity in the Pathology of Neurodegenerative Diseases. *Neural Plasticity, 2020*, 1–2. https://doi.org/10.1155/2020/4245821
14. Du, L., Xi, H., Zhang, S., Zhou, Y., Tao, X., Lv, Y., Hou, X., & Yu, L. (2024). Effects of exercise in people with multiple sclerosis: a systematic review and meta-analysis. *Frontiers in Public Health, 12*. https://doi.org/10.3389/fpubh.2024.1387658
15. Hirsch, M. A., van Wegen, E. E. H., Newman, M. A., & Heyn, P. C. (2018). Exercise-induced increase in brain-derived neurotrophic factor in human Parkinson's disease: a systematic review and meta-analysis. *Translational Neurodegeneration*, *7*, 7. https://doi.org/10.1186/s40035-018-0112-1
16. Horiba, M., Ueki, Y., Nojima, I., Shimizu, Y., Sahashi, K., Itamoto, S., Suzuki, A., Yamada, G., Matsukawa, N., & Wada, I. (2019). Impaired Motor Skill Acquisition Using Mirror Visual Feedback Improved by Transcranial Direct Current Stimulation (tDCS) in Patients With Parkinson's Disease. *Frontiers in Neuroscience, 13*. https://doi.org/10.3389/fnins.2019.00602
17. Hortobágyi, T., Vetrovsky, T., Balbim, G. M., Sorte Silva, N. C. B., Manca, A., Deriu, F., Kolmos, M., Kruuse, C., Liu-Ambrose, T., Radák, Z., Váczi, M., Johansson, H., dos Santos, P. C. R., Franzén, E., & Granacher, U. (2022). The impact of aerobic and resistance training intensity on markers of neuroplasticity in health and disease. *Ageing Research Reviews, 80*. https://doi.org/10.1016/j.arr.2022.101698
18. Johansson, H., Hagströmer, M., Grooten, W. J. A., & Franzén, E. (2020). Exercise-Induced Neuroplasticity in PD's Disease: A Metasynthesis of the Literature. *Neural Plasticity, 2020*, 1–15. https://doi.org/10.1155/2020/8961493

19. Kaagman, D. G. M., van Wegen, E. E. H., Cignetti, N., Rothermel, E., Vanbellingen, T., & Hirsch, M. A. (2024). Effects and Mechanisms of Exercise on Brain-Derived Neurotrophic Factor (BDNF) Levels and Clinical Outcomes in People with Parkinson's Disease: A Systematic Review and Meta-Analysis. *Brain Sciences*, *14*(3). https://doi.org/10.3390/brainsci14030194
20. Kadkhodaie, M., Sharifnezhad, A., Ebadi, S., Marzban, S., Habibi, S. A., Ghaffari, A., & Forogh, B. (2020). Effect of eccentric-based rehabilitation on hand tremor intensity in Parkinson disease. *Neurological Sciences*, *41*(3), 637–643. https://doi.org/10.1007/s10072-019-04106-9
21. Kambestad, O. B., Sirevåg, K., Mrdalj, J., Hovland, A., Trygve, †, Endal, B., Andersson, E., Sjøbø, T., & Stavestrand, S. H. (2023). Physical Exercise and Serum BDNF Levels: Accounting for the Val66Met Polymorphism in Older Adults. *Cognitive and Behavioral Neurology*, *36*(4), 219–227. https://doi.org/10.1097/WNN.0000000000000349
22. Khan, S., Barve, K. H., & Kumar, M. S. (2020). Recent Advancements in Pathogenesis, Diagnostics and Treatment of Alzheimer's Disease. *Current Neuropharmacology*, *18*(11), 1106–1125. https://doi.org/10.2174/1570159X18666200528142429
23. Koch, G., & Spampinato, D. (2022). *Alzheimer disease and neuroplasticity* (pp. 473–479). https://doi.org/10.1016/B978-0-12-819410-2.00027-8
24. Kovacs, G. G. (2019). Molecular pathology of neurodegenerative diseases: principles and practice. *Journal of Clinical Pathology*, *72*(11), 725–735. https://doi.org/10.1136/jclinpath-2019-205952
25. Kuhlmann, T., Moccia, M., Coetzee, T., Cohen, J. A., Correale, J., Graves, J., Marrie, R. A., Montalban, X., Yong, V. W., Thompson, A. J., Reich, D. S., Amato, M. P., Banwell, B., Barkhof, F., Chataway, J., Chitnis, T., Comi, G., Derfuss, T., Finlayson, M., … Waubant, E. (2023). Multiple sclerosis progression: time for a new mechanism-driven framework. *The Lancet Neurology*, *22*(1), 78–88. https://doi.org/10.1016/S1474-4422(22)00289-7
26. León Ruiz, M., Sospedra, M., Arce Arce, S., Tejeiro-Martínez, J., & Benito-León, J. (2022). Current evidence on the potential therapeutic applications of transcranial magnetic stimulation in multiple sclerosis: a systematic review of the literature. *Neurología (English Edition)*, *37*(3), 199–215. https://doi.org/10.1016/j.nrleng.2020.05.004

27. London: National Institute for Health and Care Excellence. (2022). *Multiple sclerosis in adults: management NICE guideline*. www.nice.org.uk/guidance/ng220
28. Marzola, P., Melzer, T., Pavesi, E., Gil-Mohapel, J., & Brocardo, P. S. (2023). Exploring the Role of Neuroplasticity in Development, Aging, and Neurodegeneration. *Brain Sciences*, *13*(12), 1610. https://doi.org/10.3390/brainsci13121610
29. McGinley, M. P., Goldschmidt, C. H., & Rae-Grant, A. D. (2021). Diagnosis and Treatment of Multiple Sclerosis. *JAMA*, *325*(8). https://doi.org/10.1001/jama.2020.26858
30. Mey, G. M., Mahajan, K. R., & DeSilva, T. M. (2023). Neurodegeneration in multiple sclerosis. *WIREs Mechanisms of Disease*, *15*(1). https://doi.org/10.1002/wsbm.1583
31. Monteiro, A. R., Barbosa, D. J., Remião, F., & Silva, R. (2023). Alzheimer's disease: Insights and new prospects in disease pathophysiology, biomarkers, and disease-modifying drugs. *Biochemical Pharmacology*, *211*. https://doi.org/10.1016/j.bcp.2023.115522
32. Monteiro, F., Carvalho, Ó., Sousa, N., Silva, F. S., & Sotiropoulos, I. (2022). Photobiomodulation and visual stimulation against cognitive decline and Alzheimer's disease pathology: A systematic review. *Alzheimer's & Dementia: Translational Research & Clinical Interventions*, *8*(1). https://doi.org/10.1002/trc2.12249
33. Moretti, J., Poh, E. Z., & Rodger, J. (2020). rTMS-Induced Changes in Glutamatergic and Dopaminergic Systems: Relevance to Cocaine and Methamphetamine Use Disorders. *Frontiers in Neuroscience*, *14*. https://doi.org/10.3389/fnins.2020.00137
34. Orban, A., Garg, B., Sammi, M. K., Bourdette, D. N., Rooney, W. D., Kuehl, K., & Spain, R. I. (2019). Effect of High-Intensity Exercise on Multiple Sclerosis Function and Phosphorous Magnetic Resonance Spectroscopy Outcomes. *Medicine & Science in Sports & Exercise*, *51*(7), 1380–1386. https://doi.org/10.1249/MSS.0000000000001914
35. Ozben, T., & Ozben, S. (2019). Neuro-inflammation and anti-inflammatory treatment options for Alzheimer's disease. *Clinical Biochemistry*, 72, 87–89. https://doi.org/10.1016/j.clinbiochem.2019.04.001
36. Palasz, E., Wysocka, A., Gasiorowska, A., Chalimoniuk, M., Niewiadomski, W., & Niewiadomska, G. (2020). BDNF is a promising therapeutic agent in Parkinson's disease. *International Journal of Molecular Sciences*, *21*(3). https://doi.org/10.3390/ijms21031170

37. Piscopo, P., Crestini, A., Carbone, E., Rivabene, R., Ancidoni, A., Lo Giudice, M., Corbo, M., Vanacore, N., & Lacorte, E. (2022). A systematic review on drugs for synaptic plasticity in the treatment of dementia. *Ageing Research Reviews*, *81*. https://doi.org/10.1016/j.arr.2022.101726
38. Popescu, B. O., Batzu, L., Ruiz, P. J. G., Tulbă, D., Moro, E., & Santens, P. (2024). Neuroplasticity in Parkinson's disease. *Journal of Neural Transmission*. https://doi.org/10.1007/s00702-024-02813-y
39. Sandroff, B. M., Jones, C. D., Baird, J. F., & Motl, R. W. (2020). Systematic Review on Exercise Training as a Neuroplasticity-Inducing Behavior in Multiple Sclerosis. *Neurorehabilitation and Neural Repair*, *34*(7), 575–588. https://doi.org/10.1177/1545968320921836
40. Sasmita, A. O., Kuruvilla, J., & Ling, A. P. K. (2018). Harnessing neuroplasticity: modern approaches and clinical future. *International Journal of Neuroscience*, *128*(11), 1061–1077. https://doi.org/10.1080/00207454.2018.1466781
41. Savšek, L., Stergar, T., Strojnik, V., Ihan, A., Koren, A., Špiclin, & Jazbec, S. (2021). Impact of aerobic exercise on clinical and magnetic resonance imaging biomarkers in persons with multiple sclerosis: An exploratory randomized controlled trial. *Journal of Rehabilitation Medicine*, *53*(4). https://doi.org/10.2340/16501977-2814
42. Schaeffer, E., Roeben, B., Granert, O., Hanert, A., Liepelt-Scarfone, I., Leks, E., Otterbein, S., Saraykin, P., Busch, J., Synofzik, M., Stransky, E., Bartsch, T., & Berg, D. (2022). Effects of exergaming on hippocampal volume and brain-derived neurotrophic factor levels in Parkinson's disease. *European Journal of Neurology*, *29*(2), 441–449. https://doi.org/10.1111/ene.15165
43. Schootemeijer, S., Darweesh, S. K. L., & De Vries, N. M. (2022). Clinical Trial Highlights - Aerobic Exercise for Parkinson's Disease. *Journal of Parkinson's Disease*, *12*(8), 2297–2306. https://doi.org/10.3233/JPD-229006
44. Schuepbach, W. M. M., Rau, J., Knudsen, K., Volkmann, J., Krack, P., Timmermann, L., Hälbig, T. D., Hesekamp, H., Navarro, S. M., Meier, N., Falk, D., Mehdorn, M., Paschen, S., Maarouf, M., Barbe, M. T., Fink, G. R., Kupsch, A., Gruber, D., Schneider, G.-H., … Deuschl, G. (2013). Neurostimulation for Parkinson's Disease with Early Motor Complications. *New England Journal of Medicine*, *368*(7), 610–622. https://doi.org/10.1056/NEJMoa1205158

45. Sciacca, G., Mostile, G., Disilvestro, I., Donzuso, G., Nicoletti, A., & Zappia, M. (2023). Long-Duration Response to Levodopa, Motor Learning, and Neuroplasticity in Early Parkinson's Disease. *Movement Disorders*, *38*(4), 626–635. https://doi.org/10.1002/mds.29344
46. Shen, X., Wong-Yu, I. S. K., & Mak, M. K. Y. (2016). Effects of Exercise on Falls, Balance, and Gait Ability in Parkinson's Disease. *Neurorehabilitation and Neural Repair*, *30*(6), 512–527. https://doi.org/10.1177/1545968315613447
47. Tavazzi, E., Cazzoli, M., Pirastru, A., Blasi, V., Rovaris, M., Bergsland, N., & Baglio, F. (2021). Neuroplasticity and Motor Rehabilitation in Multiple Sclerosis: A Systematic Review on MRI Markers of Functional and Structural Changes. *Frontiers in Neuroscience*, *15*. https://doi.org/10.3389/fnins.2021.707675
48. Trempe, J.-F., & Gehring, K. (2023). Molecular Mechanisms of Neurodegeneration in Parkinson's Disease. *Journal of Molecular Biology*, *435*(12), 168131. https://doi.org/10.1016/j.jmb.2023.168131
49. Wang, Y., Liu, H., Zhang, B.-S., Soares, J. C., & Zhang, X. Y. (2016). Low BDNF is associated with cognitive impairments in patients with Parkinson's disease. *Parkinsonism & Related Disorders*, *29*, 66–71. https://doi.org/10.1016/j.parkreldis.2016.05.023
50. Yuan, T.-F., Li, W.-G., Zhang, C., Wei, H., Sun, S., Xu, N.-J., Liu, J., & Xu, T.-L. (2020). Targeting neuroplasticity in patients with neurodegenerative diseases using brain stimulation techniques. *Translational Neurodegeneration*, *9*(1), 44. https://doi.org/10.1186/s40035-020-00224-z

Exploring the Brain-Gut Axis: the Microbiota Connection

Harun Dere

1. Introduction

The brain-gut axis connects the central and enteric nervous systems, with gut microbiota as a key player (Mayer et al., 2022). Research shows it helps with gut balance and brain function, including neuroplasticity (Wenger & Kühn, 2021).

Gut bacteria produce substances that affect learning and memory, highlighting their importance for brain health. An imbalance in gut bacteria can lead to brain and mood disorders, showing the need for more research (Pan et al., 2023). This chapter will explore the brain-gut axis and its link to microbiota and detail how microbiota support neuroplasticity.

2. Introduction to the Brain-Gut Axis

2.1. Historical Background and Exploration

The brain-gut axis has drawn the attention of scientists and doctors since the nineteenth century. They discovered that intestines do more than digest; they interact with the nervous system. In 1986, German physiologist Leopold Auerbach suggested that the enteric nervous system could control the intestine independently. This sparked further study into the brain-gut axis (Miller, 2018). By the twentieth century, it was clear that these two systems communicated bi-directionally. This communication impacts neurological and psychological health. Now, research shows that gut microbiota is key in this axis. It highlights the need to explore the link between microbiota and neuroplasticity (Mayer et al., 2022).

2.2. Anatomical and Functional Overview

The brain-gut axis is a complex network. It connects the central nervous system, enteric nervous system, immune system, endocrine system, and gastrointestinal system. Information flows between the brain and intestines via hormones and neural pathways (Chaudhry et al., 2023). The vagus nerve is crucial, transmitting signals from the gut to the brain. Additionally, gut microbiota plays a key role in this communication. These elements regulate functions such as digestion, learning, stress management, and mental health (Han et al., 2022).

2.3. Basic Components of the Brain-Gut Axis

The brain-gut axis includes several interconnected systems. The central nervous system (CNS), enteric nervous system, gut microbiota, immune system, and neuroendocrine system are key parts. The CNS, made up of the brain and spinal cord, manages the body. The enteric nervous system allows the gut to function independently. Both systems communicate via the vagus nerve and neurotransmitters (Martin et al., 2018). Gut microbiota, with trillions of microorganisms, affects gut and brain health. The immune system protects the body and manages inflammation in the gut. The neuroendocrine system uses hormones to communicate these components. This complex network impacts physical and mental health (Ruan et al., 2020).

2.4. Role of the Central Nervous System

The central nervous system (CNS), which includes the brain and spinal cord, controls most body functions. It links with the gut through two main pathways. First, the vagus nerve allows direct communication. Second, signaling molecules enable indirect communication (Fleming et al., 2020). The vagus nerve carries gut information to the brain and constantly updates it on gut health. Meanwhile, the CNS sends hormones to manage gut functions. For instance, the CNS boosts stress hormones like cortisol in stressed people. This can alter gut movement and balance bacteria. Thus, the CNS affects not just thinking and feelings but also gut health (Martin et al., 2018).

2.5. Interaction of the Brain-Gut Axis with the Enteric Nervous System and Central Nervous System

The enteric nervous system (ENS) is often called the "second brain." It autonomously manages intestinal functions. The ENS controls motility, enzyme release, blood flow, and mucus production. It communicates directly with the central nervous system (CNS), mainly through the vagus nerve. This nerve carries signals from the intestines to the brain and vice versa. Though the ENS can work alone, it often interacts with the CNS. For instance, gut pain is relayed to the CNS, potentially causing stress or anxiety. Similarly, the CNS can influence bowel movements and secretions (Fleming et al., 2020).

3. Microbiota

3.1. Overview of Intestinal Microbiota

Gut microbiota includes microorganisms like bacteria and viruses in our intestines (Turroni et al., 2018). It is unique to each person and shaped by genetics, birth method, diet, environment, and medications (Nagai et al., 2016). Recent studies suggest microbiota development starts in the womb, not just after birth (Cryan et al., 2019). A healthy gut microbiota typically contains Bacteroidetes, Actinobacteria, Firmicutes, Fusobacteria, Cyanobacteria, Proteobacteria, and Verrucomicrobia. Bacteroidetes and Firmicutes comprise about 90% of the population (Jandhyala et al., 2015). These microorganisms interact to form a complex system. The Firmicutes/Bacteroidetes (F/B) ratio is a key health indicator for gut microbiota (Fujisaka et al., 2023).

3.2. Diversity and Stability of Intestinal Microbiota

The gut microbiota's diversity includes various microorganisms in different amounts. More diversity typically means a better-developed gut microbiome. Different microorganisms boost efficiency and block harmful ones. They also aid in nutrient absorption (Danneskiold-Samsøe et al., 2019). Moreover, diversity supports the immune system. It helps manage responses, lowering inflammation and autoimmune risks. Conversely, low diversity is tied to health issues like digestive problems, obesity, type 2 diabetes, and inflammatory bowel diseases.(Laitinen & Mokkala, 2019).

Microbiota stability can be defined as the resistance to changes in the microbiota ecosystem. A high level of stability indicates that the microbiota ecosystem is resistant to external and internal stress factors and has the ability to maintain its structural integrity. A stable microbiota almost always means a more organized and in good-condition intestinal flora. Factors that play a role in shaping microbiota stability include eating habits, antibiotic use, stress, hereditary characteristics, and environmental factors. Dietary habits, especially the use of fibrous foods and fermentable carbohydrates, can strengthen microbiota stability. Antibiotics can lead to a disturbance of the microbiota balance as they kill both beneficial and pathogenic bacteria in the human body. Stress and hereditary factors may also affect the microbiome content (Olsson et al., 2022).

A diverse gut microbiota is more stable. This diversity helps adapt to changes in the environment. In contrast, low diversity leads to low stability and gut issues. Thus, diversity and stability are key to a healthy gut (Ruan et al., 2020).

3.3. Factors Affecting Microbiota Composition

Genetics, age, birth method, diet, and antibiotics influence gut bacteria (Patangia et al., 2022). Gut microbiota starts at birth and evolves with environmental exposure. Vaginally born babies share microbiota with their mothers, while cesarean babies have different bacteria influenced by skin contact (Arrieta et al., 2014). Bacteria increase after birth, especially with breastfeeding. Breastfed babies have healthier gut flora. In contrast, formula-fed babies show reduced diversity (Catassi et al., 2024).

Dietary habits impact microbiota directly or indirectly. Long-term dietary patterns are crucial for gut microbiota. Specifically, the Mediterranean diet fosters beneficial gut bacteria. Conversely, diets high in fat, sugar, and processed foods disrupt microbiota balance and promote harmful bacteria (Leeming et al., 2019).

The environment significantly impacts an individual's microbiota. Thus, different societies have unique microbiota. People in rural areas often have more diverse microbiota. This is due to their closer contact with nature. Activities like working with soil, drinking natural water, and eating organic food boost gut microbiota. Conversely, urban life, industrialization, and modern habits harm microbiota diversity. Factors such as air pollution, overusing cleaning products, taking antibiotics, and living in sterile conditions reduce beneficial microorganisms. Notably, excessive antibiotic use is particularly harmful. It cuts microbial diversity, disrupts gut balance, and may cause long-term health issues (Ruan et al., 2020).

The gut microbiota and brain are directly linked, known as the brain-gut line. Stress impacts the microbiota via this line, with effects lasting from short to long-term (Dere et al., 2023). Chronic stress boosts intestinal permeability. This allows harmful bacteria to enter the body, disrupting microbiota balance and causing dysbiosis. Conversely, a healthy microbiota mitigates stress's brain impact. It regulates stress hormones like cortisol and maintains the body's stress response (Olsson et al., 2022).

3.4. Microbiota and Metabolic Functions

The microbiota is vital for energy balance and metabolic processes in our body. It breaks down complex carbohydrates to produce short-chain fatty acids. These acids are key energy sources that regulate fat storage and insulin sensitivity. Additionally, the microbiota impacts energy intake and use. Some bacteria help extract more energy, while others limit absorption. Dysbiosis, an imbalance in the microbiota, can disrupt metabolism. This leads to obesity, insulin resistance, and type 2 diabetes (Rueb et al., 2021).

3.5. Microbiota and Immune System Modulation

The immune system directly interacts with gut microbiota. It helps develop and regulate immune cells (Choden & Cohen, 2022). Microbiota influences immune responses through interactions with intestinal cells. They are vital for gut balance and the mucosal immune system. Moreover, they protect against harmful microorganisms by maintaining gut integrity (Yoo et al., 2020).

Beneficial bacteria like Lactobacillus and Bifidobacterium support this balance. They prevent excessive immune reactions and aid in tolerance development. Additionally, these bacteria produce short-chain fatty acids (SCFAs), which are crucial for the immune system. SCFAs strengthen the gut barrier and reduce inflammation (Rueb et al., 2021). The microbiota links innate and adaptive immune systems. Innate cells use it to detect and fight pathogens. It also manages the activation and tolerance of T and B cells. Dendritic cells and macrophages in the gut respond to microbiota signals (Ratajczak et al., 2019).

3.5.1. Dysbiosis and Immune System

A balanced microbiota is crucial for a healthy immune system. Imbalances, or dysbiosis, can cause the immune system to overreact, leading to inflammation. This condition is linked to chronic diseases, autoimmune disorders, allergies, and inflammatory bowel diseases (IBD) (Lau et al., 2017). For instance, a dominance of harmful bacteria can trigger "leaky gut" syndrome. This increases intestinal permeability, causing immune overstimulation and systemic inflammation. Thus, the microbiota is key in developing and maintaining immune function. It is vital for gut and overall immune health (Christovich & Luo, 2022).

4. Molecular Mechanisms Linking the Gut Microbiota to the Brain

The gut microbiota produces substances like short-chain fatty acids and neurotransmitters. It also sends signals that impact the central nervous system (CNS). These signals travel through the vagus nerve and the immune system. Understanding this could lead to new treatments for neurological and psychiatric disorders (Chen et al., 2021).

The brain-gut interaction is due to the direct effects of gut microorganisms and their influence on immune and neurological processes. Dysbiosis, or an imbalance in gut bacteria, can harm CNS function and lead to neuropsychiatric disorders (Christovich & Luo, 2022).

4.1. Microbial Metabolites and Neurotransmitter Production

Gut bacteria can produce neurotransmitters and metabolites that affect the CNS. These substances help nerve cells communicate. They also influence emotion, thinking, and behavior (Suganya & Koo, 2020).

4.1.1. Microbial Metabolites and Their Effects on the Nervous System

Gut microorganisms turn food into short-chain fatty acids and other metabolites. SCFAs like butyrate, propionate, and acetate impact the gut–brain connection. Butyrate nourishes gut cells, reduces brain inflammation, and boosts brain adaptability. Meanwhile, propionate and acetate influence brain communication and immune responses. These compounds are key in managing gut and brain functions (O'Riordan et al., 2022).

4.1.2. Neurotransmitter Production and Modulation

The gut microbiota directly affects neurotransmitter production. For instance, it metabolizes tryptophan, which is crucial for serotonin synthesis. Serotonin, made in the gut, regulates mood, sleep, and appetite. Gut microorganisms also influence dopamine and GABA levels. Bacteria like Lactobacillus and Bifidobacterium boost GABA, easing anxiety and stress (Chen et al., 2021).

These mechanisms show how gut microbes affect the central nervous system (CNS). A lack of diversity or harmful microbes can disrupt neurotransmitter production. This, in turn, may lead to neurological and psychiatric diseases (Ullah et al., 2023).

4.1.3. Clinical Importance of Microbiota-Neurotransmitter Interactions

Microbial metabolites and neurotransmitters are crucial in medicine. For instance, gut levels of serotonin, dopamine, and GABA are linked to depression, anxiety, autism, and Parkinson's disease. Specifically, serotonin fluctuations can lead to mood and depressive symptoms. Research is ongoing to understand how microbiota influence these neurotransmitters and their potential for disease treatment (O'Riordan et al., 2022).

4.2. Role of Short-Chain Fatty Acids

Gut bacteria turn complex carbs, like fiber, into short-chain fatty acids (SCFAs). The key ones are butyrate, propionate, and acetate. SCFAs provide energy to gut cells and impact the central nervous system (CNS). Notably, butyrate protects the brain. It strengthens the blood-brain barrier, reduces inflammation, and aids brain adaptability (Rasouli-Saravani et al., 2023).

SCFAs also manage inflammation in the body by influencing gut cell responses. Propionate and acetate help reduce brain inflammation by affecting microglia cells, which is vital for treating neurodegenerative diseases and depression. Moreover, SCFAs play a crucial role in brain function through the gut–brain axis. Understanding this helps link microbial metabolites to CNS effects (Ratajczak et al., 2019).

4.3. Toxins from Gut Bacteria and Their Impact on the Brain

An imbalance in gut bacteria (dysbiosis) can produce harmful substances. These substances can negatively affect the central nervous system (CNS). Notably, key examples are lipopolysaccharides (LPS) and toxins from certain harmful bacteria. LPS, an endotoxin from Gram-negative bacteria, can cause a "leaky gut." When it breaches the intestinal barrier, it enters the bloodstream. This triggers inflammatory cytokines, increasing CNS inflammation (Christovich & Luo, 2022).

4.4. Vagus Nerve: A Key Communication Link

The vagus nerve is vital for communication between the brain and the gut. It is the longest nerve in the parasympathetic system. It carries signals from the gut to the CNS and vice versa. This nerve sends gut sensory information to the brainstem. Additionally, it influences digestion, immune responses, and neurological functions (Han et al., 2022). For more on the vagus nerve, see *Chapter 10: Neuromodulation Techniques and Their Impact on Brain Plasticity*.

4.5. The Immune System's Role

The immune system links gut microbiota and the CNS. Intestinal cells interact with immune cells and microbes. The microbiota controls immune responses, preventing or starting neuroinflammation (O'Riordan et al., 2022). Microbial metabolites, especially short-chain fatty acids, reduce inflammation in the CNS. They act on immune cells. Moreover, gut microbes influence brain functions by altering cytokine production. These cytokines enter the CNS through the bloodstream. Conditions like depression, multiple sclerosis, and autism spectrum disorder are linked to microbiota-induced inflammation (Rasouli-Saravani et al., 2023).

5. The Concept of Neuroplasticity and the Effect of Microbiota on Neuroplasticity

Neuroplasticity is the nervous system's ability to adapt. It changes in response to the environment, learning, and experiences. This process occurs in both children and adults. It helps the brain adjust for learning, memory, and recovery. Neuroplasticity involves more than just strengthening or weakening connections between neurons. It also includes creating new neurons (neurogenesis), forming new connections (synaptogenesis), and long-term changes in synaptic strength. First identified in the 1950s, neuroplasticity showed that the brain is not fixed. It is central to learning and memory, with connections forming and changing throughout life (Murciano-Brea et al., 2021).

Gut microbiota significantly affects brain function and neuroplasticity. Research is growing on its impact on brain development, learning, and cognitive functions. The microbiota–brain axis is key. It uses substances produced by gut bacteria to regulate brain health (Damiani et al., 2023).

5.1. The Effect of Intestinal Microbiota on Brain Development

Gut microbiota influences brain development, both directly and indirectly. This begins in the womb. The microbiota–brain axis is crucial here. It affects brain structure and function by influencing the immune and central nervous systems. For example, interaction with the microbiota before and immediately after birth can influence the morphology of the cerebral cortex and neurotransmitter balance. In the case of dysbiosis, the risk of abnormalities in brain development and neurological disorders may increase both during developmental age and later in life (Vaher et al., 2022).

5.2. Microbiota and Adult Neurogenesis

Adult neurogenesis refers to the regeneration of brain cells and the formation of new neurons in adult individuals. Gut microbiota has also been shown to have important effects on adult neurogenesis. Microbiota can influence the release of neurotrophic factors and the survival of nerve cells. In particular, short-chain fatty acids produced by microbiota have been found to promote the process of neurogenesis in brain regions such as the hippocampus and support the formation of new neurons. This indicates that gut microbiota is crucial for neurogenesis (Sarubbo et al., 2022).

5.3. Microbiota and Synaptic Plasticity

Synaptic plasticity involves strengthening or weakening nerve connections. It's vital for learning and memory. Microbiota plays a key role in this process. Clinical studies show gut microbiota affects synaptic connections. It does so through short-chain fatty acids (SCFAs) and other products (Damiani et al., 2023).

Microbiota influences synaptic transmission by altering synapse numbers and density. This is crucial for learning and memory. Metabolites from microbiota can directly affect synaptic plasticity. They support the adaptation of neural networks (Glinert et al., 2022).

5.4. Impact on Learning and Memory

Gut microbiota significantly affects learning, memory, and cognitive functions. This is part of the microbiota–brain axis. Microbiota produces metabolites and signaling molecules that influence brain function. For instance, SCFAs and neurotrophic factors support learning and memory pathways (Socała et al., 2021).

Toxins and inflammation from microbiota can harm cognitive functions. This is linked to Alzheimer's and other neurological diseases. Adjusting the microbiota–brain axis could be a therapeutic strategy to enhance cognitive functions (Wiatrak et al., 2022).

5.5. Microbiota and BDNF

Brain-derived neurotrophic factor (BDNF) is vital for nerve cell health, synaptic plasticity, and new neuron growth. Microbiota can influence BDNF levels and activity. SCFAs and other metabolites boost BDNF expression, supporting synaptic transmission and neurogenesis (Agnihotri & Mohajeri, 2022).

Dysbiosis, or an imbalance in gut microbiota, can lower BDNF levels. It is linked to inflammation, stress, and metabolic disorders. Inflammation can inhibit BDNF production, harming the brain. For example, inflammatory cytokines can reduce BDNF expression, affecting nerve cell health (Colucci-D'Amato et al., 2020).

6. Dysbiosis and Neuroplasticity Disorders

Dysbiosis disrupts gut microbiota balance and can lead to neuroplasticity disorders. It negatively affects the brain-gut axis. This disruption can harm nervous system functions and contribute to neurodegenerative and psychiatric diseases. This chapter explores dysbiosis's role in these diseases and its effects on the gut microbiota (Mitrea et al., 2022).

6.1. Dysbiosis and Its Effects on the Brain

Dysbiosis disrupts gut microbiota and harms neurological function. The gut microbiota communicates with the brain, producing essential compounds. In dysbiosis, this balance is lost, reducing nervous system regulation. This leads to neuroinflammation and hampers learning and memory. It may also impair the vagus nerve, affecting brain-intestine communication and causing long-term issues (Pan et al., 2023).

6.2. Role in Neurodegenerative Diseases

Dysbiosis is linked to neurodegenerative diseases like Alzheimer's and Parkinson's. Those with these diseases show significant gut microbiota changes. Such disturbances can trigger inflammation and damage the brain's protective barrier. They also disrupt metabolic processes, contributing to disease progression. Thus, regulating gut microbiota could slow down these diseases (Han et al., 2022).

6.3. Gut Microbiota in Mental Disorders

Dysbiosis is closely linked to mental disorders such as depression and anxiety. It disrupts communication in the gut–brain axis, harming the nervous system. This affects the production of crucial neurotransmitters like serotonin. Increased neuroinflammation and reduced neurotransmitter levels can trigger these disorders. Treatments that regulate gut microbiota, like probiotics, show promise. Ongoing trials are exploring this (Mitrea et al., 2022).

6.4. Autism Spectrum Disorders and Microbiota

Research has spotlighted gut microbiota's role in autism spectrum disorder (ASD). Individuals with ASD show significant gut microbiota differences. These changes can disrupt brain development and neuroplasticity. They may also lead to behaviors associated with autism. New treatments aimed at modifying gut microbiota could ease ASD symptoms (Murciano-Brea et al., 2021).

6.5. Link Between Gut Health and Cognitive Decline

Cognitive decline is linked to disturbances in gut microbiota. Aging changes gut microbiota, which can speed up cognitive decline. Dysbiosis increases neuroinflammation and reduces neuroplasticity. It also lowers the production of beneficial compounds, affecting learning and memory. These findings highlight gut health's importance for cognitive function and suggest targets for treatment (Liu et al., 2022).

7. Therapeutic Interventions for the Brain-Gut Axis

Research on treatments for the brain-gut-microbiota axis has surged. Clinical trials show these can help with neurological and psychiatric issues. Understanding gut microbiota's role in brain health has spurred new therapies (Fleming et al., 2020). Methods include probiotics, prebiotics, fecal transplants, and drugs, all aimed at boosting brain function (Colucci-D'Amato et al., 2020).

7.1. Probiotics and Prebiotics in Neuroplasticity

Probiotics and prebiotics are vital for the brain-gut axis. Probiotics aid intestinal health and benefit neurological health. Prebiotics, on the other hand, nourish probiotics. They also enhance brain function and support the nervous system by improving gut health. These substances balance gut microbiota, reduce inflammation, and aid neurological functions. Thus, they are natural options for boosting neurological health and neuroplasticity (Ansari et al., 2023).

7.2. Diet, Nutrition, and the Brain-Gut-Microbiota Axis

Nutrition is key in managing the brain-gut axis. Diets high in fiber, omega-3s, and polyphenols are particularly beneficial. These foods promote gut balance and support neurological functions. Fiber-rich foods boost short-chain fatty acids (SCFAs), enhancing gut barrier health and reducing inflammation. SCFAs are linked to increased neuroplasticity (Romaní-Pérez et al., 2021).

Omega-3 fatty acids, crucial for reducing inflammation, enhance brain cell health. They improve brain cell flexibility and aid in preventing neurological disorders. Meanwhile, polyphenols, with their antioxidant properties, protect brain function and slow cognitive decline (Laitinen & Mokkala, 2019).

7.3. Fecal Microbiota Transplantation (FMT)

FMT involves transferring healthy microbiota from a donor to a recipient's intestine. It's used to fix severe gut microbiota imbalances and treat related neurological and psychiatric disorders. FMT has proven effective against Clostridium difficile infections (Chinna Meyyappan et al., 2020).

Recent studies suggest FMT might also help with neurological disorders like depression, anxiety, autism, and Parkinson's disease. It quickly alters gut microbiota, reduces neuroinflammation, and boosts nervous system function. This method opens new avenues for microbiota-based therapies and may enhance neuroplasticity (Settanni et al., 2021).

7.4. Pharmacological Approaches Targeting Microbiota

New treatments targeting gut health and brain function focus on the microbiota. These treatments use psychobiotics to alleviate depression, anxiety, and stress. Psychobiotics boost gut microbiota and support cognitive functions by reducing brain inflammation (Socała et al., 2021).

Antibiotics and prebiotics can balance gut bacteria and treat brain disorders. They improve brain function, protect nerve cells, and aid learning. Psychobiotics, which affect gut bacteria, are key in treating brain and mental health issues (Feng et al., 2020).

7.5. Future Directions in Therapeutics

Future brain-gut treatments will benefit from personalized medicine and genetic advances. Targeted therapies using modified probiotics and microbiota are on the horizon. Also, treatments tailored to a person's genetics and gut bacteria will be more effective (Ugwu et al., 2024).

Research into how gut bacteria affect brain flexibility will spark new treatments for diseases like Alzheimer's, depression, anxiety, and autism. Drugs and biotech solutions linked to gut bacteria will be crucial in maintaining brain health and preventing cognitive disorders (Murciano-Brea et al., 2021).

8. Conclusion

The brain-gut axis and gut microbiota interact, revealing new aspects of brain function. The gut microbiota affects neural processes, such as synaptic plasticity, neurogenesis, and cognition. It does this by producing metabolites and neurotransmitters. This interaction is vital for gut health and influences brain development and behavior.

An imbalance in gut microbiota, known as dysbiosis, is linked to neuropsychiatric and neurodegenerative disorders. This highlights the need for microbiota-targeted therapies. Treatments could include probiotics, prebiotics, dietary changes, and drugs aimed at restoring balance and improving neuroplasticity.

Research is uncovering the mechanisms behind this system. It shows that gut microbiota is crucial for gut and brain health. Using this relationship in therapy could improve treatments for cognitive decline, neurodevelopmental, and psychiatric disorders. Future studies are essential to understand microbiota-neuroplasticity interactions. This knowledge will help develop therapies that support brain health throughout life.

9. Takeaways

1. The brain-gut axis links the brain and gut, influenced by gut microbiota, and aids neuroplasticity.
2. Gut microbiota boosts brain-derived neurotrophic factors (BDNF), vital for neuroplasticity and synapses.
3. Raising BDNF via microbiota could treat neurological diseases and enhance cognition. Prebiotics, probiotics, and diet can restore gut balance and BDNF.
4. Gut microbiota affects learning and memory mechanisms.
5. Fecal microbiota transplantation shows promise in restoring gut health and neuroplasticity.
6. Research on the brain-gut axis could lead to new treatments for cognitive decline, neurodevelopmental disorders, and mental illnesses.
7. Neuroplasticity is significantly influenced by gut microbiota activity.
8. Microbial metabolites can affect brain functions related to learning, memory, and emotions.
9. Dysbiosis, or gut imbalance, is linked to psychiatric disorders like depression and anxiety.
10. Changes in gut microbiota during key development stages can impact brain development and neuroplasticity.

Abbreviations

ASD: Autism Spectrum Disorders
BDNF: Brain-Derived Neurotrophic Factor
CNS: Central Nervous System
ENS: Enteric Nervous System
FMT: Fecal Microbiota Transplantation
GABA: Gamma Amino Butyric Acid
IBD: Inflammatory Bowel Diseases
LPS: Lipopolysaccharides
LTD: Long-Term Depression
LTP: Long-Term Potentiation
SCFAs: Short-Chain Fatty Acids

Bibliography

1. Agnihotri, N., & Mohajeri, M. H. (2022). Involvement of intestinal microbiota in adult neurogenesis and the expression of brain-derived neurotrophic factor. *International Journal of Molecular Sciences*, *23*(24), 15934. https://doi.org/10.3390/ijms232415934
2. Ansari, F., Neshat, M., Pourjafar, H., Jafari, S. M., Samakkhah, S. A., & Mirzakhani, E. (2023). The role of probiotics and prebiotics in modulating of the gut-brain axis. *Frontiers in nutrition*, *10*, 1173660. https://doi.org/10.3389/fnut.2023.1173660
3. Arrieta, M. C., Stiemsma, L. T., Amenyogbe, N., Brown, E. M., & Finlay, B. (2014). The intestinal microbiome in early life: health and disease. *Frontiers in immunology*, *5*, 427. https://doi.org/10.3389/fimmu.2014.00427
4. Catassi, G., Aloi, M., Giorgio, V., Gasbarrini, A., Cammarota, G., & Ianiro, G. (2024). The role of diet and nutritional interventions for the Infant Gut Microbiome. *Nutrients*, *16*(3), 400. https://doi.org/10.3390/nu16030400
5. Chaudhry, T. S., Senapati, S. G., Gadam, S., Mannam, H. P. S. S., Voruganti, H. V., Abbasi, Z., … & Arunachalam, S. P. (2023). The impact of microbiota on the gut–brain axis: Examining the complex interplay and implications. *Journal of clinical medicine*, *12*(16), 5231. https://doi.org/10.3390/jcm12165231
6. Chen, Y., Xu, J., & Chen, Y. (2021). Regulation of neurotransmitters by the gut microbiota and effects on cognition in neurological disorders. *Nutrients*, *13*(6), 2099. https://doi.org/10.3390/nu13062099
7. Chinna Meyyappan, A., Forth, E., Wallace, C. J., & Milev, R. (2020). Effect of fecal microbiota transplant on symptoms of psychiatric disorders: a systematic review. *BMC psychiatry*, *20*, 1–19. https://doi.org/10.1186/s12888-020-02654-5
8. Choden, T., & Cohen, N. A. (2022). The gut microbiome and the immüne system. *Exploration of Medicine*, *3*(3), 219–233. https://doi.org/10.37349/emed.2022.00087
9. Christovich, A., & Luo, X. M. (2022). Gut microbiota, leaky gut, and autoimmune diseases. *Frontiers in immunology*, *13*, 946248. https://doi.org/10.3389/fimmu.2022.946248
10. Colucci-D'Amato, L., Speranza, L., & Volpicelli, F. (2020). Neurotrophic factor BDNF, physiological functions and therapeutic potential in depression, neurodegeneration and brain cancer. *International journal of molecular sciences*, *21*(20), 7777. https://doi.org/10.3390/ijms21207777

11. Cryan, J. F., O'Riordan, K. J., Cowan, C. S., Sandhu, K. V., Bastiaanssen, T. F., Boehme, M., … & Dinan, T. G. (2019). The microbiota-gut-brain axis. *Physiological reviews*. https://doi.org/10.1152/physrev.00018.2018
12. Damiani, F., Cornuti, S., & Tognini, P. (2023). The gut-brain connection: Exploring the influence of the gut microbiota on neuroplasticity and neurodevelopmental disorders. *Neuropharmacology*, *231*, 109491. https://doi.org/10.1016/j.neuropharm.2023.109491
13. Danneskiold-Samsøe, N. B., Barros, H. D. D. F. Q., Santos, R., Bicas, J. L., Cazarin, C. B. B., Madsen, L., … & Júnior, M. R. M. (2019). Interplay between food and gut microbiota in health and disease. *Food research international*, *115*, 23–31. https://doi.org/10.1016/j.foodres.2018.07.043
14. Dere, H., Bayraktar, M., & Özden, A. V. (2023). Effect of aerobic exercise on stress and the brain-gut axis. *International Journal of Sport Culture and Science*, *11*(2), 71–78. https://doi.org/10.14486/IntJSCS.2023.675
15. Feng, W., Liu, J., Ao, H., Yue, S., & Peng, C. (2020). Targeting gut microbiota for precision medicine: Focusing on the efficacy and toxicity of drugs. *Theranostics*, *10*(24), 11278. https://doi.org/10.7150%2Fthno.47289
16. Fleming, M. A., Ehsan, L., Moore, S. R., & Levin, D. E. (2020). The enteric nervous system and its emerging role as a therapeutic target. *Gastroenterology research and practice*, *2020*(1), 8024171. https://doi.org/10.1155/2020/8024171
17. Fujisaka, S., Watanabe, Y., & Tobe, K. (2023). The gut microbiome: a core regulator of metabolism. *Journal of Endocrinology*, *256*(3). https://doi.org/10.1530/JOE-22-0111
18. Glinert, A., Turjeman, S., Elliott, E., & Koren, O. (2022). Microbes, metabolites and (synaptic) malleability, oh my! The effect of the microbiome on synaptic plasticity. *Biological Reviews*, *97*(2), 582–599. https://doi.org/10.1111/brv.12812
19. Han, Y., Wang, B., Gao, H., He, C., Hua, R., Liang, C., … & Xu, J. (2022). Vagus nerve and underlying impact on the gut microbiota-brain axis in behavior and neurodegenerative diseases. *Journal of inflammation research*, 6213–6230. https://doi.org/10.2147/JIR.S384949
20. Jandhyala, S. M., Talukdar, R., Subramanyam, C., Vuyyuru, H., Sasikala, M., & Reddy, D. N. (2015). Role of the normal gut microbiota. *World journal of gastroenterology: WJG*, *21*(29), 8787. https://doi.org/10.3748%2Fwjg.v21.i29.8787

21. Laitinen, K., & Mokkala, K. (2019). Overall dietary quality relates to gut microbiota diversity and abundance. *International journal of molecular sciences*, *20*(8), 1835. https://doi.org/10.3390/ijms20081835
22. Lau, K., Srivatsav, V., Rizwan, A., Nashed, A., Liu, R., Shen, R., & Akhtar, M. (2017). Bridging the gap between gut microbial dysbiosis and cardiovascular diseases. *Nutrients*, *9*(8), 859. https://doi.org/10.3390/nu9080859
23. Leeming, E. R., Johnson, A. J., Spector, T. D., & Le Roy, C. I. (2019). Effect of diet on the gut microbiota: rethinking intervention duration. *Nutrients*, *11*(12), 2862. https://doi.org/10.3390/nu11122862
24. Liu, P., Gao, M., Liu, Z., Zhang, Y., Tu, H., Lei, L., … & Zhang, K. (2022). Gut microbiome composition linked to inflammatory factors and cognitive functions in first-episode, drug-naive major depressive disorder patients. *Frontiers in neuroscience*, *15*, 800764. https://doi.org/10.3389/fnins.2021.800764
25. Martin, C. R., Osadchiy, V., Kalani, A., & Mayer, E. A. (2018). The brain-gut-microbiome axis. *Cellular and molecular gastroenterology and hepatology*, *6*(2), 133–148. https://doi.org/10.1016/j.jcmgh.2018.04.003
26. Mayer, E. A., Nance, K., & Chen, S. (2022). The gut–brain axis. *Annual review of medicine*, *73*(1), 439–453. https://doi.org/10.1146/annurev-med-042320-014032
27. Miller, I. (2018). The gut–brain axis: historical reflections. *Microbial ecology in health and disease*, *29*(2), 1542921. https://doi.org/10.1080/16512235.2018.1542921
28. Mitrea, L., Nemeş, S. A., Szabo, K., Teleky, B. E., & Vodnar, D. C. (2022). Guts imbalance imbalances the brain: a review of gut microbiota association with neurological and psychiatric disorders. *Frontiers in medicine*, *9*, 813204. https://doi.org/10.3389/fmed.2022.813204
29. Murciano-Brea, J., Garcia-Montes, M., Geuna, S., & Herrera-Rincon, C. (2021). Gut microbiota and neuroplasticity. *Cells*, *10*(8), 2084. https://doi.org/10.3390/cells10082084
30. Nagai, M., Obata, Y., Takahashi, D., & Hase, K. (2016). Fine-tuning of the mucosal barrier and metabolic systems using the diet-microbial metabolite axis. *International immunopharmacology*, *37*, 79–86. https://doi.org/10.1016/j.intimp.2016.04.001
31. Olsson, L. M., Boulund, F., Nilsson, S., Khan, M. T., Gummesson, A., Fagerberg, L., … & Bäckhed, F. (2022). Dynamics of the normal gut microbiota: a longitudinal one-year population study in Sweden. *Cell host & microbe*, *30*(5), 726–739. https://doi.org/10.1016/j.chom.2022.03.002

32. O'Riordan, K. J., Collins, M. K., Moloney, G. M., Knox, E. G., Aburto, M. R., Fülling, C., … & Cryan, J. F. (2022). Short chain fatty acids: microbial metabolites for gut-brain axis signalling. *Molecular and Cellular Endocrinology*, *546*, 111572. https://doi.org/10.1016/j.mce.2022.111572
33. Pan, I., Issac, P. K., Rahman, M. M., Guru, A., & Arockiaraj, J. (2023). Gut-brain axis a key player to control gut Dysbiosis in neurological diseases. *Molecular Neurobiology*, 1–19. https://doi.org/10.1007/s12035-023-03691-3
34. Patangia, D. V., Anthony Ryan, C., Dempsey, E., Paul Ross, R., & Stanton, C. (2022). Impact of antibiotics on the human microbiome and consequences for host health. *Microbiologyopen*, *11*(1), e1260. https://doi.org/10.1002/mbo3.1260
35. Rasouli-Saravani, A., Jahankhani, K., Moradi, S., Gorgani, M., Shafaghat, Z., Mirsanei, Z., … & Mirzaei, R. (2023). Role of microbiota short-chain fatty acids in the pathogenesis of autoimmune diseases. *Biomedicine & Pharmacotherapy*, *162*, 114620. https://doi.org/10.1016/j.biopha.2023.114620
36. Ratajczak, W., Rył, A., Mizerski, A., Walczakiewicz, K., Sipak, O., & Laszczyńska, M. (2019). Immunomodulatory potential of gut microbiome-derived short-chain fatty acids (SCFAs). *Acta Biochimica Polonica*, *66*(1), 1–12. https://doi.org/10.18388/abp.2018_2648
37. Romaní-Pérez, M., Bullich-Vilarrubias, C., López-Almela, I., Liébana-García, R., Olivares, M., & Sanz, Y. (2021). The microbiota and the gut–brain axis in controlling food intake and energy homeostasis. *International journal of molecular sciences*, *22*(11), 5830. https://doi.org/10.3390/ijms22115830
38. Ruan, W., Engevik, M. A., Spinler, J. K., & Versalovic, J. (2020). Healthy human gastrointestinal microbiome: composition and function after a decade of exploration. *Digestive diseases and sciences*, *65*, 695–705. https://doi.org/10.1007/s10620-020-06118-4
39. Rueb, A. M., Tsakmaklis, A., Gräfe, S. K., Simon, M. C., Vehreschild, M. J., & Wuethrich, I. (2021). Biomarkers of human gut microbiota diversity and dysbiosis. *Biomarkers in Medicine*, *15*(2), 139–150. https://doi.org/10.2217/bmm-2020-0353
40. Sarubbo, F., Cavallucci, V., & Pani, G. (2022). The influence of gut microbiota on neurogenesis: evidence and hopes. *Cells*, *11*(3), 382. https://doi.org/10.3390/cells11030382
41. Settanni, C. R., Ianiro, G., Bibbò, S., Cammarota, G., & Gasbarrini, A. (2021). Gut microbiota alteration and modulation

in psychiatric disorders: Current evidence on fecal microbiota transplantation. *Progress in Neuro-Psychopharmacology and Biological Psychiatry*, *109*, 110258. https://doi.org/10.1016/j.pnpbp.2021.110258
42. Socała, K., Doboszewska, U., Szopa, A., Serefko, A., Włodarczyk, M., Zielińska, A., … & Wlaź, P. (2021). The role of microbiota-gut-brain axis in neuropsychiatric and neurological disorders. *Pharmacological research*, *172*, 105840. https://doi.org/10.1016/j.phrs.2021.105840
43. Suganya, K., & Koo, B. S. (2020). Gut–brain axis: role of gut microbiota on neurological disorders and how probiotics/prebiotics beneficially modulate microbial and immune pathways to improve brain functions. *International journal of molecular sciences*, *21*(20), 7551. https://doi.org/10.3390/ijms21207551
44. Turroni, S., Brigidi, P., Cavalli, A., & Candela, M. (2018). Microbiota–host transgenomic metabolism, bioactive molecules from the inside: Miniperspective. *Journal of medicinal chemistry*, *61*(1), 47–61. https://doi.org/10.1021/acs.jmedchem.7b00244
45. Ugwu, O. P. C., Alum, E. U., Okon, M. B., & Obeagu, E. I. (2024). Mechanisms of microbiota modulation: Implications for health, disease, and therapeutic interventions. *Medicine*, *103*(19), e38088. https://doi.org/10.1097/MD.0000000000038088
46. Ullah, H., Arbab, S., Tian, Y., Liu, C. Q., Chen, Y., Qijie, L., … & Li, K. (2023). The gut microbiota–brain axis in neurological disorder. *Frontiers in neuroscience*, *17*, 1225875. https://doi.org/10.3389/fnins.2023.1225875
47. Vaher, K., Bogaert, D., Richardson, H., & Boardman, J. P. (2022). Microbiome-gut-brain axis in brain development, cognition and behavior during infancy and early childhood. *Developmental Review*, *66*, 101038. https://doi.org/10.1016/j.dr.2022.101038
48. Wenger, E., & Kühn, S. (2021). Neuroplasticity. *Cognitive training: An overview of features and applications*, 69–83.
49. Wiatrak, B., Balon, K., Jawień, P., Bednarz, D., Jęśkowiak, I., & Szeląg, A. (2022). The role of the microbiota-gut-brain axis in the development of Alzheimer's disease. *International Journal of Molecular Sciences*, *23*(9), 4862. https://doi.org/10.3390/ijms23094862
50. Yoo, J. Y., Groer, M., Dutra, S. V. O., Sarkar, A., & McSkimming, D. I. (2020). Gut microbiota and immune system interactions. *Microorganisms*, *8*(10), 1587. https://doi.org/10.3390/microorganisms8101587

Notes on Contributors

ECENUR ÖZKUL ERDOĞAN
*Bahçeşehir University, Faculty of Health Sciences, Nutrition and Dietetics Deparment
Istanbul, Türkiye
ecenur.ozkul@bau.edu.tr

YONCA SEVİM
*Bahçeşehir University, Faculty of Health Sciences, Nutrition and Dietetics Deparment
Istanbul, Türkiye
yonca.sevim@bau.edu.tr

İDİL ÖZKOÇ
* Bahçeşehir University, Graduate School, Physiotherapy and Rehabilitation Doctoral Program, Istanbul, Türkiye
idilozkoc@gmail.com

AYGÜL KÖSEOĞLU KURT
* Bahçeşehir University, Graduate School, Physiotherapy and Rehabilitation Doctoral Program, Istanbul, Türkiye
aygulkoseoglu@gmail.com

Mehmet Kaan Altunok
* Selçuk University, Faculty of Health Sciences, Department of Physiotherapy and Rehabilitation, Konya, Türkiye
**Bahçeşehir University, Graduate School, Physiotherapy and Rehabilitation Doctoral Program, Istanbul, Türkiye
mehmetkaan.altunok@selcuk.edu.tr

Pelin Pişirici
* Bahçeşehir University, Faculty of Health Sciences, Physiotherapy and Rehabilitation Department, Istanbul, Türkiye
pelin.pisirici@bau.edu.tr
** California State University, Long Beach, Department of Physical Therapy, California, USA
Pelin.Pisirici@csulb.edu

Büşra Tamgüç
* Bahçeşehir University, Graduate School, Physiotherapy and Rehabilitation Doctoral Program, Istanbul, Türkiye
busra.tamguc@bahcesehir.edu.tr
** İstanbul Nişantaşı University, Faculty of Health Sciences, Department of Physiotherapy and Rehabilitation, Istanbul, Türkiye
busra.tamguc@nisantasi.edu.tr

Öznur Şimşek
* Bahçeşehir University, Graduate School, Neuroscience Master Program, Istanbul, Türkiye
oznurksimsek@gmail.com

Selen Gür Özmen
*Bahçeşehir University, Faculty of Health Sciences, Physiotherapy and Rehabilitation Department, Istanbul, Türkiye
**Bahçeşehir University, Graduate School of Educational Sciences, Lecturer in The Programs of Neuroscience, Istanbul, Türkiye
selen.gurozmen@bau.edu.tr

Hande Besna Göçen
*Bahçeşehir University, Graduate School, Physiotherapy and Rehabilitation Doctoral Program, Istanbul, Türkiye
** Istanbul Gelişim University, Faculty of Health Sciences, Department of Physiotherapy and Rehabilitation, Istanbul, Türkiye
handegocen@outlook.com

Okan Şahin
* Istanbul Health and Technology University, Faculty of Health Sciences, Department of Physiotherapy and Rehabilitation, Istanbul, Türkiye
**Bahçeşehir University, Graduate School, Physiotherapy and Rehabilitation Doctoral Program, Istanbul, Türkiye
okan.sahin@istun.edu.tr

Berkay Eren Pehlivanoğlu
* Bahçeşehir University, Faculty of Health Sciences, Physiotherapy and Rehabilitation Deparment, Istanbul, Türkiye
berkayeren.pehlivanoglu@bau.edu.tr

Ömer Faruk Candal
* Bahçeşehir University, Graduate School, Neuroscience Doctoral Program, Istanbul, Türkiye
omerfarukcandal@gmail.com

Adnan Kuşman
* Bahçeşehir University, Graduate School, Neuroscience Doctoral Program, Istanbul, Türkiye

Suzan Aydin
* Istanbul Gelişim University, Faculty of Health Sciences, Department of Physiotherapy and Rehabilitation, Istanbul, Türkiye
**Bahçeşehir University, Graduate School, Physiotherapy and Rehabilitation Doctoral Program, Istanbul, Türkiye
suzanaltnkeser@gmail.com

Öznur Kaya Sağlam
*Bahçeşehir University, Graduate School, Physiotherapy and Rehabilitation Doctoral Program, Istanbul, Türkiye
** Istanbul Gelişim University, Faculty of Health Sciences, Department of Physiotherapy and Rehabilitation, Istanbul, Türkiye
oznurkayasaglam@gmail.com

Yusuf Yaşasin
* Bahçeşehir University, Graduate School, Physiotherapy and Rehabilitation Doctoral Program, Istanbul, Türkiye
yusufyasasin@gmail.com

Cansu Demir
* Bahçeşehir University, Graduate School, Physiotherapy and Rehabilitation Doctoral Program, Istanbul, Türkiye
**Istanbul Rumeli University, Faculty of Health Sciences, Department of Physiotherapy and Rehabilitation, Istanbul, Türkiye
fztcansudemir@gmail.com

Harun Dere
*Bahçeşehir University, Graduate School, Physiotherapy and Rehabilitation Doctoral Program, Istanbul, Türkiye
**Private Clinical Practice, Manchester, United Kingdom
doctorharundere@gmail.com

www.ingramcontent.com/pod-product-compliance
Ingram Content Group UK Ltd.
Pitfield, Milton Keynes, MK11 3LW, UK
UKHW021829100926
8556IPUK00004B/231

9 783631 930182